Radar Sea Clutter: Modelling and target detection

Other volumes in this series:

Volume 1	**Optimised Radar Processors** A. Farina (Editor)
Volume 3	**Weibull Radar Clutter** M. Sekine and Y. Mao
Volume 4	**Advanced Radar Techniques and Systems** G. Galati (Editor)
Volume 7	**Ultra-Wideband Radar Measurements: Analysis and processing** L. Yu. Astanin and A.A. Kostylev
Volume 8	**Aviation Weather Surveillance Systems: Advanced radar and surface sensors for flight safety and air traffic management** P.R. Mahapatra
Volume 10	**Radar Techniques Using Array Antennas** W. Wirth
Volume 11	**Air and Spaceborne Radar Systems: An introduction** P. Lacomme (Editor)
Volume 13	**Introduction to RF Stealth** D. Lynch
Volume 14	**Applications of Space-Time Adaptive Processing** R. Klemm (Editor)
Volume 15	**Ground Penetrating Radar, 2nd Edition** D. Daniels
Volume 16	**Target Detection by Marine Radar** J. Briggs
Volume 17	**Strapdown Inertial Navigation Technology, 2nd Edition** D. Titterton and J. Weston
Volume 18	**Introduction to Radar Target Recognition** P. Tait
Volume 19	**Radar Imaging and Holography** A. Pasmurov and S. Zinovjev
Volume 20	**Sea Clutter: Scattering, the K distribution and radar performance** K. Ward, R. Tough and S. Watts
Volume 21	**Principles of Space-Time Adaptive Processing, 3rd Edition** R. Klemm
Volume 22	**Waveform Design and Diversity for Advanced Radar Systems** F. Gini, A. De Maio and L.K. Patton
Volume 23	**Tracking Filter Engineering: The Gauss–Newton and polynomial filters** N. Morrison
Volume 25	**Sea Clutter: Scattering, the K distribution and radar performance, 2nd Edition** K. Ward, R. Tough and S. Watts
Volume 33	**Radar Automatic Target Recognition (ATR) and Non-Cooperative Target Recognition** D. Blacknell and H. Griffiths (Editor)
Volume 26	**Radar Techniques Using Array Antennas, 2nd Edition** W. Wirth
Volume 101	**Introduction to Airborne Radar, 2nd Edition** G.W. Stimson
Volume 534	**New Methodologies for Understanding Radar Data** A.K. Mishra and S. Brüggenwirth
Volume 537	**Ocean Remote Sensing Technologies: High frequency, marine and GNSS-based radar** W. Huang and E.W. Gill (Editor)

Radar Sea Clutter

Modelling and target detection

Luke Rosenberg and Simon Watts

The Institution of Engineering and Technology

Published by SciTech Publishing, an imprint of The Institution of Engineering and Technology, London, United Kingdom

The Institution of Engineering and Technology is registered as a Charity in England & Wales (no. 211014) and Scotland (no. SC038698).

First published 2022

The Institution of Engineering and Technology
Michael Faraday House
Six Hills Way, Stevenage
Herts, SG1 2AY, United Kingdom

www.theiet.org

British Library Cataloguing in Publication Data
A catalogue record for this product is available from the British Library

ISBN 978-1-78561-913-7 (hardback)
ISBN 978-1-78561-914-4 (PDF)

Typeset in India by MPS Limited
Printed in the UK by CPI Group (UK) Ltd, Croydon

Contents

About the authors xi
Acknowledgements xiii
List of symbols xv
List of acronyms xxix

1 Introduction **1**
1.1 Prologue 1
1.2 Maritime surveillance radars 1
1.3 The sea surface 3
1.4 Clutter characteristics 5
1.5 Sea clutter modelling 6
1.5.1 Electromagnetic scattering models 7
1.5.2 Empirical modelling 8
1.6 Data sets 9
1.6.1 Ingara X-band data set 9
1.6.2 CSIR X-band data set 10
1.7 Book outline 10
References 11

2 Monostatic sea clutter **15**
2.1 Introduction 15
2.2 Mean backscatter 17
2.2.1 Anomalous propagation 20
2.2.2 Multipath scattering 21
2.2.3 GO-SSA mean backscatter models 24
2.3 Amplitude distributions 27
2.3.1 Distribution models 31
2.3.2 Example fits 39
2.3.3 Effective shape relationships 42
2.4 Sea-spike characteristics 42
2.5 Doppler characteristics 44
2.5.1 Coherent amplitude statistics 49
2.5.2 Mean Doppler spectra 52
2.5.3 Time and range varying spectra 53
2.6 Texture correlation 58
References 60

3 Bistatic sea clutter **67**
3.1 Introduction 67
3.2 Bistatic geometry 68
3.3 Active bistatic radar 71
3.3.1 Sea clutter measurements 72
3.3.2 In-plane reflectivity 74
3.3.3 Out-of-plane reflectivity 75
3.3.4 Amplitude statistics 77
3.3.5 Correlation 78
3.3.6 Doppler spectra 80
3.3.7 Sea spikes 82
3.4 Passive bistatic radar 84
3.4.1 Sea clutter measurements 84
3.4.2 Data processing 89
3.4.3 Delay/Doppler amplitude statistics 91
3.4.4 Time domain amplitude statistics and correlation 93
3.4.5 Doppler spectra 98
3.4.6 Summary 101
3.5 Discussion on bistatic sea clutter 102
References 102

4 Parametric modelling **107**
4.1 Introduction 107
4.2 Monostatic mean backscatter 107
4.2.1 Royal Radar Establishment model 108
4.2.2 Sittrop's model 109
4.2.3 Georgia Institute of Technology model 109
4.2.4 The hybrid model 111
4.2.5 The Technology Service Corporation model 112
4.2.6 The Masuko models 114
4.2.7 The Imaging Radar Systems Group model 115
4.2.8 Naval Research Laboratory model 118
4.2.9 DSTG continuous mean backscatter model 119
4.3 Bistatic mean reflectivity 122
4.3.1 In-plane scattering 122
4.3.2 Out-of-plane scattering 126
4.4 K-Distribution shape 127
4.4.1 Low grazing angle models 127
4.4.2 DSTG continuous shape model 129
4.5 Doppler spectrum 132
4.5.1 Mean Doppler spectrum models 132
4.5.2 Temporal decorrelation model 136
4.5.3 Evolving Doppler spectrum model 136
4.6 Spatial correlation 143
References 147

5 Sea clutter simulation **151**
5.1 Introduction 151
5.2 Uniform Doppler spectrum 152
5.2.1 Memoryless non-linear transform 153
5.3 Evolving Doppler spectrum 157
5.3.1 Frequency domain simulation 157
5.3.2 Time domain simulation 160
5.3.3 Multiple phase centres 161
5.3.4 Scanning radar scenario 163
5.3.5 Simulation examples 165
5.4 Other simulation algorithms 171
5.4.1 Spatially and temporally limited clutter model 172
5.4.2 Compound G2 model 175
5.4.3 Algorithm comparison 178
5.4.4 Summary 185
References 189

6 Performance prediction modelling **193**
6.1 Introduction 193
6.2 Radar range equation 195
6.3 Non-coherent processing 197
6.3.1 Constant target 200
6.3.2 Fluctuating target 202
6.3.3 Compound clutter 203
6.3.4 Correlated returns 205
6.4 Coherent processing 210
6.4.1 Doppler filtering 210
6.4.2 Adaptive processing 213
6.5 Assessing end-to-end radar performance 218
6.5.1 Radar performance examples 218
6.5.2 Simulated performance prediction 225
6.5.3 Discussion 229
References 229

7 Maritime target detection techniques **235**
7.1 Introduction 235
7.2 Non-coherent detection 236
7.2.1 Clutter statistics 237
7.2.2 Constant false alarm rate detectors 240
7.2.3 CA-CFAR performance in spatially correlated clutter 243
7.2.4 Scan-to-scan processing 244
7.2.5 Alternative methods 247
7.3 Pareto CFAR detectors 248
7.3.1 Type I transformed CFAR 249
7.3.2 Type I complete sufficient statistics 250

7.3.3 Type II transformed CFAR 251
7.3.4 Type II Bayesian CFAR 252
7.3.5 Detection results 253
7.3.6 Summary 257
7.4 Coherent detection 257
7.4.1 Multichannel radar 258
7.4.2 Coherent detection strategies 260
7.4.3 Pre-Doppler whitening 260
7.4.4 Post-Doppler detection 263
7.4.5 Post-Doppler whitening 263
7.4.6 Model-based detection 264
7.4.7 Doppler spectrum aliasing 265
7.5 Detection in heterogeneous environments 267
7.5.1 The single data set detector 267
7.5.2 Fixed scale hybrid 269
7.5.3 Variable scale hybrid 269
7.5.4 Texture estimation and normalisation hybrid 270
7.5.5 Detection results 270
References 275

8 New approaches for target detection 281
8.1 Time/frequency target separation 282
8.1.1 Sub-band isolation and reconstruction 283
8.1.2 Sub-band selection 285
8.1.3 Detection performance 289
8.1.4 Summary and future directions 290
8.2 Sparse signal separation 291
8.2.1 Tuned *Q* wavelet transform 292
8.2.2 Short-time Fourier transform 293
8.2.3 Detection approaches 296
8.2.4 Detection performance 301
8.2.5 Summary and further developments 303
8.3 Dictionary learning 304
8.3.1 Dictionary learning techniques 305
8.3.2 Performance evaluation 306
8.3.3 Detection performance 309
8.3.4 Summary and further developments 311
8.4 Track before detect 311
8.4.1 Mathematical models 312
8.4.2 The Bernoulli track before detect filter 315
8.4.3 Simulation details 317
8.4.4 Detection and tracking performance 318
8.4.5 Summary and further developments 321
References 321

Appendix A: Goodness of fit metrics **329**
A.1 Chi-square test 329
A.2 Kolmogorov–Smirnov test 330
A.3 Bhattacharyya distance 330
A.4 Threshold error 330
References 331

Appendix B: Bimodal parameters for the evolving Doppler spectrum **333**
References 338

Appendix C: Atmospheric and precipitation losses **339**
References 341

Index **343**

About the authors

Luke Rosenberg is a research specialist at the Defence Science and Technology Group in Australia and an adjunct associate professor with the University of Adelaide. His work covers a range of areas related to airborne radars. Dr Rosenberg has received the prestigious Defence Science and Technology Achievement Award for Science and Engineering Excellence in 2016 and the IEEE AESS Fred Nathanson award in 2018 for 'Fundamental Experimental and Theoretical Work in Characterizing Radar Sea Clutter'.

Simon Watts MBE FREng was a deputy scientific director and a technical fellow in Thales UK until 2013 and is a visiting professor in the Department of Electronic and Electrical Engineering at University College London. He has over 40 years' experience of researching the design of airborne maritime radar systems and the modelling of radar sea clutter.

Acknowledgements

Much of the work described in this book was supported by the Australian Defence Science Technology Group over the past 10 years. The NATO SET-185 group on sea-clutter modelling from high grazing angles also played a big role with a number of significant developments achieved in sea clutter modelling.

There are a number of people who helped make this book possible. We would like to thank our families for being patient with us over many the many weekends, early mornings and late nights spent writing. We would also like to thank the many authors who have contributed to the body of knowledge presented here.

List of symbols

α	Bragg weighting factor in the bimodal Doppler spectrum models
α_{bp}	Loss factor due to non-rectangular antenna beamshape
α_{mf}	Loss factor due to pulse compression mismatch
α_{s}	Energy ratio test statistic
$\tilde{a}, \tilde{\mathbf{a}}$	3MD-distribution discrete texture parameters
$a_{\mathrm{az}}(\cdot)$	Azimuth two-way beampattern in the time domain
$a_{\mathrm{cont}}, \tilde{a}_{\mathrm{cont}}$	DSTG continuous model parameter
a_{D}	Parameter used in Domville's bistatic parametric model
a_{gen}	Generic model parameter
a_{ext}	Extended Doppler model coefficient
a_{GIT}	GIT mean backscatter model parameter
$a_{\mathrm{GO\text{-}SSA}}(\cdot)$	Azimuth term used in the extended GO-SSA model
a_{h}	Relative humidity
a_{hyb}	Hybrid mean backscatter model parameter
a_{IRSG}	IRSG mean backscatter model parameter
a_{Mas}	Masuko's mean backscatter model coefficient
a_{NRL}	NRL mean backscatter model coefficient
a_{o}	Oxygen factor
a_{r}	Briggs' rain model coefficient
a_{RRE}	Royal Radar Establishment mean backscatter model coefficient
a_{Sit}	Sittrop's mean backscatter model coefficient
a_{Sitv}	Sittrop's RMS velocity model coefficient
a_{temp}	Temporal decorrelation model coefficient
a_{tex}	Texture weighting component
a_{TSC}	TSC mean backscatter model parameter
a_{w}	Water content (g/m^3)
a_{WW}	Watts and Wicks shape model coefficient
a_{v}	Water absorption factor
A	Target amplitude
$A_{\mathrm{az}}(\cdot)$	Azimuth two-way beampattern in the frequency domain
A_{Br}	Bragg magnitude used in mean Doppler spectrum model
A_{c}	Parameter used in the Doppler spectrum model
A_{cl}	Clutter cell area (m^2)
$A_{\mathrm{mp}}(\cdot)$	Multipath effect term used in the extended GO-SSA model
A_q	Wavelet approximation sub-band, level q

$\tilde{A}_q$	Reconstructed approximation wavelet sub-band, level q
A_0	Clutter cell area reference (m^2)
A_{S}	Sea spike magnitude used in mean Doppler spectrum model
A_{TSC}	TSC mean backscatter model parameter
A_{W}	Whitecap magnitude used in mean Doppler spectrum model
β	Bistatic angle (rad)
$\tilde{\beta}$	Slope of the angle/Doppler space
$\beta_0(\cdot)$	Beta function
$\beta_{\mathrm{inc}}(\cdot)$	Incomplete beta function
β_{TSC}	TSC mean backscatter model parameter
b	K-distribution scale parameter
$\tilde{b}(\cdot)$	Target birth PDF used in Bernoulli TBD
b_{Br}	Bragg scattering scale parameter in the G2 model
$b_{\mathrm{cont}}, \tilde{b}_{\mathrm{cont}}$	DSTG continuous model parameter
b_{ext}	Extended Doppler model coefficient
b_{F}	Fast scattering scale parameter in the G2 model
b_{gen}	Generic model parameter
b_{GIT}	GIT mean backscatter model parameter
$b_{\mathrm{GO\text{-}SSA}}(\cdot)$	Azimuth term used in the extended GO-SSA model
b_{hyb}	Hybrid mean backscatter model parameter
b_{IRSG}	IRSG mean backscatter model parameter
b_{K}	KA-distribution scale parameter
b_{Mas}	Masuko's mean backscatter model coefficient
b_{NRL}	NRL mean backscatter model coefficient
b_{p}	Pareto distribution scale parameter
b_{r}	Briggs' rain model coefficient
b_{RRE}	Royal Radar Establishment mean backscatter model coefficient
b_{Sit}	Sittrop's mean backscatter model coefficient
b_{Sitv}	Sittrop's RMS velocity model coefficient
b_{t}	Gamma distribution scale parameter
b_{temp}	Temporal decorrelation model coefficient
b_{TSC}	TSC mean backscatter model parameter
b_{w}	Weibull distribution scale parameter
b_{WW}	Watts and Wicks shape model coefficient
B	Radar bandwidth (Hz)
B_{a}	Frequency agile bandwidth (Hz)
B_{c}	Parameter used in the Doppler spectrum model
$\chi(\cdot)$	Delay/Doppler bistatic radar signal
χ_{c}	Curvature of the contour in the Laplace domain
$\mathbf{c}$	Clutter vector
$\tilde{c}, \tilde{\mathbf{c}}$	3MD-distribution discrete texture weightings
c_0	Speed of light (m/s)
c_{B}	Constant used in Domville's bistatic parametric model

$\mathbf{c}_{\mathrm{Br}}$	Bragg component vector
$c_{\mathrm{cont}}, \tilde{c}_{\mathrm{cont}}$	DSTG continuous model parameter
c_{ext}	Extended Doppler model coefficient
$\mathbf{c}_{\mathrm{F}}$	Fast scattering vector
c_{gen}	Generic model parameter
c_{GIT}	GIT mean backscatter model parameter
$c_{\mathrm{GO\text{-}SSA}}(\cdot)$	Azimuth term used in the extended GO-SSA model
c_{hyb}	Hybrid mean backscatter model parameter
c_{IRSG}	IRSG mean backscatter model parameter
c_{Mas}	Masuko's mean backscatter model coefficient
c_{NRL}	NRL mean backscatter model coefficient
c_{RRE}	Royal Radar Establishment mean backscatter model coefficient
c_{Sit}	Sittrop's mean backscatter model coefficient
c_{Sitv}	Sittrop's RMS velocity model coefficient
c_{temp}	Temporal decorrelation model coefficient
c_{TSC}	TSC mean backscatter model parameter
c_{WW}	Watts and Wicks shape model coefficient
$\mathcal{C}$	Clutter to noise ratio
$\mathbb{C}$	Contour
$C(\cdot)$	Clutter spectrum
$\mathcal{C}_{\mathrm{r}}$	Clutter to noise plus Rayleigh ratio
$\mathcal{CN}(\cdot)$	Complex normal distributed random variable
$\mathcal{C}_{\mathrm{av}}$	Average clutter-to-noise ratio
$\delta(\cdot)$	Delta function
δ_{B}	Target bisector angle (rad)
δ_{cont}	DSTG continuous shape model parameter
ΔR	Slant range resolution (m)
ΔR_{B}	Bistatic range resolution (m)
ΔR_{BG}	Bistatic range resolution projected onto the ground (m)
$\Delta\theta$	Angular resolution (rad)
Δ_z	Intensity spacing between PDF samples
d	Distance between two spatial antenna channels (m)
d_{cont}	DSTG continuous model coefficient
d_{ext}	Extended Doppler model coefficient
d_{gen}	Generic model parameter
d_{hyb}	Hybrid mean backscatter model parameter
d_{IRSG}	IRSG mean backscatter model coefficient
d_{Mas}	Masuko's mean backscatter model coefficient
d_{NRL}	NRL mean backscatter model coefficient
d_{p}	Constant used in the Pareto method of moments estimator
d_{r}	Constant used in the K+Rayleigh $z \log z$ estimator
d_{Sit}	Sittrop's mean backscatter model coefficient
d_{Sitv}	Sittrop's RMS velocity model coefficient
d_{TSC}	TSC mean backscatter model parameter

d_{sp}	Mean spike duration (s)
d_z	Constant used in the $z \log z$ estimator
D	Number of components in the 3MD-distribution
D_{BD}	Bhattacharyya distance metric
D_{CS}	Chi-square metric
D_{KS}	Kolmogorov–Smirnov metric
D_{max}	Maximum order used for the 3MD method of moments estimator
D_{MBD}	Modified Bhattacharyya distance metric
D_{MCS}	Modified chi-square metric
D_{MKS}	Modified Kolmogorov–Smirnov metric
D_q	Wavelet detail sub-band, level q
$\tilde{D}_q$	Reconstructed detail wavelet sub-band, level q
D_{TE}	Threshold error metric
ε	Binary random variable for target existence used in Bernoulli TBD
ε_{dev}	Model deviation used in sparse signal separation
ε_{GIP}	Confidence level used for the variable scale hybrid detector
ε_{lim}	Magnitude limit used in multichannel clutter simulation
ε_{rec}	Threshold used in dictionary learning reconstruction
η	Threshold multiplier
η_{cont}	DSTG continuous shape model parameter
η_{DL}	Diagonal loading
$\eta_p(\cdot)$	Function used to define the Pareto CDF
η_s	Penalty parameter multiplier used in sparse signal separation
e_{cont}	DSTG continuous model coefficient
e_{ext}	Extended Doppler model coefficient
e_{gen}	Generic model parameter
e_{IRSG}	IRSG mean backscatter model coefficient
e_{NRL}	NRL mean backscatter model coefficient
e_p	Constant used in the Pareto method of moments estimator
$E(\cdot)$	Entropy metric used in SWT detector
$E_\nu(\cdot)$	Generalised exponential integral function
f	Doppler frequency (Hz)
f_0	Target Doppler frequency (Hz)
f_B	Target bistatic Doppler (Hz)
f_{bi}	Fluctuation around centre Doppler frequency (bimodal model) (Hz)
f_{Br}	Bragg Doppler frequency (Hz)
$\bar{f}_{Br}$	Mean Bragg Doppler frequency (Hz)
f_{cont}	DSTG continuous model coefficient
f_F	Fast scattering frequency (Hz)
f_{gen}	Generic model coefficient
f_{IRSG}	IRSG mean backscatter model coefficient
f_{lin}	Fluctuation around centre Doppler frequency (linear model) (Hz)
f_r	Pulse repetition frequency (Hz)
f_{RF}	Radar carrier frequency (Hz)

f_s	Spatial target frequency (Hz)
$f_\theta(\cdot)$	Central azimuth frequency for each Bragg event (Hz)
$\bar{f}_W$	Mean whitecap Doppler frequency (Hz)
$F(\cdot), F_-(\cdot)$	Cumulative distribution function
$\bar{F}(\cdot), F_+(\cdot)$	Complementary cumulative distribution function
${}_1F_1(\cdot)$	Confluent hypergeometric function
F_a	Normalising factor for covariance analysis
F_{dat}	Data CDF
F_e	Normalising factor under mismatched conditions
F_m	Normalising factor under matched conditions
F_n	Receiver noise figure
F_{mod}	Model CDF
F_p	Propagation factor
$\bar{F}_{\zeta_0}$	Complementary cumulative distribution function at the point ζ_0
γ	Detection threshold
γ_{cont}	DSTG continuous model parameter
γ_t	Reporting threshold used in Bernoulli TBD
γ_X	Maximum entropy of the reconstructed sub-band X
$\Gamma(\cdot)$	Gamma function
g	Gravitational acceleration (m/s^2)
$\tilde{\mathbf{g}}$	Received radar signal vector used in the SDS algorithm
$g_a(\cdot)$	Analysis high-pass filter
$g_{az}(\cdot)$	Clutter impulse response scaled by azimuth beampattern
g_{gen}	Generic model coefficient
$g_s(\cdot)$	Synthesis high-pass filter
$G(\cdot)$	Doppler spectrum model
$G_2(\cdot)$	G2 Doppler spectrum model
$G_a(\cdot)$	Analysis high-pass filter response
$G_{av}(\cdot)$	Averaged Doppler spectrum model
$G_{az}(\cdot)$	Doppler spectrum model convolved with azimuth beampattern
$G_{Br}(\cdot)$	Bragg Doppler spectrum model
G_c	Number of guard cells
$G_F(\cdot)$	Fast scattering Doppler spectrum model
$G_{FP}(\cdot)$	Fourier power spectrum
$G_{MV}(\cdot)$	Minimum variance optimal power spectrum
G_r	One-way antenna receive gain
$G_s(\cdot)$	Synthesis high-pass filter response
G_t	One-way antenna transmit gain
$G_{uni}(\cdot)$	Uniform Doppler spectrum model
$h, \mathbf{h}$	Received radar signal
$\bar{\mathbf{h}}$	Received radar signal vector used in the SDS algorithm
$h_a(\cdot)$	Analysis low-pass filter
$h_d(\cdot)$	Received direct path signal
h_{gen}	Generic model coefficient

$\mathbf{h}_{\mathrm{G2}}$	Compound G2 radar signal vector
h_{n}	Normalising factor
h_{r}	Received surveillance signal
$h_{\mathrm{s}}(\cdot)$	Synthesis low-pass filter
$\mathbf{h}_{\mathrm{STLC}}$	Spatially and temporally limited clutter radar signal vector
$H(\cdot)$	Received radar power spectrum
H_0	Null hypothesis
H_1	Alternate hypothesis
$H_{1/3}$	Significant wave height (m)
$H_{\mathrm{a}}(\cdot)$	Analysis low-pass filter response
H_{avg}	Average wave height (m)
$H_{\mathrm{e}}(\cdot)$	Entropy
H_{noise}	Noise power spectrum
$H_{\mathrm{p}}(\cdot)$	Hermite polynomial
H_{rms}	Root-mean-square wave height (m)
$H_{\mathrm{s}}(\cdot)$	Synthesis low-pass filter response
I	Number of patches modelled in a clutter patch
$\mathbf{I}_N$	Identity matrix of size $N \times N$
J	Range cell used for estimating the clutter mean in the OS-CFAR
J_0	Number of successive two-channel filter banks used in the TQWT
k_{B}	Boltzmann's constant
k_{c}	Parameter used in out-of-plane bistatic model
$k_{\zeta 0}$	Intensity bin corresponding to intensity threshold, ζ_0
k_{e}	Earth radius scaling factor
k_{r}	K+Rayleigh distribution Rayleigh to clutter ratio
k_{rat}	KK-distribution weighting factor
k_{s}	Swell wavenumber
k_{sr}	Swell wavenumber component parallel to radar beam
k_{WW}	Watts and Wicks shape model polarisation coefficient
$\mathcal{K}(\cdot)$	Modified Bessel function of the second kind
K	Number of range bins
K_{a}	Maximum number of sub-bands in SWT
K_{atom}	Number of atoms in a dictionary
K_{cont}	Continuous mean backscatter model parameter
K_{int}	Number of bins used to form discrete PDF
K_{s}	Number of range bins used to average clutter backscatter
K_{t}	Target fluctuation parameter
K_{T}	Number of snapshots in the CUT for the SDS algorithm
$K_{\mathrm{T_t}}$	Number of temporal snapshots in the CUT for the SDS algorithm
$K_{\mathrm{T_s}}$	Number of spatial snapshots in the CUT for the SDS algorithm
$K_{\mathcal{T}}$	Number of rows in the training data
λ	Radar wavelength (m)
λ_0	Penalty parameter offset used in sparse signal separation
λ_G	Gravity wavelength (m)

λ_n	The *n*th eigenvalue
λ_s	Penalty parameter used in sparse signal separation
$\mathbf{\Lambda}$	Parameter in singular value decomposition
ℓ_n	Vector norm function of order *n*
l_c	Clear air attenuation per km
l_f	Fog attenuation per km
l_r	Rain attenuation per km
l_s	Snow attenuation per km
$\mathcal{L}(\cdot)$	Likelihood ratio function
L	Number of spatial channels
L_0	Filter order used in clutter simulation
L_a	Atmospheric losses
L_q	Overlap factor for the interference STFT dictionary
L_{RF}	RF hardware and radome losses in the radar
L_s	Overlap factor for the target STFT dictionary
L_{sp}	Signal-processing losses in the radar
L_t	Total losses in the radar
μ	Normalised received signal
μ_0	Constant used in target ambiguity function
μ_c	Pulse compression gain
μ_k	*k*th-order sample moment
μ_{sim}	Mean of the similarity measure
μ_{TSC}	TSC mean backscatter model parameter
$m_{bi}(\cdot)$	Centre frequency in the bimodal Doppler spectrum model
m_B	Constant used in Domville's bistatic parametric model
m_c	Parameter used in out-of-plane bistatic model
$m_f(\cdot)$	Centre frequency in the Doppler spectrum model
m_{LN}	Mean of the log–normal distribution
m_w	Doppler spectrum width mean
$\mathcal{M}$	Moment generating function
M	Number of range cells used in the clutter range profile
$\bar{M}$	Mean number of spikes in a range cell for the KA-distribution
M_a	Filter length used in SWT
M_b	Number of scans used for binary integrator
ν	K-distribution shape parameter
$\tilde{\nu}$	K-distribution effective shape parameter
ν_B	Bistatic K-distribution shape parameter
ν_{Br}	Shape parameter for the Bragg scattering component in the G2 model
ν_{CS}	Watts and Wicks cross-swell shape model parameter
ν_{diff}	Watts and Wicks shape model parameter
ν_F	Shape parameter for the fast scattering component in the G2 model
ν_K	KA-distribution shape parameter
ν_{occ}	Shape parameter for the probability of occurrence in the STLC model

ν_{p}	Pareto distribution shape parameter
$\tilde{\nu}_{\mathrm{p}}$	Pareto distribution effective shape parameter
ν_{r}	K+Rayleigh distribution shape parameter
ν_{sum}	Watts and Wicks shape model parameter
ν_{t}	Gamma distribution shape parameter
ν_{US}	Watts and Wicks shape model parameter
ν_{w}	Weibull distribution shape parameter
n_{c}	Parameter used in out-of-plane bistatic model
$\mathcal{N}(\cdot)$	Normal distributed random variable
N	Number of pulses
N_{b}	Number of scans used for binary integrator
$N_{\mathrm{eff}}, \tilde{N}_{\mathrm{eff}}$	Effective number of independent pulses
N_{Mas}	Masuko's mean backscatter model parameter
N_{p}	Number of pulses in a scan
N_{P}	Number of samples in the PDF, $P(\cdot)$
N_{q}	Number of pulses used in the interference STFT dictionary
N_{Q}	Number of samples in the PDF, $Q(\cdot)$
N_{s}	Number of pulses used in the target STFT dictionary
N_{sc}	Number of scatterers
N_{t}	Number of particles used in the Bernoulli TBD
N_{targ}	Number of pulses that the target is in the radar beam
N_{tex}	Number of texture realisations generated in a radar scan
N_{σ}	Number of Bragg spectral widths to include in the STLC model
N_{wave}	Number of STLC points mapped between Doppler and azimuth
$\omega_{\mathrm{i}}, \omega_{\mathrm{s}}$	Process noise intensities used in Bernoulli TBD
$\boldsymbol{\Omega}, \boldsymbol{\omega}$	Process noise matrices used in Bernoulli TBD
$\boldsymbol{\Omega}_k$	Clutter range profile for the kth range bin
$\boldsymbol{\Omega}_{\mathrm{p}}$	Set of parameters used for simulation
$\boldsymbol{\Pi}$	Transition probability matrix used in Bernoulli TBD
ϕ	Grazing angle (rad)
$\tilde{\phi}$	Speckle phase (rad)
ϕ_0	Grazing angle reference (rad)
ϕ_{11}, ϕ_{21}	Incidence and reflected angles in Domville's parametric model (rad)
$\phi_{3\mathrm{dB}}$	Elevation 3-dB beamwidth (rad)
$\phi_{\mathrm{b}}, \phi_{\mathrm{d}}$	Angles used in Domville's bistatic parametric model (rad)
ϕ_{p}	Parameter used in the Pareto P_{fa} model
$\phi_{\mathrm{n}}, \phi_{\mathrm{s}}$	Angles used in Domville's bistatic parametric model (rad)
ϕ_{t}	Transition grazing angle (rad)
φ	Spatial phase shift (rad)
$\Phi(\cdot)$	Phase function used for evaluating contour integral
$\boldsymbol{\Phi}_{\mathrm{c}}$	Clutter dictionary
$\boldsymbol{\Phi}_{\mathrm{q}}$	Interference dictionary
$\boldsymbol{\Phi}_{\mathrm{s}}$	Target dictionary

ψ	Target angle used in bistatic geometry
ψ_{B}	Azimuth scattering angle for bistatic radar (rad)
p_{c}	Clutter mean power
$\bar{p}_{\mathrm{c}}$	Normalised clutter mean power
p_{mf}	Received power after matched filtering
p_{n}	Noise mean power
$\bar{p}_{\mathrm{n}}$	Normalised noise mean power
p_{r}	Rayleigh mean power
p_{rec}	Received power in the receiver front end
p_{mf}	Received power after matched filtering
p_{s}	Target mean power
p_{t}	Transmitted average power
$P(\cdot)$	Probability density function
P_0	Probability of no target present
P_1	Probability of target present
P_{AR}	Autoregressive order number
P_{b}	Probability of target birth used in Bernoulli TBD
P_{Br}	Number of Bragg events in a single range cell in the STLC model
P_{d}	Probability of detection
P_{dat}	Data PDF
P_{dist}	Desired distribution in the MNLT
P_{fa}	Probability of false alarm
$P_{\mathrm{fa,ideal}}$	Ideal probability of false alarm
P_{GIP}	Generalised inner product statistic
P_{in}	Input probability
P_{m}	Probability of miss
P_{mod}	Model PDF
P_{occ}	Probability of a fast scattering event occurring in the STLC model
P_{out}	Output probability
P_{s}	Probability of target survival used in Bernoulli TBD
P_{sp}	Percentage of sea spikes
P_{T}	Number of temporal elements used in the SDS algorithm
P_{wave}	PDF for the large wave slopes in the GO-SSA model
$\tilde{q}$	Posterior PDF of target presence used in Bernoulli TBD
$\mathbf{q}_{\mathrm{I}}$	Interference signal
q_{sim}	Similarity measure between two vectors
$\mathbf{Q}$	Covariance estimate used in the SDS algorithm
$Q(\cdot)$	Mismatched filter output
Q_0	Q-factor for the TQWT
$Q_{\mathrm{c}}(\cdot)$	Complementary quantile function
Q_{e}	Whitening filter output under mismatched conditions
Q_{m}	Whitening filter output under matched conditions
$Q_N(\cdot)$	Marcum Q function

Q_T	Number of spatial elements used in the SDS algorithm
$Q_w(\cdot)$	Window function
Q_{wave}	Fraction of surface covered by breaking waves in the GO-SSA model
$\rho(\cdot)$	Correlation function
$\bar{\rho}(\cdot)$	Normalised correlation function
$\tilde{\rho}$	Correlation value
ρ_G	Normalised Gaussian autocorrelation function
ρ_L	Spatial decorrelation length for short ranges (bins)
ρ_r	Rain rate (mm/h)
$\rho_{spat}(\cdot)$	Spatial autocorrelation function
$\rho_{temp}(\cdot)$	Temporal autocorrelation function
$r, \mathbf{r}$	Magnitude
r_0	Redundancy factor for the TQWT
r_e	Earth radius (m)
r_{eff}	Effective Earth radius (m)
$\mathbf{r}_s$	Residual component used in sparse signal separation
R	Slant range (m)
$\mathbf{R}$	Speckle covariance matrix
R_0	Target range (m)
$\mathbf{R}_{av}$	Covariance matrix long-term average
$\mathbf{R}_{Br}$	Bragg covariance
R_e	Range extent (m)
$\mathbf{R}_F$	Fast scattering covariance
R_G	Spatial decorrelation length associated with the gravity wave (m)
R_L	Spatial decorrelation length for short ranges (m)
$\mathbf{R}_m$	Covariance matrix of the clutter, target and noise
R_{max}	Maximum instrumented range (m)
R_{res}	Residual component after using the saddlepoint approximation
R_{Rx}	Slant range between receiver and ground patch (m)
$\mathbf{R}_s$	Target covariance matrix
$\mathbf{R}_{SCM}$	Sample covariance matrix
$\mathbf{R}_{NSCM}$	Normalised sample covariance matrix
$\mathbf{R}_\theta$	Spatial component of the covariance
$\mathbf{R}_t$	Temporal component of the Bragg covariance
$\mathbf{R}_T$	Temporal component of the fast scattering covariance
R_{Tx}	Slant range between transmitter and ground patch (m)
ς	Log magnitude
σ	Radar cross section
$\tilde{\sigma}$	Speckle radar cross section
σ^0	Monostatic mean backscatter
σ_B^0	Bistatic mean backscatter
σ_c^0	Crosswind mean backscatter
σ_d^0	Downwind mean backscatter

σ^0_{GIT}	GIT mean backscatter
σ^0_{HH}	Horizontally polarised mean backscatter
σ^0_{Mas}	Masuko's model mean backscatter
σ^0_{med}	Continuous mean backscatter model component
$\sigma^0_{\mathrm{quasi}}$	Monostatic quasi-specular mean backscatter
σ^0_{u}	Upwind mean backscatter
σ^0_{VV}	Vertically polarised mean backscatter
σ_0	Reference mean backscatter
σ_1, σ_2	Normalised backscatter from the Domville parametric model
σ_{c}	Clutter radar cross section
σ_{CPI}	CPI length Doppler spectrum standard deviation (Hz)
$\sigma_{f_{\mathrm{Br}}}$	Bragg standard deviation of the Doppler spectrum mean (Hz)
$\sigma_{f_{\mathrm{F}}}$	Fast component standard deviation of the Doppler spectrum mean (Hz)
σ_{GIT}	GIT mean backscatter model parameter
σ_{h}	Sea surface height standard deviation (m)
σ_{L}	Root mean square large scale wave height in the GO-SSA model
σ_{LN}	Standard deviation of the log–normal distribution
σ_{plat}	Platform motion Doppler spectrum standard deviation (Hz)
$\sigma_{\mathrm{s}}, \tilde{\sigma}_{\mathrm{s}}$	Backscatter standard deviation and averaged standard deviation
σ_{S}	Root mean square small scale wave height in the GO-SSA model
σ_{scan}	Scanning motion Doppler spectrum standard deviation (Hz)
σ_{sp}	KA-distribution sea spike intensity
σ_{TSC}	TSC mean backscatter model parameter
σ_{w}	Standard deviation of the Doppler spectrum width (Hz)
σ_{wave}	RCS of a breaking wave
σ_z	Standard deviation of the intensity
$\mathbf{\Sigma}$	Compound clutter covariance
$\tilde{\mathbf{\Sigma}}$	Scaled compound clutter covariance
$\mathbf{\Sigma}_{\mathrm{FS}}$	Fixed-scale covariance
$\mathbf{\Sigma}_{\mathrm{TN}}$	Texture estimation and normalisation covariance
$\mathbf{\Sigma}_{\mathrm{VS}}$	Variable scale covariance
s	Complex frequency used in Laplace transform
$\mathbf{s}$	Target signal vector
$\tilde{s}(\cdot)$	Posterior spatial PDF used in Bernoulli TBD
s_0	Random variable for the normalised target fluctuation
$\mathrm{s}_0^{\pm}$	Saddlepoint locations
$\mathbf{s}_\theta(\cdot)$	Spatial steering vector
$\mathbf{s}_{\mathrm{t}}(\cdot)$	Temporal steering vector
$\mathcal{S}$	Signal-to-interference ratio
$\mathcal{S}_0$	Local signal-to-interference ratio
$S(\cdot), S_{\mathrm{I}}(\cdot)$	Target point spread function in amplitude/intensity
S_{ss}	Sea state
$S_{\mathrm{t}}(\cdot)$	Target slow-time point spread function

$S_R(\cdot)$	Target range point spread function
θ	Azimuth angle relative to up-wind (rad)
θ_0	Target azimuth angle (rad)
θ_{3dB}	Two-way azimuth 3 dB antenna beamwidth (rad)
$\theta_{3dB,Rx}$	One-way azimuth 3 dB antenna beamwidth – receiver (rad)
$\theta_{3dB,Tx}$	One-way azimuth 3 dB antenna beamwidth – transmitter (rad)
$\dot{\theta}_{az}$	Azimuth scan rate (rad/s)
θ_{plat}	Relative azimuth angle between the radar platform and radar look direction (rad)
θ_{sw}	Azimuth swell angle relative to upswell (rad)
θ_t	Target steering angle (rad)
ϑ	Transition probability used in Bernoulli TBD
ϑ_L, ϑ_U	Lower and upper bounds for the variable scale hybrid detector
$\boldsymbol{\Theta}$, $\boldsymbol{\vartheta}$	Transition matrices used in Bernoulli TBD
τ	Mean speckle power
$\bar{\tau}$	Normalised mean speckle power
τ_{bi}	Threshold used in the bimodal Doppler spectrum model
τ_{Br}	Mean power of the Bragg component
$\bar{\tau}_{Br}$	Normalised mean power of the Bragg component
τ_F	Mean power of the fast scattering component
τ_r	Mean speckle power used in the K+Rayleigh distribution
t	Slow time
t_0	Target time offset (s)
t_D	Time lag for passive radar cross-correlation
t_e	Time extent (m)
t_f	Fast time
T	Pulse width (s)
T_0	Receiver temperature (K)
T_1, T_2	Number of cells trimmed in the TM-CFAR
$T_{a,C}$	Atmospheric temperature (K)
T_d	Temporal decorrelation time for the texture (s)
T_I	Coherent integration time(s)
T_p	Compressed pulse width (s)
T_r	Pulse repetition interval (s)
T_s	Swell period (s)
T_{sc}	Scan time (s)
T_{sp}	Mean interval between spikes (s)
T_{targ}	Time interval when target is in the radar beam (s)
T_{temp}	Temporal decorrelation time for the speckle (s)
$\mathcal{T}$	Training data set
$\mathbf{u}$	State vector used in Bernoulli TBD
u_{Br}	Bragg component uniform Doppler spectrum standard deviation (Hz)
u_F	Fast component uniform Doppler spectrum standard deviation (Hz)

U	Wind speed (m/s)
$\mathbf{U}$	Component in singular value decomposition
U_0	Wind speed reference (m/s)
v_0	Target radial velocity (m/s)
v_{HH}	Velocity of the horizontal polarised scatterers
v_{max}	Maximum target velocity (m/s)
v_{p}	Radar platform velocity (m/s)
v_{RMS}	Root-mean-square velocity (m/s)
v_{Rx}	Radar receiver velocity (m/s)
v_{Tx}	Radar transmitter velocity (m/s)
v_{VV}	Velocity of the vertical polarised scatterers
$\mathbf{V}$	Component in singular value decomposition
V_{f}	Visibility in fog (m)
w	Doppler spectrum standard deviation (Hz)
$\tilde{w}$	Weighting factor used in Bernoulli TBD
w_{bi}	Bimodal Doppler spectrum standard deviation (Hz)
w_{RMS}	Root-mean-square width (Hz)
w_{av}	Averaged Doppler spectrum standard deviation (Hz)
w_{Br}	Bragg component Doppler spectrum standard deviation (Hz)
$\bar{w}_{\mathrm{Br}}$	Mean Bragg spectral width (Hz)
$\mathbf{w}_{\mathrm{c}}$	Clutter weight used in sparse signal separation
$w_{\mathrm{D}}(\cdot)$	Window function
w_{F}	Fast component Doppler spectrum standard deviation (Hz)
w_{HP}	Half power width (Hz)
$\mathbf{w}_{\mathrm{q}}$	Interference weight used in sparse signal separation
$\mathbf{w}_{\mathrm{s}}$	Target weight used in sparse signal separation
$\mathbf{w}_{\mathrm{sp}}$	Width of the Doppler spectrum including broadening
$\bar{w}_{\mathrm{S}}$	Mean sea-spike spectral width (Hz)
$\bar{w}_{\mathrm{W}}$	Mean whitecap spectral width (Hz)
w_x, w_y	Standard deviation for the target point spread function in range/azimuth
W	Target spectrum
$W_{\mathrm{af}}(\cdot)$	Wind aspect angle scaling term used in the extended GO-SSA model
$\mathbf{W}_{\mathrm{c}}$	Matrix of clutter weights used in sparse signal separation
W_{ext}	Extended Doppler model component
x	Geometry direction
$x, \mathbf{x}$	Speckle component
$x_{\mathrm{c}}, \mathbf{x}_{\mathrm{c}}$	Correlated speckle component
$\mathbf{x}_{\mathrm{F}}$	Complex Gaussian realisation used to model fast scattering component
$X(\cdot)$	Generic parameter model
$X_{\mathrm{ext}}(\cdot)$	Extended Doppler parameter model
$\boldsymbol{\xi}$	Noise vector
y	Geometry direction
$y, \mathbf{y}$	Compound Gaussian model

y_s	Variable used in saddlepoint approximation
Y	Global detection threshold
Y_0	Normalised detection threshold
Y_{cont}	Parameter used in the continuous model
ζ	Detection statistic
ζ_0	Intensity threshold used to define distribution tail region
ζ_{TN}	Normalisation term used for the TN hybrid detector
z	Geometry direction
$\mathrm{z_p}$	Radar platform altitude
$\mathrm{z_t}$	Target altitude
$z, \mathbf{z}$	Intensity
$\bar{z}$	Intensity offset by Pareto scale parameter
Z	Multi-look intensity

List of acronyms

3MD	Tri-modal distribution
ACF	Autocorrelation function
AMF	Adaptive matched filter
AR	Autoregressive
AREPS	Advanced refractive effects prediction system
ARPDD	Automatic radar periscope detection and discrimination
BD	Bhattacharyya distance
BPD	Basis pursuit de-noising
CA	Cell-averaging
CAMP	Complex approximate message passing
CFAR	Constant false alarm rate
CCDF	Complementary cumulative distribution function
CDF	Cumulative distribution function
CM	Censored mean
CNR	Clutter-to-noise ratio
CPI	Coherent processing interval
CRP	Clutter range profile
CS	Chi-square
CSW	Cross-swell
CSIR	Council of Scientific and Industrial Research
CUT	Cell under test
CW	Continuous wave
CWT	Continuous wavelet transform
DL	Dictionary learning
DOMINODL	Drop off mini-batch online dictionary learning
DSTG	Defence Science and Technology Group
DWT	Discrete wavelet transform
EM	Electromagnetic
FFT	Fast Fourier transform
FISTA	Fast iterative shrinking thresholding algorithm
FRF	Field research facility
GLRT	Generalised likelihood ratio test
GO	Greater-of
GPU	Graphical processing unit
GM	Geometric mean

KS	Kolmogorov–Smirnov
HF	High frequency
HH	Horizontal transmit, horizontal receive
HP	Half power
H-PMHT	Histogram-probabilistic multi-hypothesis tracker
HV	Horizontal transmit, vertical receive
IEEE	Institute of Electrical and Electronic Engineers
ISTA	Iterative shrinking thresholding algorithm
IRSG	Imaging Radar Systems Group
KSVD	K-means singular value decomposition
GIT	Georgia Institute of Technology
GO-SSA	Geometric optics-small slope approximation
LASSO	Least absolute shrinkage and selection operator
LARS	Least-angle regression
LN	Log–normal
LSD	Log–normal standard deviation
MCA	Morphological component analysis
MCS	Modified chi-square
MGF	Moment generating function
MKS	Modified Kolmogorov–Smirnov
MNLT	Memoryless non-linear transform
NAAWS	NATO anti-air warfare systems
NAMF	Normalised adaptive matched filter
NATO	North Atlantic Treaty Organisation
NetRAD	Netted radar
NM	Nautical miles
NRL	Naval Research Laboratory
NSCM	Normalised sample covariance estimate
ODL	Online dictionary learning
OMP	Orthogonal matching pursuit
OS	Ordered statistic
OSPA	Optimal sub-pattern assignment
PDF	Probability density function
PSD	Power spectral density
PSF	Point spread function
PRF	Pulse repetition frequency
RCS	Radar cross section
RFI	Radio frequency interference
RMS	Root mean square
RRE	Royal Radar Establishment
SALSA	Split augmented Lagrangian shrinkage algorithm
SAR	Synthetic aperture radar
SCM	Sample covariance estimate

SDS	Single data set
SIRP	Spherically invariant random process
SIRV	Spherically invariant random vector
SIR	Signal-to-interference ratio
SNR	Signal-to-noise ratio
SO	Smaller of
SS	Sea state
STAP	Space–time adaptive processing
STFT	Short-time Fourier transform
STLC	Spatially and temporally limited clutter
SWT	Stationary wavelet transform
TBD	Track before detect
TE	Threshold error
TN	Texture estimation and normalisation
TM	Trimmed mean
TQWT	Tuned *Q*-wavelet transform
TSC	Technology Services Corporation
UCL	University College London
USW	Upswell
VH	Vertical transmit, horizontal receive
VV	Vertical transmit, vertical receive
WT	Wavelet transform
WTW	Ward, Tough and Watts

Chapter 1
Introduction

1.1 Prologue

Radar for maritime surveillance has been essential capability since its first development in the 1930s. The original requirements were for military applications, but today there are also many important paramilitary and civil uses. The first maritime surveillance radars in World War II quickly discovered that returns from the sea, soon to be known as sea clutter, were often the limiting factor when attempting to detect small targets while controlling false alarms. This remains true for modern radars, where the detection of small, slow moving targets on a rough sea surface remains one of the main drivers for maritime radar design, particularly in the development of detection processing.

The design, development and testing of radar signal processing for maritime surveillance requires a very detailed understanding of the characteristics of radar sea clutter and of the combined target and clutter returns. This book provides an updated and comprehensive review of the latest research into radar sea clutter and detection methods for targets in sea clutter, building on the earlier book by Ward, Tough and Watts [1]. The emphasis is on understanding the characteristics of radar sea clutter as observed with different radars, viewing geometries and environmental conditions. This understanding is assisted by the development of mathematical models that are used in the radar design process.

The work reported here has mainly been driven from the requirements of airborne maritime reconnaissance, although many of the ideas and models can be equally applied to ground-based or ship-borne radars. Since the publication of [1] there has been an increased interest in operating at higher altitudes, resulting in the sea surface being illuminated with larger grazing angles than used in traditional airborne surveillance platforms or ground-based systems. There has also been significant research into bistatic operation, including passive radars using illuminators of opportunity. The use of coherent and multi-aperture systems in a maritime radar is also of increasing interest and these new application areas are covered in this book.

1.2 Maritime surveillance radars

Maritime surveillance radars can be ground, air or space based. For ground-based radar systems, the most common are navigation radars that monitor the coastline, or on naval ships providing situational awareness. The radars on military ships are typically more advanced and used for multiple applications ranging from navigation to air defence. All these radars generally operate at S- or X-band frequencies and scan either mechanically or electronically to provide 360° coverage. These radars may typically be 20–30 m above the sea surface, resulting in low grazing angles over most of the surveyed sea surface. This can result in lower backscatter and spikier amplitude statistics compared with higher grazing angles, complicating the detection of slow moving small targets.

Airborne maritime surveillance radars are primarily concerned with the detection of targets on the sea surface. These radars may be used in military applications, such as anti-submarine warfare and surface surveillance of shipping, or in paramilitary or civil applications, looking for smugglers or refugee boats. They are typically X-band radars scanning over 360° but may also include radars on fast jets with forward looking antennas or fixed side looking radars, suitable for radar imaging. When attempting to detect small stationary and slow moving targets on the sea surface, airborne radars have traditionally operated at low altitudes, typically below 3,000 ft (914 m). This is because of the higher signal-to-clutter power ratio that results from operation at low grazing angles. However, there is a push for current and future airborne platforms, including unmanned aerial systems (UASs), to operate at higher altitudes. This extends the radar horizon and provides a much larger surveillance region. The consequence is that the grazing angle also increases, particularly for close-in ranges, and the strength of the sea clutter increases, making the detection of small targets difficult. Two example maritime platforms include the P-8A Poseidon and MQ-4C Triton UAS shown in Figures 1.1 and 1.2, respectively.

Figure 1.1 P-8A Poseidon aircraft. © Australian Department of Defence 2021

Figure 1.2 MQ-4C Triton unmanned aerial system. © Northrop Grumman 2021

Radars have long been used in space for oceanographic measurements, following on from the pioneering flight of the Seasat satellite in 1978 [2]. Oceanographic satellites use a radar scatterometer to estimate wind speed and direction by sensing the variation of average backscatter power with look direction. High-resolution imaging radars can map ocean currents, the movement of sea ice and so on. Persistent maritime surveillance is difficult from these satellites due to their low earth orbit and long revisit time. One exception is the RADARSAT Constellation Mission [3] which comprises three satellites and is primarily focussed on the surveillance of ships.

1.3 The sea surface

In developing an understanding of sea clutter characteristics, it is important to realise that the sea surface is not a random rough surface but has a significant structure determined by the wind and currents. As the wind blows, the surface is roughened in the form of small ripples which grow and transfer their energy to longer waves [1].

At some point the waves cannot sustain their shape and break, which redistributes the energy further. Ultimately, a point of equilibrium is reached between the energy input from the wind and the energy dissipated in the waves. As the wind speed increases, the time required to reach equilibrium increases, as does the wavelength of the resulting sea surface. The result is a spectrum of waves propagating in the sea. There may also be a swell due to gravity waves that continue to propagate when the wind ceases and will be added to the wave spectrum.

There are various terms used to describe the sea surface characteristics [4]:

Wind wave: A wave resulting from the action of the wind on the water surface.

Gravity wave: A wave whose velocity of propagation is controlled by gravity and typically has a wavelength greater than 5 cm.

Capillary wave (ripple): A wave whose velocity of propagation is controlled by the surface tension of the liquid in which the wave is travelling. Water waves with a wavelength of less than 2.5 cm are considered capillary waves.

Fetch: This is the generating area of the sea surface produced by wind having a constant direction and speed. The length of the fetch area is measured in the direction of the wind from where the seas are generated.

Duration: The length of time the wind blows in essentially the same direction over the fetch.

Fully developed (arisen) sea: The maximum height ocean waves that can be generated by a given wind force blowing over sufficient fetch, regardless of duration. It is the result of all possible wave components in the spectrum being present with their maximum amount of spectral energy.

Sea state: The numerical or written description of ocean-surface roughness. Ocean sea state may be defined more precisely as the significant wave height, which is the mean peak-to-trough height of the highest one-third of the waves observed in a wave train.

In the specification of radar system performance, the Douglas sea state has traditionally been used. Table 1.1 shows the Douglas sea state definition with the

Table 1.1 Douglas sea state and associated wind speeds for a fully developed sea

Sea state	Description	Sig. wave height (m)	Wind speed (m/s)	Fetch (NM)	Duration (h)
0	Glassy	no waves	0		
1	Calm	0–0.1	0–3		
2	Smooth	0.1–0.5	3–6	50	5
3	Slight	0.5–1.25	6–7.5	120	20
4	Moderate	1.25–2.5	7.5–10	150	23
5	Rough	2.5–4.0	10–12.5	200	25
6	Very rough	4.0–6.0	12.5–17	300	27
7	High	6.0–9.0	17–25	500	30
8	Very high	9.0–14.0	>25		
9	Phenomenal	14.0+	$\gg$25		

approximate wind speeds that would produce the wave heights assuming a fully developed sea. This is an approximation that requires the wind to blow continuously for a minimum duration over the fetch distance. The fetch and duration linking wind speeds with sea states are also shown in Table 1.1. Note that the common definition of Douglas sea state does not agree exactly with the definition given in [5] and other older text books. In addition to the Douglas sea state scale, there is the World Meteorological Organisation sea state and Beaufort wind speed scales that are slightly different. This emphasises that great care must be taken when using such scales to define performance. Also, the definition of sea state covers a range of conditions and the boundary between one sea state and another will be very difficult to determine in practice. In addition, the accurate estimation of significant wave height by an observer is very difficult as conditions may vary over a local area.

1.4 Clutter characteristics

The backscattered radar signals from the sea surface are known as clutter. In order to capture the clutter characteristics that will affect radar performance, models are built that represent typical behaviour. The general sea clutter characteristics of interest to the modern radar designer include the following:

- the average magnitude (mean backscatter) of the returns at a given point on the sea surface;
- the fluctuations in amplitude around the local mean backscatter level;
- the temporal characteristics of these fluctuations in amplitude;
- the spatial variation in mean backscatter over a larger area.

All these characteristics will be dependent on the sea and weather conditions, the radar viewing geometry and the radar parameters such as frequency and polarisation. The principle characteristics and features that are used to describe the clutter returns include the following:

- reflectivity (normalised mean backscatter);
- amplitude distribution;
- clutter spikes;
- Doppler spectra;
- spatial and temporal correlation.

The received radar signal comprises a combination of clutter, target and noise, with the latter usually dominated by the thermal noise generated within the receiver. The strength of the sea clutter is determined by the radar transmitter power and antenna gain together with the area illuminated by the radar pulse and the local reflectivity of the sea surface. The clutter will also fluctuate in amplitude due to the combination of many scatterers within the radar field of view. It is represented by an amplitude distribution or probability density function (PDF). With coarse range resolution, the PDF of coherent sea clutter may have a complex Gaussian distribution [1]. However, as the range resolution becomes finer, the variation of the sea surface becomes better resolved and the effect of breaking waves and other discrete events (sea spikes) are more pronounced. These returns then have a higher probability of large magnitudes relative to the overall mean level and the PDFs of amplitude or intensity are described as having longer 'tails'.

Along slow time (pulse-to-pulse), the clutter amplitude fluctuations are not totally random but are correlated. This correlation is typically characterised in the frequency domain, in terms of the power spectral density or Doppler spectrum. In addition to the amplitude fluctuations, large returns known as sea spikes may sometimes be observed. These can be either short-lived, existing for only a few tens of milliseconds or persistent over seconds. As a result, they can influence both the amplitude distribution and the Doppler spectrum, making target detection more challenging. Studying the characteristics of sea spikes is therefore essential to understand their effect on radar detection performance. Just as the amplitude of the returns may be correlated from pulse-to-pulse, so the fluctuations in clutter intensity over range or over a long-time scale may also be correlated. This may reflect the physical structure of the sea waves and swell and appear as a spatial or temporal correlation of clutter intensity.

1.5 Sea clutter modelling

The clutter characteristics described before need to be captured using models, which can then be used to predict and assess radar performance. Models of sea clutter are used throughout the development life cycle of a radar and include [1]:

Performance requirement definition: Specifying the radar performance requires an understanding of how the radar will behave in different operational environments and can be assisted by using a model that relates the performance requirements to changes in the environment.

Modelling of potential performance: Initial modelling is based on the radar range equation. Detection performance in sea clutter requires knowledge of the clutter characteristics and the radar signal processing.

System and algorithm development: As the radar design is refined, the detection performance is reassessed through modelling. New detection algorithms should be benchmarked against idealised models of performance.

Performance assessment and acceptance trials: Once the radar is built and integrated into a working system, its overall performance needs to be assessed. This is best achieved by measuring the environment in which the trials are performed and then comparing the trial results with the modelled performance.

In-service tactics and training: There are typically a number of operator controls and settings that need tuning to achieve optimal detection performance in different conditions. A radar operator requires both on-the-job training and practice in a radar simulator that accurately reflects the real-world behaviour. Modelling of the environment can also be used to aid the radar operator. For example, the modelled detection results can be used to define the optimal altitude for detecting a certain sized target in a particular mode and sea state.

In-service upgrades: As user requirements change or the radar is upgraded with newer technology, radar modelling can be used to define the expected improvement in the detection performance.

Radar modelling can broadly be subdivided into physical models, based on calculations of electromagnetic (EM) scattering from modelled representations of the sea surface, and empirical models based on observations of the clutter in different conditions. The EM modelling of scattering is still a very difficult problem, especially at low grazing angles. For that reason most radar development is still undertaken using empirical models.

1.5.1 Electromagnetic scattering models

Advances in oceanography over many years have resulted in good models of ocean waves that can be used to develop realistic models of the sea surface in different conditions. Ideally such models of the sea surface might provide the foundation for EM scattering calculations and reduce the need for empirical models.

Sea water has a high dielectric constant and hence absorbs very little radar energy. Moreover, radio-frequency energy has very little penetration into the ocean surface. In calm conditions, the ocean surface is flat and radar reflections are highly specular. For this case, most of the transmitted energy is scattered away from the radar transmitter, and the measured backscatter levels are low. However, as winds increase and the ocean surface becomes rougher, the backscatter becomes more diffuse and measured backscatter levels increase. The qualitative degree of roughness depends on the radar wavelength with the same ocean surface generating specular reflections at high frequencies and diffuse reflections at microwave frequencies.

The perturbation theory for EM scattering was developed by Rice [6] and later applied by Peake [7] to compute the radar cross section of land. Wright [8] then extended this theory to water and classified it as a 'slightly rough' surface. He showed that the scattering elements of primary importance for small grazing angles are capillary or short-gravity waves. Bragg resonance occurs when the distances travelled by the radar waves to successive capillary wave crests are half the radar wavelength. In this case, the reflections add together in phase and resonate to produce strong backscatter. A theoretical backscatter model based on Bragg resonance is presented by Ulaby *et al.* [9]. They note, however, that since their model uses only first-order theory, the magnitude of the cross-polarised channels is predicted to be zero, which is not realistic.

While the Bragg theory does not always match measured backscatter data well, it does provide some insight into the properties of measured backscatter data. For instance, it is commonly observed that backscatter measured with the vertical polarisation is larger than that measured with the horizontal. Bragg theory explains this difference due to Fresnel reflection coefficients [9] and also proposes a reason for the sinusoidal-like variation in ocean backscatter with an azimuth angle.

To extend the first-order Bragg scattering theory, two-scale composite scattering theories were proposed to divide the rough surface into large- and small-scale components [10–14]. The physical motivation for the model comprises a large-scale component which is modelled by geometric or physical optics and a small-scale component described by the existing perturbation theory.

Neither the simple Bragg scattering model nor the two-scale composite model can fully account for what is observed, especially at low grazing angles. More recent work has introduced empirical terms to account for breaking waves and the multipath reflections and shadowing effects that are encountered at low grazing angles. These results can give good correspondence with measured data [15] and with the established empirical models of reflectivity (see Chapter 4). There is a large body of literature based on extensions to the composite model, applied to the retrieval of wind and sea conditions from synthetic aperture radar and wind-scatterometer measurements made from space. Kudryavtsev *et al.* [16] describe the use of multi-scale composite models based on Bragg scattering, extended to include the effects of breaking waves. Work was also undertaken by the same authors [17] to infer the wave radar modulation transfer function, which relates the shape of long waves to the wave-induced variation of radar returns. An alternative approach to using the small slope approximation is to solve the Stratton–Chu equations numerically with the 'forward backward' method. This is able to model the effects of multipath and shadowing at low grazing angles, without resorting to empirical correction factors, although breaking waves must still be treated in a statistical manner [1,18,19].

Some work has been done to investigate the mean Doppler characteristics of sea backscatter using EM modelling[18–21]. This requires time-evolving models of the sea surface and has still not been wholly successful, particularly in relation to the time- and range-varying characteristics discussed in Chapter 2. However, Johnson *et al.* [19] have used EM scattering modelling to demonstrate an approximately linear relationship between the clutter intensity and the mean Doppler shift. In addition to the difficulties in modelling Doppler spectra, it has not yet been possible to develop EM modelling methods that predict the amplitude statistics of sea clutter, especially at low grazing angles.

1.5.2 Empirical modelling

To have value in the radar design process, it is necessary to assign values to the different clutter characteristics that are modelled. In the absence of reliable physical models, it is usual for models to be developed on the basis of observations. For example, the original idea for the compound Gaussian model for clutter amplitude statistics was developed from observations of clutter characteristics [22].

The most developed empirical models are those for clutter reflectivity (see Chapter 4) and these have been derived from many thousands of measurements. However, even with the very large data sources that have been used, none of these models are identical. This diversity reflects the different geographic areas used to collect the data, the difficulties in quantifying the prevailing conditions under which the measurements were made (wind speed and direction, wave height and direction, swell, etc.), the overall variability of clutter characteristics and the difficulty in calibrating the radar systems to make absolute measurement of backscatter power. The propagation conditions due to variations in atmospheric refractive index can also have a marked effect on clutter measurements [4,5,23]. This great variability in the models must be understood if they are to be used effectively. A common aphorism, generally attributed to George E. P. Box, a British statistician, is that 'all models are wrong, but some are useful'. This is particularly true of sea clutter models. If a radar trial is undertaken on a particular day and measurements are carefully made of the sea and weather conditions, the viewing geometry, the radar parameters and so on, it is still most unlikely that the observed clutter will correspond exactly with any modelled predictions. The purpose of these models is not to predict exactly the values of clutter characteristics on a given trial. Rather, since they are developed from many measurements made on different occasions and in different locations, these models are very good at predicting average characteristics and the likely spread of values over different conditions. In this way, models are invaluable in radar design and for providing the means to understand the range of clutter signals that a radar may encounter. In addition, the more complex models, such as those for amplitude statistics and Doppler spectra described in this book, can provide guidance to the radar designer on the efficacy of different signal-processing strategies (such as pulse-to-pulse integration, the effects of frequency agility, and Doppler processing), when attempting to detect targets and control false alarms.

1.6 Data sets

Good quality radar data sets are essential for the understanding of sea clutter and the development of empirical models. In recent years, some high quality data sets have been made available to the wider research community. An early example is the coherent X-band data set, collected by the McMaster IPIX radar [24]. This data set has been widely accessed and analysed. Some of the newer models reported here have made use of data collected by two groups. These include the Defence Science Technology Group (DSTG), Australia and the Council for Scientific and Industrial Research (CSIR), South Africa. These data sets are described in more detail here.

1.6.1 Ingara X-band data set

In many of the examples in this book, the X-band Ingara medium grazing angle sea-clutter data set will be used [25]. This data was collected by DSTG in 2004 and 2006 with the goal of characterising as wide a parameter space as possible. The airborne radar had a 200 MHz bandwidth (0.75 m range resolution) centred at 10.1 GHz and a nominal pulse repetition frequency (PRF) of 600 Hz. At a slant range of 3.4 km and with a two-way 3 dB (half-power) azimuth beamwidth of 1°, the azimuth resolution was approximately 63 m. The sea clutter was collected over 12 days with differing sea states ranging from calm to very rough. As shown in Figure 1.3, the radar platform flew in a circular spotlight collection mode with at least six full orbits around the same

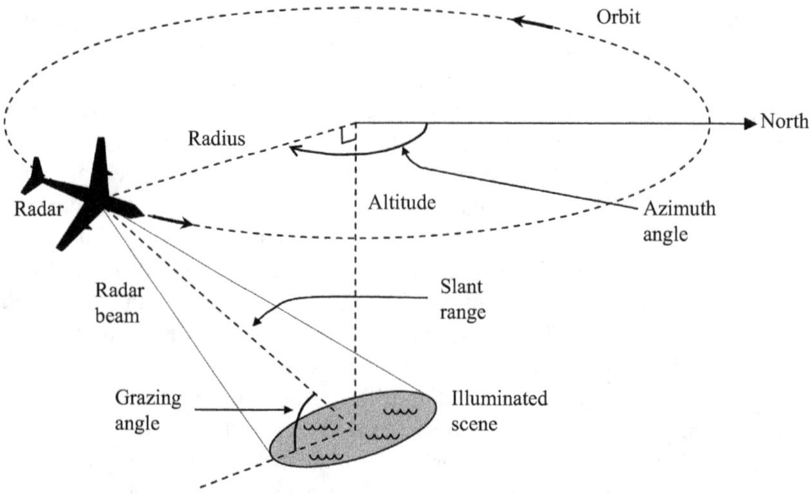

Figure 1.3 Ingara circular spotlight collection geometry [26]. © Commonwealth of Australia 2014

patch of ocean to cover a large portion of grazing angles between 15° and 45°. Both dual and fully polarised data were collected, with horizontal and vertical polarisations transmitted alternately in the latter case. Note that as the Ingara data was collected in a side looking collection geometry from a moving platform, the observed Doppler spectra are broader than would be observed from a stationary radar.

1.6.2 CSIR X-band data set

A very useful data set has been made available by CSIR in South Africa. It was collected by their Fynmeet radar [27] with a full description of the various data sets available in [28]. For the results reported in this book, the radar was vertically polarised with a frequency of 9 GHz, a PRF of 5 kHz, a bandwidth of 10 MHz (15 m range resolution) and an antenna azimuth beamwidth of 1.8°. The radar was situated on a clifftop overlooking the sea at a height of 67 m and data were typically collected at ranges between 3 and 7 km. The antenna pointing direction was fixed for the duration of each run and the data record for each range gate comprises complex time-series data gathered over intervals of 30–60 s.

1.7 Book outline

This book is designed to supplement the material in [1,5,29] with a summary of new work over the past 5–10 years. The first contribution in Chapter 2 is a study of the monostatic characteristics of sea clutter and the different modelling approaches.

The characteristics include the mean backscatter, amplitude statistics, sea spikes, Doppler spectra and texture correlation.

Bistatic radars have received considerable attention over recent years due to the extra degrees of freedom that can be used to extract information on targets and the covertness of the receive node. However, bistatic measurements are significantly more difficult than monostatic due to the increased complexity in collecting and then understanding the sea clutter statistics. In Chapter 3, the focus is on characterisation of sea clutter for both active and passive bistatic radars.

Chapter 4 then covers a number of parametric models that provide input values for the clutter models, allowing them to be used for simulation and performance prediction modelling. This includes new models for the medium grazing angle region that is motivated by the desire to use higher altitude airborne platforms for maritime surveillance. Radar data collected by these platforms have different statistical properties and a stronger clutter mean power than if it was collected from a traditional low flying radar platform. The models describe the expected characteristics of both monostatic and bistatic mean reflectivity, amplitude statistics, Doppler spectrum and the texture correlation.

Chapter 5 focuses on sea clutter simulation techniques. These can be used to stimulate radar processors during development and testing, generate realistic responses and displays in radar trainers and to evaluate radar detection algorithms. The simulation approaches vary in their complexity and cover both side looking and scanning radars. Chapter 6 describes how sea clutter models can be used for the prediction of radar performance which is an essential part of the design, development, assessment and marketing of radar systems. As many maritime radars are starting to exploit coherent processing techniques, both non-coherent and coherent detection approaches are considered, with an assessment of their end-to-end performance.

Chapter 7 then presents an overview of current approaches to target detection with detail on how the performance is affected by the sea clutter characteristics. These include both traditional non-coherent constant false alarm rate detection schemes and recent approaches that assume a specific amplitude distribution. The coherent approaches cover both conventional and adaptive 'whitening' approaches that are becoming popular in multichannel airborne radars. In Chapter 8, a number of newer techniques are presented that aim to overcome some of the problems encountered in more conventional detectors. These are based on the application of different signal-processing techniques to maritime target detection.

References

[1] K. D. Ward, R. J. A. Tough, and S. Watts, *Sea clutter: Scattering, the K-distribution and radar performance*, 2nd ed. The Institute of Engineering Technology, London, 2013.

[2] R. K. Raney, *Radar handbook*, 3rd ed. McGraw Hill, New York, 2008, ch. Space-based remote sensing radars.

[3] S. Iris, G. Kroupnik, D. D. Lisle, and M. Wierus, "RADARSAT constellation mission," *IEEE International Geoscience and Remote Sensing Symposium*, 2019, pp. 5749–5751.

[4] M. I. Skolnik, *Radar handbook*, 3rd ed. McGraw-Hill, New York, 2008.

[5] F. E. Nathanson, J. P. Reilly, and M. N. Cohen, *Radar design principles*, 2nd ed. McGraw-Hill, New York, 1991.

[6] D. O. Rice, "Reflections of electromagnetic waves from slightly rough surfaces," *Communications on Pure and Applied Maths*, vol. 4, pp. 361–378, 1951.

[7] W. H. Peake, "Theory of radar return from terrain," *IRE Convention Record*, 1959, pp. 27–41.

[8] J. W. Wright, "Backscattering from capillary waves with application to sea clutter," *IEEE Transactions on Antennas and Propagation*, vol. AP-14, no. 6, pp. 749–754, 1966.

[9] F. T. Ulaby, R. K. Moore, and A. K. Fung, *Microwave remote sensing: Active and passive, Volume II: Radar remote sensing and surface scattering and emission theory*. Addison-Wesley, Boston, MA, 1982.

[10] J. W. Wright, "A new model for sea clutter," *IEEE Transactions on Antennas and Propagation*, vol. 16, no. 2, pp. 217–223, 1968.

[11] N. W. Guinard and J. C. Daley, "An experimental study of a sea clutter model," *Proceedings of the IEEE*, vol. 58, pp. 543–550, 1970.

[12] G. R. Valenzuela, M. B. Laing, and J. C. Daley, "Ocean spectra for the high-frequency waves as determined from airborne radar measurements," *Journal of Marine Research*, vol. 29, pp. 69–84, 1971.

[13] F. G. Bass, I. M. Fuks, A. I. Kalmykov, I. E. Ostrovsky, and A. D. Rosenberg, "Very high frequency radiowave scattering by a distributed sea surface, part II: scattering from an actual sea surface," *IEEE Transactions on Antennas and Propagation*, vol. 16, no. 5, pp. 560–568, 1968.

[14] K. Hasselmann and M. Schieler, "Radar backscatter from the sea surface," *Symposium on Naval Hydrodynamics*, 1970, pp. 361–388.

[15] S. Angelliaume, V. Fabbro, G. Soriano, and C. A. Guerin, "The GO-SSA extended model for all-incidence sea clutter modelling," *IEEE Geoscience and Remote Sensing Symposium*, 2014, pp. 5017–5020.

[16] V. Kudryavtsev, D. Hauser, G. Caudal, and B. Chapron, "A semi-empirical model of the normalized radar cross-section of the sea surface, part I: The background model," *Journal of Geophysical Research*, vol. 108, pp. 2-1–2-24, 2003.

[17] V. Kudryavtsev, D. Hauser, G. Caudal, and B. Chapron, "A semi-empirical model of the normalized radar cross-section of the sea surface, part II: Radar modulation transfer function," *Journal of Geophysical Research*, vol. 108, pp. 3-1–3-16, 2003.

[18] E. A. Caponi, B. M. Lake, and H. C. Yuen, "Hydrodynamic effects in low-grazing angle backscattering from the ocean," *IEEE Transactions on Antennas and Propagation*, vol. 47, no. 2, pp. 354–363, 1999.

[19] J. T. Johnson, R. J. Burkholder, J. V. Toporkov, D. R. Lyzenga, and W. J. Plant, "A numerical study of the retrieval of sea surface height profiles from low grazing angle radar data," *IEEE Transactions on Geoscience and Remote Sensing*, vol. 47, no. 6, pp. 1641–1650, 2009.

[20] J. Houssay, N. Pinel, Y.-H. Hellouvry, *et al.*, "A physical radar simulation tool of the sea clutter," *International Radar Conference*, 2014, pp. 1–6.

[21] J. Louvigne and C. Cochin, "Maritime scene design tools for realistic radar performance predictions," *IEEE International Geoscience and Remote Sensing Symposium*, 2015, pp. 3123–3126.

[22] K. D. Ward, "Compound representation of high resolution sea clutter," *Electronic Letters*, vol. 17, no. 16, pp. 561–563, 1981.

[23] M. I. Skolnik, *Introduction to radar systems*, 3rd ed. McGraw-Hill, New York, 2001.

[24] McMaster University, "McMaster IPIX radar data." [Online]. Available: http://soma.ece.mcmaster.ca/ipix/index.html.

[25] L. Rosenberg and S. Watts, "High grazing angle sea-clutter literature review," Defence Science Technology Organisation, General Document DSTO-GD-0736, 2013.

[26] L. Rosenberg, "Characterisation of high grazing angle X-band sea-clutter Doppler spectra," *IEEE Transactions on Aerospace and Electronic Systems*, vol. 50, no. 1, pp. 406–417, 1 2014.

[27] P. L. Herselman and C. J. Baker, "Analysis of calibrated sea clutter and boat reflectivity data at C- and X-band in South African coastal waters," *IET International Radar Conference*, 2007, pp. 1–5.

[28] P. L. R. Herselman, "CSIR Fynmeet sea clutter measurement trial: Datasets," 2006. [Online]. Available: https://researchspace.csir.co.za/dspace/handle/10204/1847?show=full.

[29] M. W. Long, *Radar reflectivity of land and sea*, 3rd ed. Artech House, Northwood, 2001.

Chapter 2
Monostatic sea clutter

2.1 Introduction

This chapter reviews the main characteristics of monostatic sea clutter and their representation in mathematical models. The main characteristics of the backscattered returns which need to be considered to assess radar performance include

- the normalised mean backscatter;
- the amplitude distribution;
- the Doppler spectrum;
- the spatial and temporal variation;
- clutter spikes.

These characteristics will be dependent on the viewing geometry of the radar and the prevailing conditions, including the wind speed and the sea wave and swell heights. They will also be affected by radar parameters such as the frequency, bandwidth and polarisation. The association between parameter values of the models and the prevailing conditions, viewing geometry and radar parameters are discussed in Chapter 4. Finally, it must be remembered that any clutter returns will be observed in the radar with added thermal noise, which may dominate if the clutter amplitude is sufficiently low.

The received radar signal comprises a combination of clutter, noise and target returns with the predicted mean strength of the sea clutter for a particular set of conditions estimated using the radar range equation (see Chapter 6) with the clutter radar cross section modelled using the normalised mean backscatter. Section 2.2 describes how the mean backscatter is characterised.

The clutter from an illuminated patch of sea will also fluctuate in amplitude when observed from different directions or over time, as the sea surface moves. This is due to many individual scatterers within the patch combining randomly and can be represented by an amplitude distribution or probability density function (PDF). With coarse range resolution, a reasonable model for the PDF of the in-phase or quadrature sea clutter is the Gaussian distribution [1], with adjacent clutter cells (areas resolved by the radar) having the same amplitude distribution with fixed overall mean intensity. The Gaussian model is a result of the central limit theorem applied to the resulting scattered field from a clutter patch. However, as the range resolution becomes finer, the radar may start to resolve the sea wave and swell patterns. The local variations in sea surface slope, combined with shadowing from waves and changes in surface roughness due to the wind and breaking waves, give rise to variations in the mean intensity of the backscatter from one patch to the next. The scattering within an individual patch will still be Gaussian, but the fluctuation in mean intensity gives rise to non-Gaussian statistics when returns over many patches are considered. The average mean intensity level over a local area will still be determined by the normalised backscatter coefficient. Relative to this average mean level, the clutter returns have a higher probability of having a large magnitude compared to returns having Gaussian statistics. This has led to the development of PDF models for 'spiky' clutter, characterised by a longer distribution tail. One of the most common formulations for the sea clutter PDF is the compound Gaussian model. This model comprises a speckle component with Gaussian statistics, for modelling the small ripples on top of larger waves [1], with its mean amplitude modulated by a slowly varying texture, which represents the varying local mean level due to the wave and swell structure. A summary of PDF models for sea clutter is given in Section 2.3.

The amplitude of the returns in a clutter patch will fluctuate over time due to the movement of the sea waves and swell relative to the radar. This variation may be approximately characterised by its short-term autocorrelation function, although in practice the returns are non-stationary with the spectrum evolving over range and time. Such variation can be observed by the radar in slow-time (pulse-to-pulse) and is usually characterised in the frequency domain, in terms of the power spectral density or Doppler spectrum. Section 2.5 examines the mean characteristics of the Doppler spectra with a model for characterising their time-varying and range-varying nature.

In addition to the variation in local surface slope and roughness that gives rise to non-Gaussian amplitude statistics, additional scattering events are also observed that affect the amplitude statistics and Doppler spectra of the clutter. These events are called sea spikes and can be either short-lived, lasting for tens of milliseconds, or persistent over one or more seconds. These discrete events often result in signals to that do not conform to the standard models for PDF and Doppler spectra and may appear target-like to the radar processing, making the control of false alarms more challenging. Studying the characteristics of sea spikes is therefore essential to understand how they can be distinguished from targets. A detailed description of sea spikes and methods for modelling them is covered in Section 2.4.

In order to characterise the statistics of the interference, detection schemes need to consider multiple samples collected over range and/or long time periods. For this reason, understanding the correlation from sample to sample along both of these domains is also important. The local clutter mean level (texture) may exhibit a structured autocorrelation coefficient as a function of range (often dominated by a periodic wave or swell pattern) or over time periods on the order of seconds. Section 2.6 examines this characteristic.

2.2 Mean backscatter

The mean backscatter depends on the radar frequency and polarisation, the collection geometry, the ocean surface roughness and the viewing direction relative to the wind, wave and swell directions. It is characterised by the normalised backscatter coefficient, which is the mean radar cross section per unit area of sea surface illuminated by the radar. This is usually given as a dimensionless value, σ^0, expressed in dBm^2/m^2 (i.e. the radar cross section in m^2 per unit area in m^2, expressed in dB).

As discussed in Chapter 4, many measurements have been made of the mean backscatter over a wide range of conditions and radar parameters. The key influences on mean backscatter for a given radar frequency and polarisation will be the local grazing angle (defined as the angle between the centre of the radar beam pointing direction and the tangent to the plane of the local average sea surface), the surface roughness and the look direction relative to the wind and waves.

Although sea state is a standard measure of sea surface conditions, it may not always be the best parameter to describe clutter backscatter, which may be more strongly influenced by the local wind speed and direction. For a given wind speed, the waves take time to develop in height, but the local surface may still be very rough. There are also more breaking waves as the wave height develops and these may have high backscatter. The sea state is defined by the wave height, and estimates of sea state may also be confused by the presence of a sea swell. A large swell with no local wind will have low backscatter, while a strong local wind with poorly developed waves can result in higher backscatter.

Propagation effects will also affect the mean backscatter observed by a radar. For example, a surface duct may effectively modify the apparent grazing angle and result in a larger backscatter than might otherwise be expected. The impact of propagation is often captured by expressing the mean backscatter as $\sigma^0 F_p^4$, where F_p is the one-way propagation factor in terms of voltage. It represents the variation due to anomalous propagation (ducting) and the impact of multipath scattering. While this latter effect can significantly alter the mean backscatter at low grazing angles and for low sea states, it is less significant as the grazing angle and sea state increases. Also, it is very difficult to distinguish the impact of propagation as the radar measures the combined $\sigma^0 F_p^4$. Further detail on anomalous propagation is given in Section 2.2.1 and multipath scattering in Section 2.2.2.

Good starting points for values of the sea clutter mean backscatter are the tables given by Nathanson [2]. These cover a wide range of sea states, grazing angles and radar frequencies but are averaged over all wind directions. The general trends for the mean backscatter as it varies with grazing angle are illustrated in Figure 2.1. This shows typical results for σ^0 at X-band, for a wind speed of 7.7 m/s (15 kts). It can be seen that the horizontally polarised transmit and receive (HH) mean backscatter is generally lower than vertically polarised transmit and receive (VV) values over a range of grazing angles up to 50°. For the first 10°, the low grazing angle or interference region has backscatter that is strongly affected by multipath scattering and shadowing from waves. From 10° to 45°, the mean backscatter is approximately linear and is

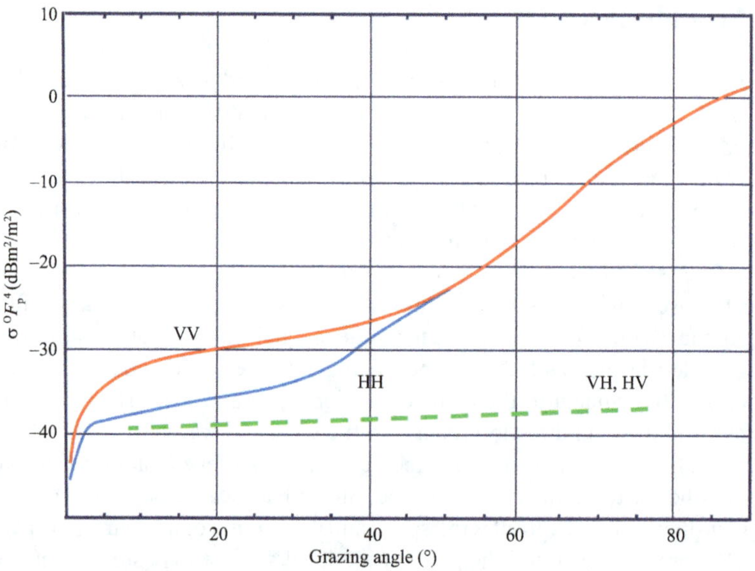

Figure 2.1 Typical variation of sea clutter mean backscatter with grazing angle at X-band with a wind speed of 7.7 m/s. Adapted from [1]

known as the medium grazing angle or 'plateau' region. Also shown in Figure 2.1 is the cross-polarised, VH or HV mean backscatter, which is significantly lower than either of the vertical or horizontal values and does not show much variation with grazing angle over the plateau region.

The variation with the wind direction is shown in Figure 2.2 for both low and medium grazing angles at X-band. At low grazing angles ($<10°$), there is only a single peak in the up-wind direction where the mean backscatter is largest, while at medium grazing angles, there is a second peak in the down-wind direction. This implies that target detection performance will be worst in the up-wind direction and best in the down-wind and/or cross wind directions depending on the grazing angle.

Figure 2.3 then shows the variation with wind speed for vertically polarised returns. A power law model is often used to model the approximate relationship between wind speed and sea state for a fully developed sea, with $U \approx 3.18 S_{ss}^{0.8}$, where U is the wind speed in m/s and S_{ss} is the sea state. Finally, the variation of backscatter with carrier frequency is shown in Figure 2.4, using a model developed by NRL [3] based on Nathanson's tables. It can be seen that at low grazing angles, there is strong dependency on radar frequency, with lower values of backscatter at lower radar frequencies. For 10° grazing angle and above, the dependency on radar frequency is much reduced.

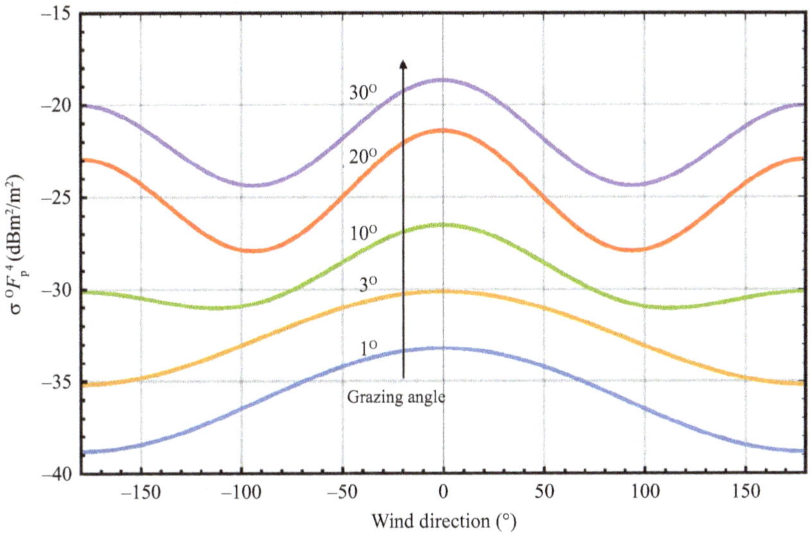

Figure 2.2 Typical variation of sea clutter mean backscatter with wind direction (0° is up-wind) at X-band, vertical polarisation and sea state 4

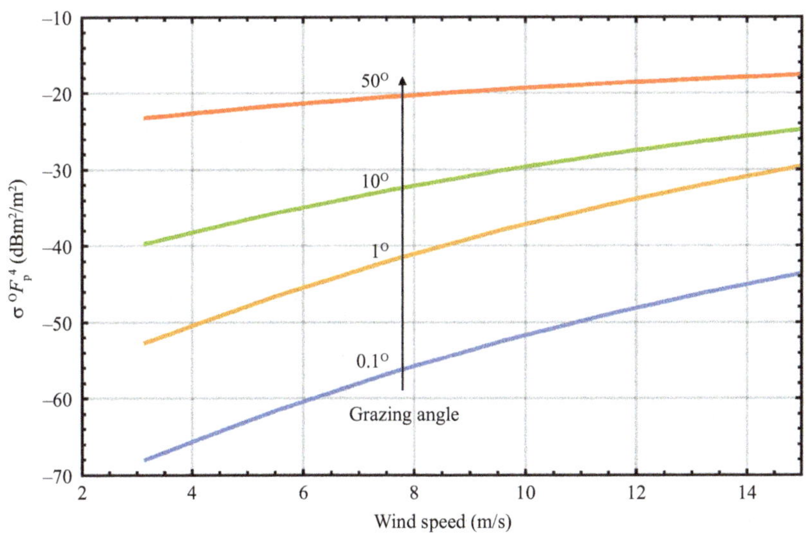

Figure 2.3 Typical variation of sea clutter mean backscatter with wind speed at X-band, vertical polarisation and different grazing angles

A number of models for the mean backscatter have been developed, as described in Chapter 4. These are mainly empirical and are based on large amounts of recorded data collected over a wide range of conditions. As such they can only represent average values for a given set of conditions. Nevertheless, the trends discussed here

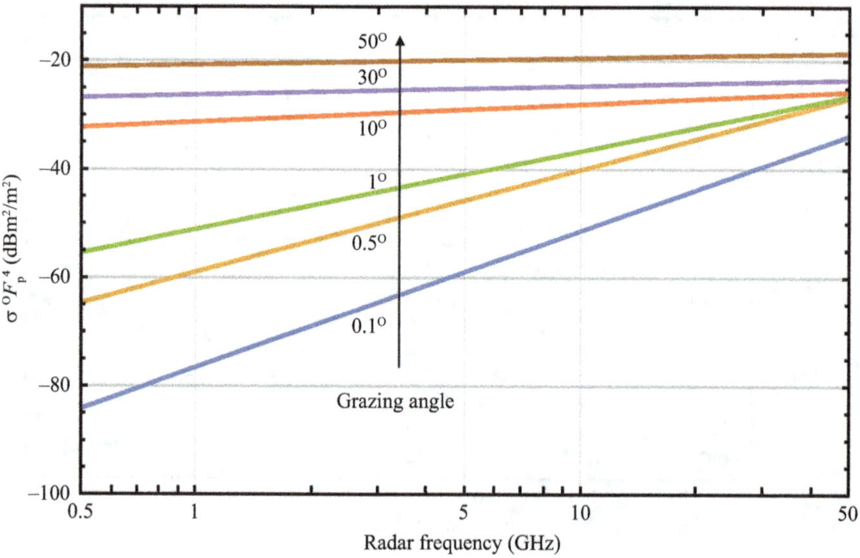

Figure 2.4 Typical variation of sea clutter mean backscatter with radar frequency and different grazing angles

match closely with those produced by electromagnetic (EM) modelling with physical modelling of the sea surface. The geometric optics-small slope approximation (GO-SSA) models [4–6] are recent examples and are further described in Section 2.2.3.

2.2.1 Anomalous propagation

In free space, EM waves will travel in a straight line. However, the refractive index of air varies with barometric pressure, water vapour pressure and temperature. Its value slowly decreases with altitude at an approximately linear rate and causes a propagating EM wave to bend downwards. This can cause errors in the apparent direction of arrival in elevation, errors in range measurement and can sometimes result in anomalous propagation with ducts that cause unexpected clutter levels and target detection ranges. Good introductions to this complex topic can be found in [2,7–9].

Clutter characteristics for otherwise constant conditions depend on the local grazing angle of the incident radar signal. In order to estimate the effective grazing angle over a curved Earth, the effects of refraction must be taken into account. If the variation of refractive index with height is linear and constant, then EM waves can be modelled as travelling in a straight line over a curved Earth. This is achieved using an effective value of the Earth radius, $r_{\text{eff}} = k_e r_e$, where r_e is the true Earth radius. A value of $k_e = 4/3$ can be used for the case of an average variation of refractive index with height (known as a standard atmosphere). In practice, values of k_e will vary with values measured between 1.24 and 1.90 [7]. Using an effective Earth radius, the local grazing angle, ϕ, of a radar signal on the sea surface at a range R from a radar at an altitude of z_p is given by

$$\phi \equiv \sin^{-1}\left(\frac{z_p}{R} + \frac{z_p^2}{2r_{eff}} - \frac{R}{2r_{eff}}\right) \tag{2.1}$$

where the radar horizon is defined when $\phi = 0$, i.e. $R \equiv \sqrt{2r_{eff}z_p}$. For a target at height z_t, the radar horizon is defined by $R \equiv \sqrt{2r_{eff}}\left(\sqrt{z_p} + \sqrt{z_t}\right)$. In a non-standard atmosphere, both the radar horizon and the grazing angles at different ranges will differ from standard modelling assumptions and there will be differences in clutter reflectivity from those predicted in a standard atmosphere. Extreme gradients of the refractive index can give rise to super-refraction with the potential for enhanced detection ranges and if the gradient becomes positive, sub-refraction will be observed with decreased detection ranges [9].

A common form of anomalous propagation over the sea is due to evaporation ducts. If the air at the sea surface is saturated with water vapour, there will be a rapid reduction in refractivity with height until normal atmospheric conditions (and so normal variation of refractive index with height) are reached. This has the effect of forming a duct that may have a height varying from about 5 m in the North Atlantic, up to 16 m in the Tropics. If a radar antenna is located in the duct, radar signals launched along the duct may become trapped. This can result in signals from the sea surface (targets and clutter) being received from beyond the expected radar horizon. In addition, the clutter reflectivity may not drop off as quickly as expected at low grazing angles (i.e. in a standard atmosphere). For an airborne radar at a height well above the duct, it is less likely that its signals will become trapped in the duct and clutter and target signals may be less affected. If the air at a few hundred metres altitude is very warm and dry compared with air at the surface, then elevated ducts may occur due to the mixing of air streams with different temperatures and pressures. Enhanced detection ranges may then be observed if the target and radar are both within the duct. Surface targets will not be within an elevated duct, but when the radar is above the duct, it is also possible for reduced coverage below the duct, compared with what may be expected in a standard atmosphere.

Knowledge of atmospheric conditions is required to understand the clutter returns of ground- and ship-based radars. For airborne radars, it may also be useful to know if elevation ducts are expected, especially for low flying aircraft. Under some circumstances, it may also be possible to infer the propagation conditions from the observed clutter returns. This technique is known as 'refractivity from clutter' [10,11].

2.2.2 Multipath scattering

Multipath scattering occurs when the signal received by the radar is reflected from the sea surface as well as directly from the target. In a rough sea, multipath scattering may also occur between adjacent parts of the sea surface, and affect the clutter return. The signals received along the direct and reflected paths will have different amplitude and phase, as well as possible differences in polarisation. When these signals combine at the radar receiver, multipath interference is observed. Simple models may be used to assess multipath propagation for point targets above a slightly rough sea surface. If the target is distributed in height or range and/or if the reflecting sea surface is changing due to waves or swell, the propagation environment is much more complex to model. A good discussion on methods for modelling multipath is given in [9] for both point and distributed targets. In general, the 'multipath factor' comprises three regions. These include the interference region at short ranges, the transition region and a diffraction region at the radar horizon. The interference region consists of the sum of the direct and reflected paths which results in both constructive and destructive interference, while the diffraction occurs when the EM signal diffracts around an object and re-radiates energy, potentially beyond the radar horizon.

There are a number of different approaches to calculating the propagation factor, including solving the parabolic equation model, waveguide models and geometric optics. The Advanced Refractive Effects Prediction System (AREPS) software [12] offers a hybrid solution based on a specified region and is used to generate the following examples. Consider an airborne radar flying at 1,000 ft (305 m) looking down at the sea surface with a 0.2° elevation angle. The one-way propagation factor for a point target as a function of height and range is shown in Figure 2.5. A slice at 5 m is then shown in Figure 2.6 for both polarisations. This shows the interference region and the transition to the diffraction region, which begins at the radar horizon or 72 km. For this example, there is little difference between the polarisations.

It is interesting to examine how the propagation factor varies with grazing angle and sea state. Consider point targets observed with local grazing angles of 1°, 5° and 10° in sea states 1 and 3 (wind speeds of 3.27 and 7.31 m/s), respectively. Figure 2.7 shows the one-way propagation factor as a function of height for both horizontal and

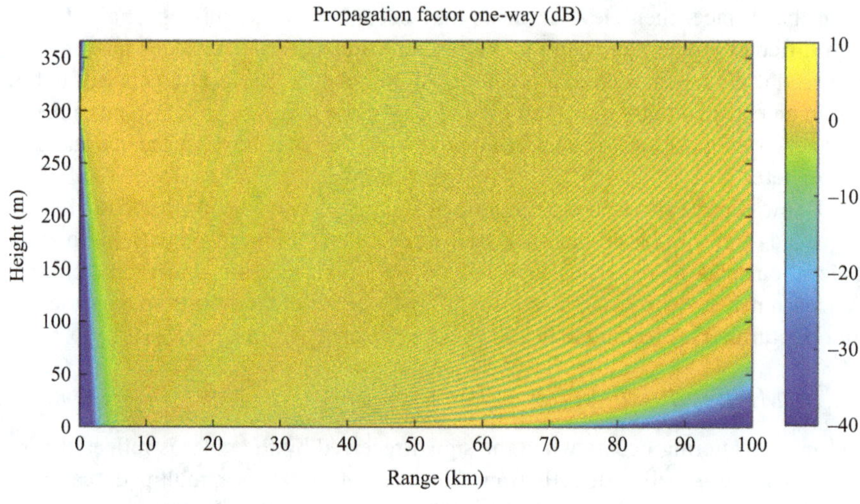

Figure 2.5 Example one-way propagation factor (dB)

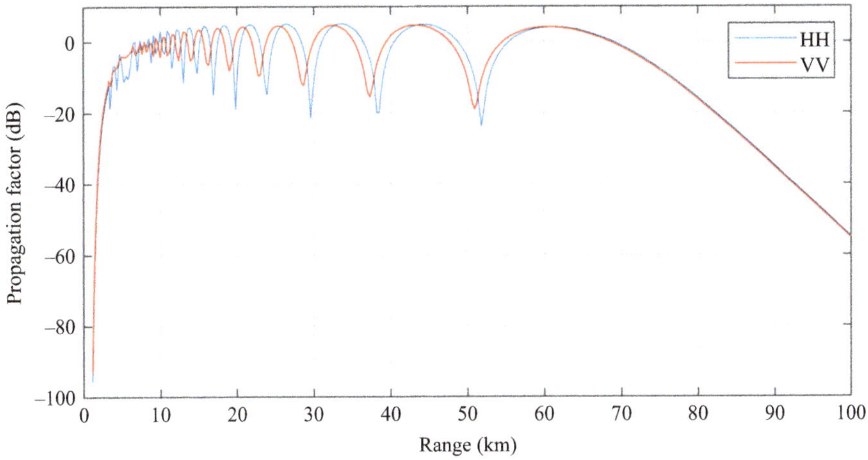

Figure 2.6 Example one-way propagation factor, slice at 5 m (dB)

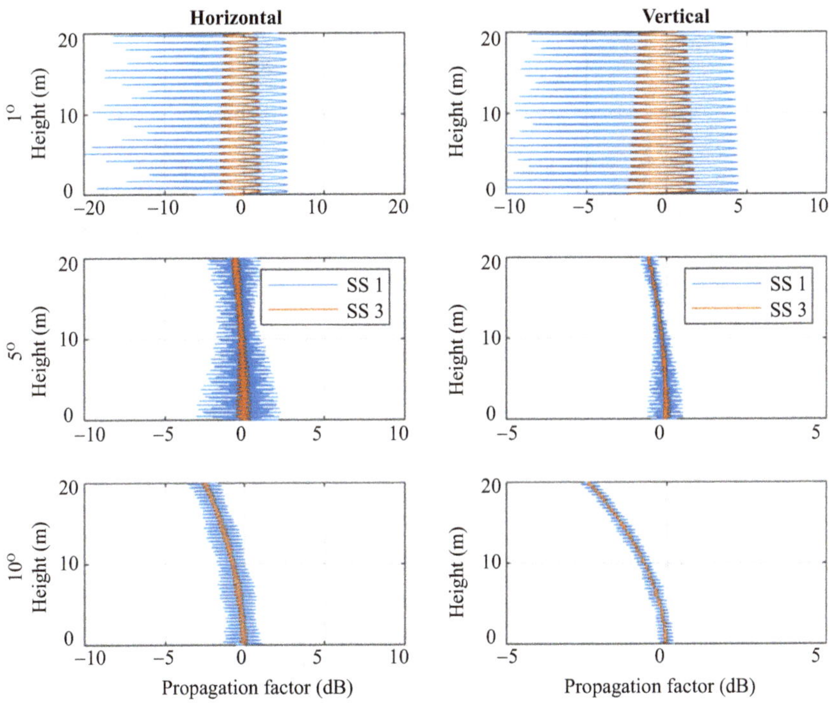

Figure 2.7 Example one-way propagation factor as a function of height (dB)

vertical polarisations. As the grazing angle increases, the fluctuation increases, with little enhancement for 5° and 10° grazing angles. Also, as the sea state increases, the fluctuation reduces. To determine the effective multipath factor, the propagation factor can be non-coherently summed from the sea surface up to the target height.

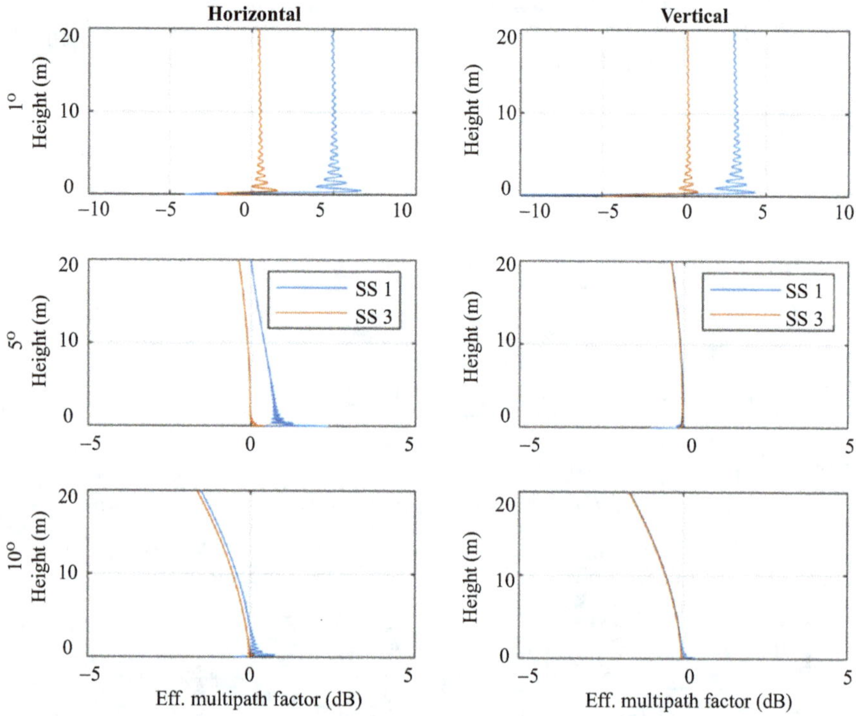

Figure 2.8 Effective multipath factor as a function of height (dB)

Figure 2.8 shows the results for these examples assuming a target that is uniformly distributed in height. It is now clear that multipath enhancement is only observed with sea state 1, horizontal polarisation and 1° grazing angle.

2.2.3 GO-SSA mean backscatter models

The GO-SSA mean backscatter models [4–6] are EM models based on physical modelling of the sea surface. Compared to empirical modelling, physical models are more universal in that they have a well-defined domain of validity with respect to the sea state and scattering environment [13]. However, issues remain in the physical modelling of backscattered signals, especially when it comes to strong wind conditions where complex phenomena such as breaking waves are present.

The original GO-SSA model [4] is based on a spectral description of the sea surface and is an improvement of the classical two-scale model. Both GO-SSA and the two-scale model represent the sea as the superposition of two independent processes describing the small- and large-scale components, respectively. Where the conventional two-scale model is a hybrid between the geometrical optics (GO) and the first-order small perturbation models, GO-SSA instead combines the GO with the first-order small slope approximation (SSA) using a function that ensures a smooth transition between components. The expression for the mean backscatter is given by

$$\sigma^0_{\text{GO-SSA}} = \sigma_{\text{GO}} \exp\left(-|\mathbf{K}_{\text{bs}} - \mathbf{K}_{\text{inc}}|^2 \sigma_S^2\right)\left(1 - \exp\left(-((\mathbf{K}_{\text{bs}} - \mathbf{K}_{\text{inc}})\hat{\mathbf{z}})^2 \sigma_L^2\right)\right) + \sigma_{\text{SSA}} P_{\text{wave}} \tag{2.2}$$

where σ_S and σ_L are the root mean square wave height for the small- and large-scale wave components, respectively, $\mathbf{K}_{\text{inc}}$ and $\mathbf{K}_{\text{bs}}$ are wave vectors of the incident and backscattered waves, $\hat{\mathbf{z}}$ is a unit vector in the vertical direction and P_{wave} is the PDF of the large wave slopes.

To take into account the contribution of breaking waves, Spiga [5] added an isotropic correction term to the GO-SSA model, based on the work by Kudryavtsev *et al.* [14,15]. This empirical term allows a significant correction of the model at low grazing angles and hence the GO-SSA breaking wave model is given by

$$\sigma^0_{\text{GO-SSA,wave}} = \sigma^0_{\text{GO-SSA}}(1 - Q_{\text{wave}}) + \sigma_{\text{wave}} Q_{\text{wave}} \tag{2.3}$$

where Q_{wave} represents the fraction of the sea surface covered by breaking waves and σ_{wave} is the RCS of a breaking wave. The main innovation of GO-SSA is its robustness with respect to the somewhat arbitrary cut-off frequency that defines the small and large scales. However, there are two key phenomena that are not taken into account in this formulation. These include multipath scattering which could have a strong effect at low grazing angles and the variation of reflectivity with the wind direction. The GO-SSA extended model [6] was developed to account for these two effects.

To include the effect of multipath scattering, a common approach is to scale the mean backscatter with an empirical multipath effect term, $A_{\text{mp}}(\phi)$ [16,17]. Figure 2.9 shows an example of this term at X-band, where the biggest impact is observed for grazing angles below 2° with a sea state 2 and below 0.3° for sea state 4.

The behaviour of sea clutter in terms of azimuth angle has been studied by various authors [18,19]. While the maximum reflectivity is always observed in the up-wind direction, at low grazing angles, the minima are observed in the down-wind direction and for medium grazing angles, the minima are observed in the cross-wind directions [20]. To include this effect, a wind aspect factor can be included as a scaling function for the GO-SSA model:

$$W_{\text{af}}(\theta, \phi) = a_{\text{GO-SSA}}(\phi) + b_{\text{GO-SSA}}(\phi)\cos\theta + c_{\text{GO-SSA}}(\phi)\cos(2\theta) \tag{2.4}$$

where θ is the azimuth angle relative to up-wind and $a_{\text{GO-SSA}}(\cdot)$, $b_{\text{GO-SSA}}(\cdot)$ and $c_{\text{GO-SSA}}(\cdot)$ are polynomial functions of the grazing angle, ϕ. For the GO-SSA extended model, these were determined using sea clutter measurements from the Defence Science and Technology Group (DSTG) X-band Ingara sea clutter data set (see Chapter 1). Figure 2.10 shows an illustration of this term for the HH polarisation and three different grazing angles, 15°, 30° and 45°.

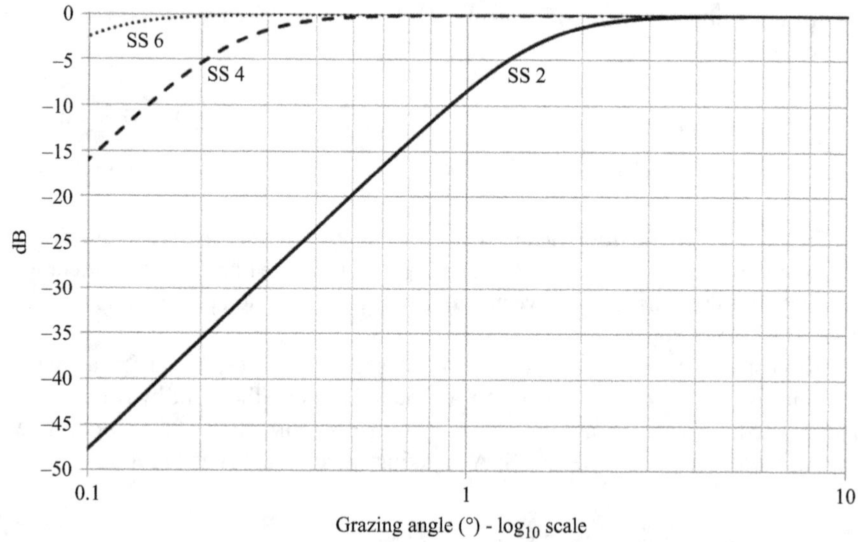

Figure 2.9 GO-SSA multipath term for Douglas sea states: (—) 2, (- -) 4, (· · ·) 6. Adapted from [6]. © IEEE 2014

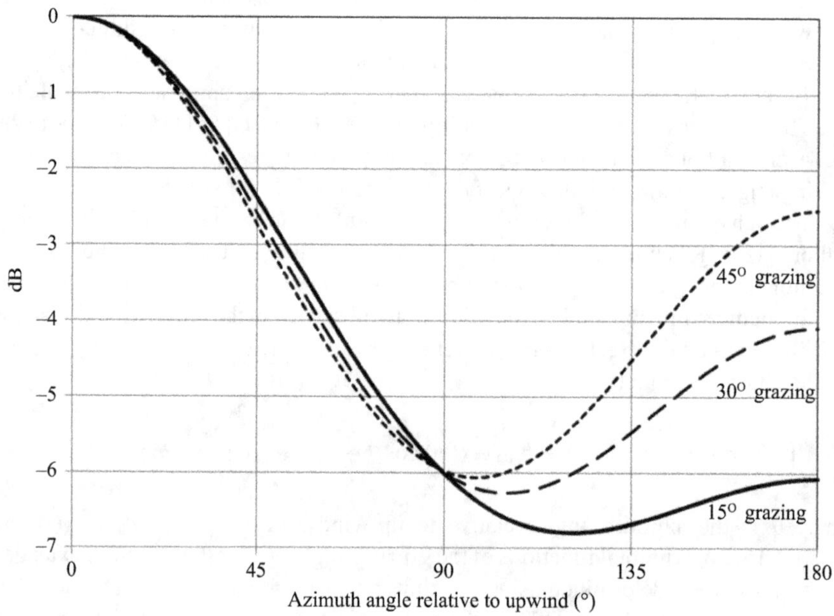

Figure 2.10 GO-SSA extended wind aspect factor at grazing angles: (—) 15°, (- -), 30°, (· · ·) 45°. Adapted from [6]. © IEEE 2014

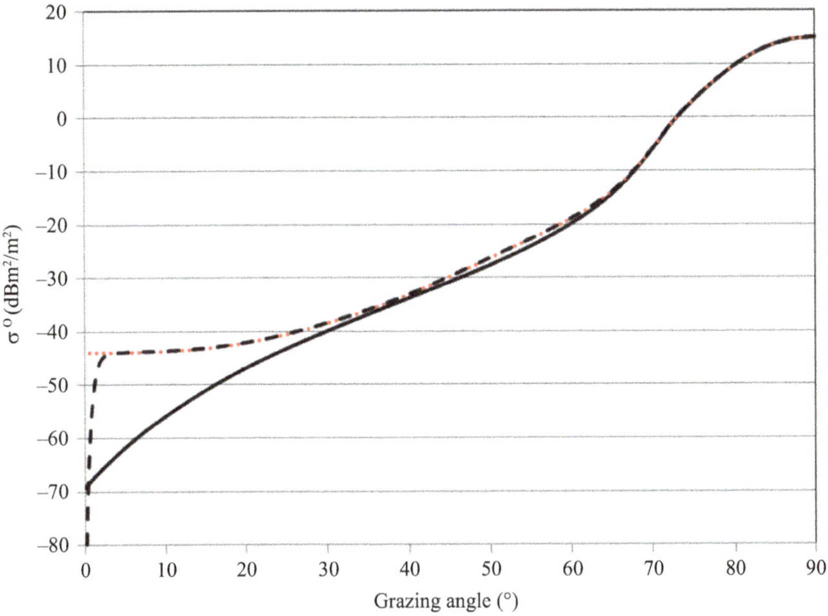

Figure 2.11 Reflectivity at X-Band, HH polarisation and sea state 2. (—) GO-SSA, (· · ·) GO-SSA breaking wave, (- -) GO-SSA extended. Adapted from [6]. © IEEE 2014

The final expression for the extended GO-SSA mean backscatter model is then given by the GO-SSA breaking wave model with the two correction terms [6]:

$$\sigma^0 = \sigma^0_{\text{GO-SSA,wave}} A_{\text{mp}}(\phi) W_{\text{af}}(\theta, \phi). \tag{2.5}$$

Figure 2.11 shows an example of the X-band horizontally polarised reflectivity for a Douglas sea state 2 using the GO-SSA, GO-SSA breaking wave and GO-SSA extended models. The results show how breaking waves cause a significant increase in the level of reflectivity even at medium grazing angles. For example, at a 10° grazing angle, the reflectivity estimated by the GO-SSA breaking wave model is more than 10 dB greater than that estimated by GO-SSA. The multipath term also affects the level of reflectivity most strongly at low grazing angles, i.e. below 2° at X-band for a sea state 2. Good agreement between the GO-SSA extended model and airborne measurement of reflectivity made at X-band was reported by Johnsen [21], for grazing angles from 5° to 55° and for wind speeds of 2 and 12 m/s.

2.3 Amplitude distributions

Models for the sea clutter amplitude distribution are usually developed empirically from measurements of real data as it is not currently possible to accurately predict the

PDF using physical models of the sea surface and EM scattering analysis. There has been a long development of PDF models used to fit data from both real aperture radar and synthetic aperture radars [22]. Early models include the Rayleigh, log–normal (LN) and Weibull, with the latter two distributions used to model the longer tails observed from non-Gaussian backscatter [23,24]. The most common model used in the literature is the K-distribution which was first introduced by Jakeman and Pusey [25] and applied to sea clutter by Ward [26]. There have since been many other models used to characterise 'spiky' sea clutter, such as the Pareto, KA and others described in this chapter.

The simplest model for backscatter from the sea assumes multiple scatterers uniformly distributed over the sea surface with the returns having a complex amplitude, x and modelled as the coherent sum of a number, N_{sc}, of individual scatterers with a random magnitude, $\tilde{\sigma}$, and phase, $\tilde{\phi}$:

$$x = \sum_{n=1}^{N_{sc}} \sqrt{\tilde{\sigma}_n} \exp\left(j\tilde{\phi}_n\right). \tag{2.6}$$

If the number of scatterers is sufficiently high, the central limit theorem applies and both the in-phase and quadrature components of the backscatter will have Gaussian statistics. The signal processing in most radars utilises either the magnitude (i.e. following a linear detector) or power (i.e. following a square law detector) of the radar returns. For the former, the magnitude, $r = |x|$, becomes Rayleigh distributed

$$P_r(r) = \frac{2r}{\tau} \exp\left(-\frac{r^2}{\tau}\right) \tag{2.7}$$

with mean, $\langle r \rangle = \sqrt{\pi\tau}/2$ and mean square, $\langle r^2 \rangle = \tau$. For the power distribution, $z = r^2$ and the speckle is represented by a negative exponential distribution,

$$P_z(z) = \frac{1}{\tau} \exp\left(-\frac{z}{\tau}\right) \tag{2.8}$$

with a mean $\langle z \rangle = \tau$ that is directly related to the mean backscatter discussed in Section 2.2. This model is only suitable for radars with a coarse spatial resolution that is greater than the sea wavelength. As the spatial resolution becomes finer, the radar is able to start resolving the structure of the sea surface and there is a higher probability of large amplitude returns (relative to the mean) being observed. This is due to variations in clutter intensity, τ, produced by modulations on the sea surface. Jakeman and Pusey [25] considered a modification to (2.6), where the number of scatterers is considered to be a random variable. The final magnitude distribution can then take different forms depending on the distribution of N_{sc}. For example, if N_{sc} has a negative binomial distribution, the magnitude, r, has a K-distribution.

From the statistics literature, the complex elliptical symmetric distributions [27,28] were proposed as means to represent a large number of distributions, including the complex normal, complex-t, generalised Gaussian and the compound Gaussian models. The latter member of this family has become a popular and widely used framework for developing clutter PDF models. It was originally proposed for use in sea clutter by Ward [26] and comprises a temporal or fast varying component known as speckle which relates to the Bragg scattering from wind-driven capillary waves (see Chapter 1). This component has Gaussian statistics with its mean power, τ, varying over range and slowly over time. This varying local intensity is called the 'texture' and captures the effect of the resolved sea surface structure due to waves or swell.

For the first-order compound Gaussian model, the texture is defined as the speckle mean power and treated as a random variable with a distribution $P_\tau(\tau)$. The overall amplitude of the clutter returns is given by $y = \sqrt{\tau}x$. This separate treatment of the speckle and texture components allows them to be modelled with different spatial and temporal correlations. For example, the pulse-to-pulse fluctuation of the speckle component in a given clutter cell can be modelled by its autocorrelation function or spectrum, with the texture assumed to be constant over the short term. The modulation of mean intensity over range is characterised by a spatial correlation of the texture, while the speckle component will decorrelate over time (typically over a few milliseconds) in a manner characterised by its spectrum or autocorrelation function. Because the multiple scatterers in a clutter patch are more or less uniformly distributed over range, it will also be possible to decorrelate the speckle component using pulse-to-pulse frequency agility. In this case, the frequency steps must at least equal the bandwidth of the radar pulses to achieve total decorrelation of the speckle. However, because there is an underlying variation of texture from range cell to range cell, the scatterers cannot be completely uniformly distributed in range and this prevents complete decorrelation of the returns [29].

Pulse-to-pulse frequency agility does not affect the correlation of the texture. This is highlighted in Figure 2.12 which shows range/time intensity plots of sea returns from an X-band radar situated on a cliff-top with a range resolution of 4.2 m and a pulse repetition frequency (PRF) of 1 kHz. The top plot shows 125 successive returns with the radar transmitting on a fixed frequency. The structure over range is clearly visible and the returns at a given range are decorrelating over a period of about 10 ms (10 pulse intervals). In the lower plot, the radar is transmitting with a repeating frequency agile sequence of 10 pulses, each spaced by 50 MHz (i.e. over a 500 MHz bandwidth in total). It can be seen that the spatial structure over range is still present, but now the speckle component appears to be decorrelating from pulse-to-pulse.

Often the radar detection processing may include non-coherent integration over N successive pulses. The effect of this integration will depend on the pulse-to-pulse correlation of the clutter speckle and texture. If pulse-to-pulse frequency agility or a sufficiently large time interval between pulses is used, the speckle component may be assumed to be independent from pulse-to-pulse, while the texture remains constant. Consider the sum of N-independent exponential random variables with a fixed mean level, τ,

$$Z = \sum_{n=1}^{N} z_n. \tag{2.9}$$

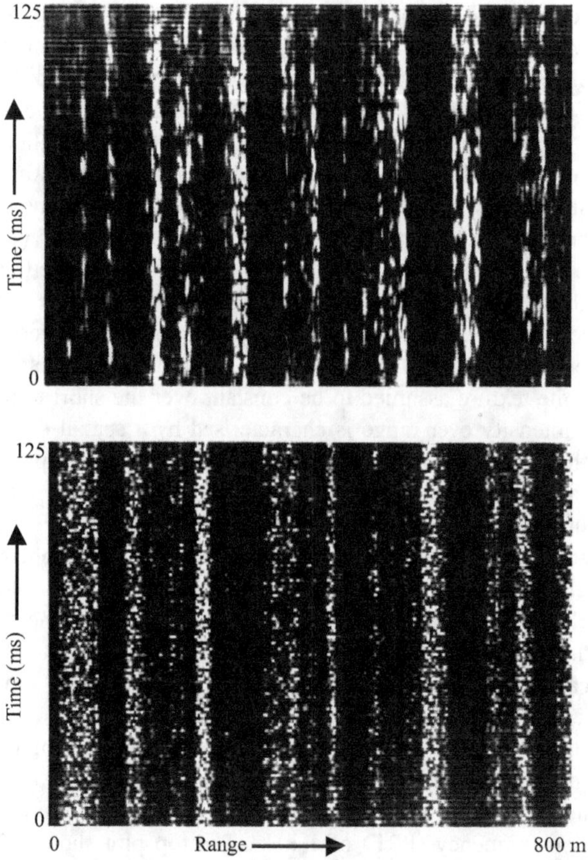

Figure 2.12 Range/time intensity plots of sea clutter. Top plot: fixed frequency; bottom plot: with pulse-to-pulse frequency agility. The data was collected at X-band with a 1 kHz PRF and range resolution of 4.2 m [1]

The PDF following the non-coherent integration of N-independent speckle samples is called a multi-look distribution, and the exponential PDF of the speckle given in (2.8) is replaced by a gamma distribution, so that

$$P_{Z|\tau}(Z|\tau) = \frac{Z^{N-1}}{\tau^N \Gamma(N)} \exp\left(-\frac{Z}{\tau}\right). \tag{2.10}$$

The compound Gaussian model of the overall intensity, Z, following the integration of N pulses with independent speckle, is found by averaging the speckle PDF, $P(Z|\tau)$, given in (2.10) over all possible values of τ. If the texture has a PDF, $P_\tau(\tau)$, then the compound Gaussian PDF is given by

$$P_Z(Z) = \int_0^\infty P_{Z|\tau}(Z|\tau)P_\tau(\tau)d\tau. \tag{2.11}$$

One common extension of the compound model is the inclusion of the noise mean power, p_n [30]. Since the noise has the same Gaussian PDF as the speckle, the mean power of the speckle-plus-noise is $\tau + p_\mathrm{n}$, with its PDF given by

$$P_{Z|\tau}(Z|\tau, p_\mathrm{n}) = \frac{Z^{N-1}}{(\tau + p_\mathrm{n})^N \Gamma(N)} \exp\left(-\frac{Z}{\tau + p_\mathrm{n}}\right). \tag{2.12}$$

The overall PDF of the clutter-plus-noise intensity is again achieved by averaging over all possible values of τ, as in (2.11). When integrating returns over successive pulses, it is important to remember that although the noise samples have the same PDF as the clutter speckle, they will also be independent from pulse-to-pulse, which may not always be the case for the speckle. The treatment of the case when the speckle is not fully decorrelated from pulse-to-pulse is discussed in Chapter 6.

In Section 2.3.1, a number of distribution models are presented. Many of these are compound models based on (2.11), with different choices for the texture distribution. Other models such as the Weibull and LN-distributions are included as they remain popular in the literature. However, they cannot easily be used to model Doppler spectra and the different correlation properties or include thermal noise. In order to characterise these distributions, it is necessary to estimate their parameters. There are many techniques outlined in the literature, including those based on maximum likelihood estimates [31], matching moments of the data with moments of the analytical distributions (method of moments) [1,31], formulations based on the mean of $z \log z$ [31,32] and others such as a least-squares model fit between the PDF or complementary cumulative distribution function (CCDF) of the model and the data [33]. This latter approach is generic in that it can be used to estimate the parameters for any distribution. It is often preferable to implement the least-squares minimisation in the $\log_{10}$ CCDF domain, to better focus on matching to the tail of the distribution.

Detection schemes may sometimes be formulated without considering the impact of thermal noise. In such cases, it may be possible to use an 'effective' distribution shape for clutter alone which can yield a similar distribution to that for the clutter-plus-noise. Details on how this can be achieved for the K and Pareto distributions are given in Section 2.3.3.

2.3.1 Distribution models

In this section, a number of popular distribution models are presented. The first three are the LN [23], Weibull [24,34] and K-distributions [26]. The subsequent models have all been designed to account for sea spikes or model longer tails in the distributions. These include the K+Rayleigh [35], Pareto [36], KK [37], KA [38] and tri-modal discrete (3MD) models [39]. Many of these were originally developed to fit to low grazing angle clutter but they are equally applicable to medium grazing angle collection geometries. There have also been other distributions used to model sea clutter. Two other compound Gaussian texture models proposed for sea clutter include the LN [40] and inverse-Gaussian [41,42].

For each model, the PDF, CDF and intensity moments are given. For the compound Gaussian models, the distributions are multilook with the assumption of non-coherent integration over N returns with a fixed local texture and independent speckle samples. The intensity moments are often used to fit data to a distribution model. In such cases, it will be normal to use single-look data, unless the speckle is well decorrelated using pulse-to-pulse frequency agility. Example model fits to real data for each distribution are given in Section 2.3.2.

2.3.1.1 Log–normal distribution

The LN distribution is a single-look distribution typically described in terms of power and arises if the logarithm of z is assumed to be normally distributed with mean m_{LN} and standard deviation σ_{LN}. It is given by [23]

$$P_z(z) = \frac{1}{z\sqrt{2\pi\sigma_{\mathrm{LN}}^2}} \exp\left(-\frac{(\ln z - m_{\mathrm{LN}})^2}{2\sigma_{\mathrm{LN}}^2}\right) \tag{2.13}$$

and has a CDF given by

$$F(z) = \frac{1}{2}\left(1 + \mathrm{erf}\left(\frac{\ln z - m_{\mathrm{LN}}}{\sigma_{\mathrm{LN}}\sqrt{2}}\right)\right) \tag{2.14}$$

where erf(·) is the error function. The kth-order moment is given by

$$\langle z^k \rangle = \exp\left(k m_{\mathrm{LN}} + \frac{k^2 \sigma_{\mathrm{LN}}^2}{2}\right). \tag{2.15}$$

In order to obtain a fit to real data, the distribution parameters must be estimated. One method is to measure the moments from the real data and apply them to the theoretical moments. For the LN distribution, the mean and variance can be estimated by

$$\begin{aligned} \hat{m}_{\mathrm{LN}} &= \ln\langle z \rangle - 0.5\ln\left(1 + \frac{\langle z^2 \rangle}{\langle z \rangle^2}\right), \\ \hat{\sigma}_{\mathrm{LN}}^2 &= \ln\left(1 + \frac{\langle z^2 \rangle}{\langle z \rangle^2}\right). \end{aligned} \tag{2.16}$$

2.3.1.2 Weibull distribution

The Weibull distribution has been used extensively for modelling both land [34] and sea clutter [24]. It is commonly represented as a single-look magnitude distribution,

$$P_r(r) = \frac{\nu_{\mathrm{w}} r^{\nu_{\mathrm{w}}-1}}{b_{\mathrm{w}}^{\nu_{\mathrm{w}}}} \exp\left(-\left(\frac{r}{b_{\mathrm{w}}}\right)^{\nu_{\mathrm{w}}}\right), \qquad \nu_{\mathrm{w}}, b_{\mathrm{w}} > 0 \tag{2.17}$$

where ν_{w} and b_{w} are the shape and scale parameters, respectively. The exponential and Rayleigh distributions can be obtained by setting $\nu_{\mathrm{w}} = 1$ or 2, respectively, with smaller ν_{w} implying a longer tailed distribution. Also for some parameter values, the Weibull distribution is equivalent to the K-distribution [43]. The CDF for the Weibull distribution is

$$F(r) = 1 - \exp\left(-\left(\frac{r}{b_{\mathrm{w}}}\right)^{\nu_{\mathrm{w}}}\right) \tag{2.18}$$

and the kth-order moment is defined by

$$\left\langle r^k \right\rangle = b_{\mathrm{w}}^k \Gamma\left(1 + \frac{k}{\nu_{\mathrm{w}}}\right) \tag{2.19}$$

where $\Gamma(\cdot)$ is the gamma function. Fitting data to a Weibull distribution by matching moments requires numerically solving the following expression for ν_{w}:

$$\frac{\left\langle r^2 \right\rangle}{\left\langle r \right\rangle^2} = \frac{\Gamma\left(1 + 2/\hat{\nu}_{\mathrm{w}}\right)}{\Gamma^2\left(1 + 1/\hat{\nu}_{\mathrm{w}}\right)} \tag{2.20}$$

with the scale then determined by

$$\hat{b}_{\mathrm{w}} = \frac{\langle r \rangle}{\Gamma\left(1 + 1/\hat{\nu}_{\mathrm{w}}\right)}. \tag{2.21}$$

2.3.1.3 K-Distribution

The most commonly used PDF model for sea clutter in both real and synthetic aperture radar is the K-distribution. This model was originally developed using a large quantity of recorded data which showed that the speckle had Gaussian statistics and the texture was a good fit to the gamma distribution [44], given by

$$P_\tau(\tau) = \frac{b^\nu}{\Gamma(\nu)} \tau^{\nu-1} \exp\left(-b\tau\right), \qquad \nu, b > 0 \tag{2.22}$$

where ν is the shape, $b = \nu / \langle \tau \rangle$ is the scale and $\langle \tau \rangle$ can be related to the mean clutter power, p_{c}. If thermal noise is present, the distribution is typically referred to as a

K+noise distribution, with no closed form solution. If thermal noise is not present ($p_n = 0$), the analytic form of the multilook distribution is given by

$$P_Z(Z) = \frac{2}{Z}(bZ)^{(N+\nu)/2}\frac{1}{\Gamma(N)\Gamma(\nu)}\mathcal{K}_{\nu-N}\left(2\sqrt{bZ}\right) \tag{2.23}$$

where $\mathcal{K}_{\nu-N}(\cdot)$ is the modified Bessel function of the second kind with order $\nu - N$. The CDF is given by

$$F(Z) = 1 - \frac{2}{\Gamma(\nu)}\sum_{n=0}^{N-1}\frac{(bZ)^{(n+\nu)/2}}{n!}\mathcal{K}_{\nu-n}\left(2\sqrt{bZ}\right) \tag{2.24}$$

with the kth-order moment for the K-distribution given by

$$\langle Z^k \rangle = \frac{\Gamma(N+k)\Gamma(\nu+k)}{b^k\Gamma(N)\Gamma(\nu)}. \tag{2.25}$$

If it is assumed that the thermal noise power is known, then the moment-based estimator for the shape parameter is given by

$$\langle \hat{\nu} \rangle = \frac{(N+1)\left(\langle Z \rangle - Np_n\right)^2}{N\langle Z^2 \rangle - (N+1)\langle Z \rangle^2}. \tag{2.26}$$

If the noise mean power is not known, the third-order moment is required and the following expressions can be derived using the single-look clutter-plus-noise moments,

$$\hat{\nu} = \frac{18\left(\langle z^2 \rangle - 2\langle z \rangle^2\right)^3}{\left(12\langle z \rangle^3 - 9\langle z \rangle\langle z^2 \rangle + \langle z^3 \rangle\right)^2}, \tag{2.27}$$

$$\hat{p}_n = \langle z \rangle - \left(\frac{\hat{\nu}}{2}\left(\langle z^2 \rangle - 2\langle z \rangle^2\right)\right)^{1/2}. \tag{2.28}$$

Another useful estimator is based on the mean of $z \log z$. This was first proposed by Blacknell and Tough [32] for the single-look case assuming no thermal noise, $p_n = 0$. More generally, the shape estimate is found by numerically solving [31]

$$\frac{\langle Z \ln Z \rangle}{\langle Z \rangle} - \langle \ln Z \rangle - \frac{1}{N} = d_z \tag{2.29}$$

with d_z given by

$$d_z = \frac{1}{1+(1/C)}\exp\left(\frac{\hat{\nu}}{C}\right)E_{\hat{\nu}+1}\left(\frac{\hat{\nu}}{C}\right) \tag{2.30}$$

where C is the clutter-to-noise ratio (CNR), and $E_\nu(\cdot)$ is the generalised exponential integral function. For the case where $p_n = 0$, the expression $d_z = 1/\hat{\nu}$. A comparison of different methods for estimating the parameters of the K+noise distribution can be found in [45].

2.3.1.4 K+Rayleigh distribution

The dominant Rayleigh component of the received backscatter has traditionally been attributed to speckle plus thermal noise. The K+Rayleigh distribution was formalised in [35] after observations by Sletten [46] and Lamont-Smith [47] who found evidence for a further Rayleigh component. It is based on the K+noise distribution with the speckle mean split into two components, $\tau = \tau_r + p_r$.

The K+Rayleigh model uses a gamma distribution for the texture,

$$P_\tau(\tau_r|\nu_r, b_r) = \frac{b_r^{\nu_r}}{\Gamma(\nu_r)}\tau_r^{\nu_r-1}\exp(-b_r\tau_r), \qquad \nu_r, b_r > 0 \tag{2.31}$$

where ν_r is the shape and $b_r = \nu_r/\langle\tau\rangle$ is the scale. To calculate the compound integral in (2.11), the integration is now performed with the modified speckle mean level, τ_r, instead of the total speckle τ. If the noise mean power is known, then the influence of the extra Rayleigh component can be measured by the ratio of the mean of the Rayleigh component to the mean of the gamma distributed clutter component, $k_r = p_r/\langle\tau\rangle$.

Parameter estimation for the K+Rayleigh distribution is based on the K-distribution methods with the shape estimate $\hat{\nu} \to \hat{\nu}_r$ and the noise mean power, $p_n \to p_n + p_r$. As the Rayleigh mean power is always unknown, the method of moments estimator in (2.27) must be used for the shape estimate with the mean noise plus Rayleigh power estimated by

$$\widehat{p_n + p_r} = \langle z\rangle - \left(\frac{\hat{\nu}}{2}\left(\langle z^2\rangle - 2\langle z\rangle^2\right)\right)^{1/2}. \tag{2.32}$$

Then if the noise mean power is known, the Rayleigh mean power can be determined by $\hat{p}_r = \widehat{p_n + p_r} - p_n$. For the $z \log z$ estimator, the expression for d_z in (2.30) is given by [31]

$$d_z = \frac{1}{1 + 1/C_r}\exp\left(\frac{\hat{\nu}_r}{C_r}\right)E_{\hat{\nu}_r+1}\left(\frac{\hat{\nu}_r}{C_r}\right) \tag{2.33}$$

where the clutter-to-noise plus Rayleigh ratio is given by

$$C_r = \frac{p_c}{p_n + p_r} = \frac{1}{1/\sqrt{\hat{\nu}_r d_r} + 1} \tag{2.34}$$

and

$$d_r = \frac{N}{N+1}\frac{\langle z^2\rangle}{\langle z\rangle^2} - 1. \tag{2.35}$$

2.3.1.5 Pareto distribution

The Pareto model was first used for sea clutter modelling by Balleri *et al.* [36], Fayard and Field [48] and later by others at the US Naval Research Laboratory (NRL) [49] and DSTG [50,51]. In SAR texture modelling, it is known as the G^0-distribution [22]. For this model, the texture has an inverse gamma distribution

$$P_\tau(\tau) = \frac{b_\text{p}^{\nu_\text{p}}}{\Gamma(\nu_\text{p})}\tau^{-\nu_\text{p}-1}\exp\left(-b_\text{p}/\tau\right), \qquad \nu_\text{p} > 1, b_\text{p} > 0 \tag{2.36}$$

where ν_p is the shape and $b_\text{p} = \langle\tau\rangle(\nu_\text{p} - 1)$ is the scale. The general solution with no thermal noise ($p_\text{n} = 0$) is

$$P_Z(Z) = \frac{Z^{N-1}b_\text{p}^{\nu_\text{p}}\Gamma(N+\nu_\text{p})}{(b_\text{p}+Z)^{N+\nu_\text{p}}\Gamma(N)\Gamma(\nu_\text{p})}, \qquad \nu_\text{p} > 1, b_\text{p} > 0 \tag{2.37}$$

which reduces to the generalised Pareto type II or Lomax distribution for $N = 1$. The CDF is given by

$$F(Z) = \frac{\beta_\text{inc}(\eta_\text{p}(Z,\nu_\text{p}))}{\beta_0(N,\nu_\text{p})} \tag{2.38}$$

where $\beta_\text{inc}(\cdot)$ is the incomplete beta function, $\beta_0(\cdot)$ is the beta function and the variable $\eta_\text{p}(\cdot)$ is given by

$$\eta_\text{p}(Z,\nu_\text{p}) = \frac{NZ}{NZ + b_\text{p}}. \tag{2.39}$$

The kth-order moment is

$$\left\langle Z^k\right\rangle = \frac{b_\text{p}^k\Gamma(N+k)\Gamma(\nu_\text{p}-k)}{\Gamma(N)\Gamma(\nu_\text{p})}, \quad k < \nu_\text{p} \tag{2.40}$$

with the restriction that moments do not exist for $k \geq \nu_\text{p}$ and hence the mean is undefined for $\nu_\text{p} \leq 1$ and the second moment for $\nu_\text{p} \leq 2$.

If the noise mean power is known, the single-look method of moments shape estimator becomes

$$\hat{\nu}_\text{p} = \frac{2\langle z\rangle^2 - 2\left\langle z^2\right\rangle + 4\langle z\rangle p_\text{n} - 2p_\text{n}^2}{2\langle z\rangle^2 - \left\langle z^2\right\rangle} \tag{2.41}$$

and if the noise mean power is not known, the third-order moment is required [31],

$$\hat{\nu}_p = 3 + \frac{6d_p(6d_p^2 + e_p)}{\left(12\langle z\rangle^3 - 9\langle z\rangle\langle z^2\rangle + \langle z^3\rangle\right)^2}, \tag{2.42}$$

$$d_p = \langle z^2\rangle - 2\langle z\rangle^2,$$

$$e_p = \sqrt{2d_p\left(\langle z^3\rangle^2 - 18\langle z\rangle\langle z^2\rangle\langle z^3\rangle + 24\langle z\rangle^3\langle z^3\rangle + 18\langle z^2\rangle^3 - 27\langle z\rangle^2\langle z^2\rangle^2\right)}.$$

This estimate is only valid if the clutter distribution has a finite third-order moment, which is the case for $\nu_p > 3$. For the $z \log z$ estimator, the shape parameter is found by numerically solving the equality in (2.29) with the following [31]:

$$d_z = \frac{1}{1 + (1/\mathcal{C})}\left(\frac{1}{\hat{\nu}_p - 1} - \exp\left((\hat{\nu}_p - 1)\mathcal{C}\right) E_{\hat{\nu}_p+1}\left((\hat{\nu}_p - 1)\mathcal{C}\right)\right) \tag{2.43}$$

where the noise-free estimate is found by substituting $d_z = 1/(\hat{\nu}_p - 1)$ into (2.29).

2.3.1.6 KK-Distribution

The KK-distribution represents a mixture of two K-distributions with the second component designed to model the long tail due to sea spikes [33,37]. To implement the KK-distribution, the texture is defined as the sum of two gamma components having shapes, $\nu_{k,1}, \nu_{k,2}$ and scales $b_{k,1}, b_{k,2}$ with the ratio of the two components given by k_{rat},

$$P_\tau(\tau) = (1 - k_{rat})P_\tau(\tau|\nu_{k,1}, b_{k,1}) + k_{rat}P_\tau(\tau|\nu_{k,2}, b_{k,2}). \tag{2.44}$$

The overall PDF then becomes

$$P_Z(Z) = (1 - k_{rat})P_Z(Z|\nu_{k,1}, b_{k,1}) + k_{rat}P_Z(Z|\nu_{k,2}, b_{k,2}). \tag{2.45}$$

This model can be simplified by equating the two shape parameters and fixing the mean of the first component. Analysis in [37] found that the ratio of means determines the degree of separation in the tail, while the ratio of the two components determines the point where the tail started diverging. Estimation of the parameters for this distribution can be implemented by a least-squares minimisation in the $\log_{10}$ CCDF domain.

2.3.1.7 KA-Distribution

The KA-distribution was originally proposed in [52] and then applied to sea clutter by Ward *et al.* [1,38,53]. The KA-distribution represents a mixture of the K-distribution with a discrete spike component. Each component of the mixture model is functionally identical to the previously discussed K+Rayleigh model, where the Rayleigh component power is now assumed to arise due to the presence of one or more Rayleigh distributed spikes. The probability of a spike being present is modelled by Poisson statistics which gives a weighting or prior probability for each mixture component. The single look KA-distribution can be written as a compound distribution with the speckle component given by

$$P_{z|\tau}(z|\tau) = \sum_{n=0}^{\infty} \frac{1}{\tau + p_{\mathrm{n}} + n\sigma_{\mathrm{sp}}} \exp\left(-\frac{z}{\tau + p_{\mathrm{n}} + n\sigma_{\mathrm{sp}}}\right) P_n(n) \tag{2.46}$$

where σ_{sp} is the mean sea-spike intensity. Note that this summation corresponds to the sum of mixture component probabilities that is typically small and can be truncated at $n = 1$. The Poisson component is described by

$$P_n(n) = \exp\left(-\bar{N}\right) \frac{\bar{N}^n}{n!} \tag{2.47}$$

where $\bar{N} \ll 1$ is the mean number of spikes in each range cell. The final PDF model is then given by

$$P_z(z) = \frac{\exp\left(-\bar{N}\right) b_{\mathrm{K}}^{\nu_{\mathrm{K}}}}{\Gamma(\nu_{\mathrm{K}})} \sum_{n=0}^{\infty} \frac{\bar{N}^n}{n!} \times \int_0^{\infty} \frac{\tau^{\nu_{\mathrm{K}}-1}}{\tau + p_{\mathrm{n}} + n\sigma_{\mathrm{sp}}} \exp\left(-b_{\mathrm{K}}\tau - \frac{z}{\tau + p_{\mathrm{n}} + n\sigma_{\mathrm{sp}}}\right) d\tau \tag{2.48}$$

where ν_{K} and b_{K} are the KA shape and scale parameters. For the KA-distribution, a least-squares fit of the $\log_{10}$ CCDF can be used to estimate the distribution parameters.

2.3.1.8 Tri-modal discrete distribution

The compound models in the literature all assume a continuous texture distribution that suggests a small probability of texture values approaching infinity. This is not physically sound as it cannot be measured by any real radar system. The 3MD model [39,54,55] instead proposes the use of a discrete texture model that assumes the sea clutter consists of a finite number of distinct modes or scatterer types, D. This implies that the scatterers in the observed scene are realisations from homogeneous clutter random variables with different texture values. In the original work, it was found that $D = 3$ modes were sufficient to model distributions from spaceborne radar imagery, hence the tri-modal in the name. However, subsequent analysis has found that up to five modes are required to accurately model bistatic sea clutter [56]. One of the consequences of this discretisation is that spatial and long-time correlation cannot be modelled as part of the texture, and hence the model is less suitable for clutter simulation. The texture PDF is given by

$$P_\tau(\tau) = \sum_{m=1}^{D} \tilde{c}_m \delta(\tau - \tilde{a}_m), \quad \sum_{m=1}^{D} \tilde{c}_m = 1, \quad \tilde{a}_m, \tilde{c}_m > 0 \tag{2.49}$$

where $\delta(\cdot)$ denotes the delta-function, $\tilde{\mathbf{a}} = [\tilde{a}_1, \ldots, \tilde{a}_D]$ are the discrete texture intensity levels and $\tilde{\mathbf{c}} = [\tilde{c}_1, \ldots, \tilde{c}_D]$ are the corresponding weightings. The continuous distribution is then given by

$$P_Z(Z) = \frac{N^N}{\Gamma(N)} Z^{N-1} \sum_{m=1}^{D} \tilde{c}_m \frac{\exp\left(-\dfrac{NZ}{\bar{p}_c \tilde{a}_m^2 + \bar{p}_n}\right)}{\left(\bar{p}_c \tilde{a}_m^2 + \bar{p}_n\right)^N} \tag{2.50}$$

with $\bar{p}_c + \bar{p}_n = \dfrac{p_c^2}{p_c^2 + p_n^2} + \dfrac{p_n^2}{p_c^2 + p_n^2} = 1$. The CDF is given by

$$F(Z) = 1 - \frac{1}{\Gamma(N)} \sum_{m=1}^{D} \Gamma\left(N, \frac{NZ}{\bar{p}_c \tilde{a}_m^2 + \bar{p}_n}\right) \tag{2.51}$$

and the kth-order moment by

$$\langle Z^k \rangle = \frac{\Gamma(N+k)}{N^k \Gamma(N)} \sum_{m=1}^{D} \tilde{c}_m \left(\bar{p}_c \tilde{a}_m^2 + \bar{p}_n\right)^k . \tag{2.52}$$

By minimising the difference between the theoretical and data moments, the $2D$ model parameters can be estimated,

$$\underset{\hat{\mathbf{a}}, \hat{\mathbf{c}}}{\operatorname{argmin}} \sum_{k=1}^{D_{\max}} \left(\langle Z^k \rangle - \mu_k\right)^2 \tag{2.53}$$

where $\mu_k = 1/N \sum_{n=1}^{N} z_n^k$ are the sample moments and the selection of the order, $D_{\max} \geq 2D$ can be chosen to balance the model accuracy and numerical complexity [39]. A number of alternative estimators have been proposed in [57] and include an analytic method of moment solution for $D \leq 3$, a maximum likelihood solution and a least-squares minimisation of the $\log_{10}$ CCDF. To determine the minimum model order, D, one proposed method in [58] first assumes a single mode ($D = 1$). The threshold error is then compared with a value of 0.5 dB and if the error is greater than this threshold, the parameters for a bimodal fit ($D = 2$) are estimated. Only if this error is greater than the threshold value, are the parameters estimated for the full 3MD model ($D = 3$). A similar approach was taken in [56] where the Bhattacharyya distance was used with an acceptance threshold of −30 dB. Further detail on the goodness of fit metrics can be found in Appendix A.

2.3.2 *Example fits*

To demonstrate the distribution models, Figures 2.13 and 2.14 show single-look fits to a block of data from the Ingara X-band data set (see Chapter 1). The data is horizontally polarised from the up-wind direction and has a 30° grazing angle. For

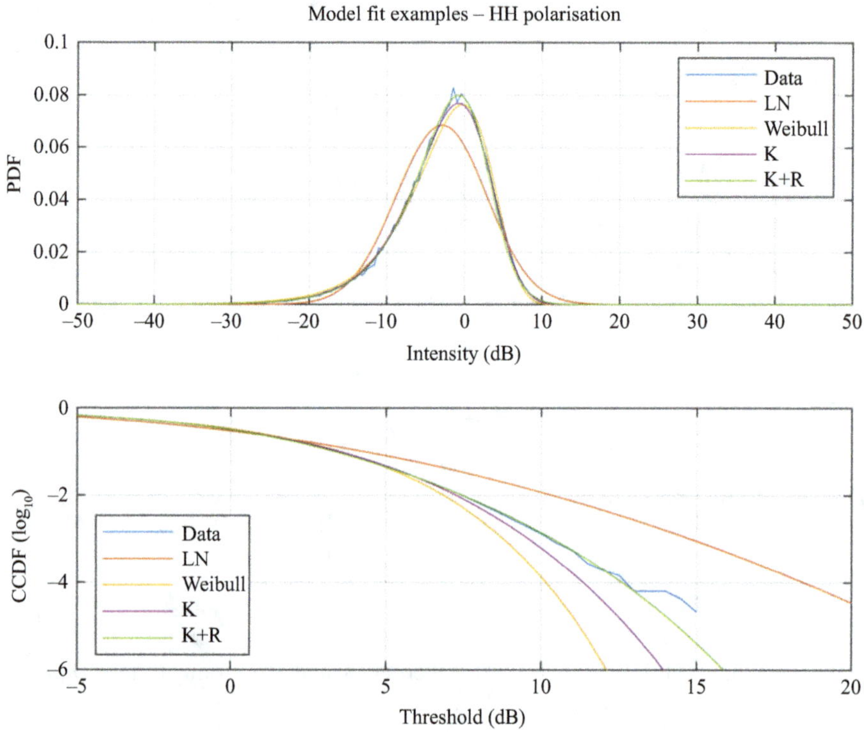

Figure 2.13 Example CCDF fits for horizontally polarised Ingara X-band data in the up-wind direction with a 30° grazing angle

these examples, the sea clutter histograms were formed by first normalising the data to have unity mean. The intensity values were then defined with uniform bin spacing in the dB domain due to the wide dynamic range of sea clutter. Both the PDF and logarithm of the CCDF are shown, with the latter used to better assess the accuracy of the distribution tail. The model fits in Figure 2.13 include the LN, Weibull, K and K+Rayleigh (K+R) distributions. They show a large mismatch in the PDF for the LN-distribution, while the CCDF results show that only the K+R-distribution matches the CCDF accurately. Figure 2.14 then presents model fits for the Pareto, KK, KA and 3MD-distributions which all show a good fit to the data.

To assess the goodness of fit between a distribution model and the histogram data, it is common to use a statistical test. A number of metrics are given in Appendix A with modifications to better assess the fit in the distribution tail. These include the chi-square (CS) and Kolmogorov–Smirnov (KS) tests, the Bhattacharyya distance (BD) and the threshold error (TE). The CS and KS tests assess the accuracy of the PDF and CDF, respectively, producing a probability value (p-value) that can be used to determine if the null hypothesis should be accepted (i.e. the model matches the data if the p-value is above 0.05). The BD is a metric for assessing the fit to the PDF, while

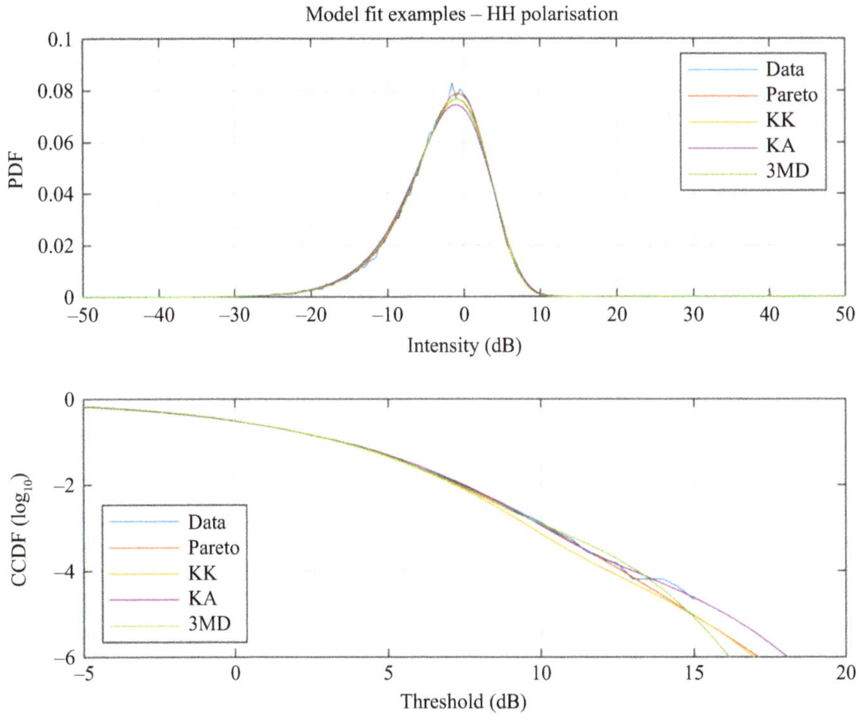

Figure 2.14 Example CCDF fits for horizontally polarised Ingara X-band data in the up-wind direction with a 30° grazing angle

Table 2.1 Goodness of fit metrics for the example in Figures 2.13 and 2.14

	LN	**Weibull**	**K**	**K+R**	**Pareto**	**KK**	**KA**	**3MD**
CS p-value	0	0	0	0.67	0.45	0	0	0
KS p-value	0	0	0.02	1.0	0.83	0	0	0
MCS p-value	0	0	0	0.9	0.9	0	0	0
MKS p-value	0	1	1	1	1	1	1	1
BD (dB)	−18.5	−27.1	−32.2	−34.1	−34.0	−32.0	−30.5	−32.3
MBD (dB)	−14.3	−22.1	−28.9	−33.0	−32.9	−29.6	−30.7	−29.4
TE (dB)	5.7	2.6	1.4	0.1	0.09	0.8	0.3	0.4

the TE measures the accuracy of the CCDF. A 'good fit' using these tests is more arbitrary with BD values less than −20 dB and TE values less than 1 dB considered acceptable [59].

Table 2.1 shows the goodness of fit metrics for the examples in Figures 2.13 and 2.14. The first three rows show p-values for the CS and KS tests with the modified chi-square (MCS) and modified Kolmogorov–Smirnov (MKS) tests only considering intensity values above 3.5 dB. This corresponds to a CCDF of 0.1 and provides a better indication of the accuracy in fitting to the distribution tail. The results show that the K+R and Pareto distributions match the data for all four tests, while the MKS test shows a good fit for all distributions except the LN. The BD metric is focussed on the PDF and shows a good fit for all distributions except the LN. The TE was measured at a $\log_{10}$(CCDF) of −4 and shows that only the K+R, Pareto, KK, KA and 3MD-distributions can accurately model the distribution tail.

2.3.3 Effective shape relationships

The effect of thermal noise on the K-distribution model has been studied by a number of authors. Watts [30] first looked at thermal noise from a target detection point of view. He formulated a modified distribution and showed that estimating the K-distribution parameters with thermal noise present leads to an 'effective' shape which can be used to achieve acceptable predictions of target detection performance. Further studies by Gini *et al.* [60] then extended the K-distribution model to include correlation and detailed analysis showed better target detection results when thermal noise is included in the model, at the expense of greater computation. A slightly different focus by Lombardo *et al.* [61] looked at the effect of thermal noise on parameter estimation for K-distributed clutter. A modified shape estimate was derived which is less sensitive to thermal noise.

Commonly in the literature, thermal noise is not explicitly modelled and the shape is really an 'effective' shape. A key relationship can be derived by matching moments of the PDF with and without noise [1]. For the K-distribution, this is given by

$$\tilde{\nu} = \nu\left(1 + \frac{1}{C}\right)^2 \tag{2.54}$$

where $\tilde{\nu}$ is the effective shape without accounting for noise, and ν is the underlying distribution shape. This approximation has been studied in [1], where it was noted that if ν and C are both small, using an effective shape parameter may not give a good fit in the distribution tail.

There have also been a number of effective shape relationships derived for the Pareto distribution [51]. Similar to the K-distribution, the relationship based on matching moments is given by

$$\tilde{\nu}_{\rm p} = (\nu_{\rm p} - 2)\left(1 + \frac{1}{C}\right)^2 + 2. \tag{2.55}$$

2.4 Sea-spike characteristics

Sea spikes are relatively rare events that result in the radar backscatter having large amplitudes compared to the surrounding clutter and distinct Doppler excursions.

The continuing evolution of the sea surface, combined with transient effects such as breaking waves, mean that some scattering events are quite short-lived but may be confused with real target returns, giving rise to false alarms.

The principal scattering mechanism for sea clutter is called 'Bragg' scattering and is primarily associated with resonant scattering from wind-induced capillary waves. As discussed in Chapter 1, the local mean power of the Bragg scattering component may be modulated over range and time by the effects of wind and changing wave slopes. This gives rise to the so-called spiky characteristics of sea clutter, where there are non-Gaussian amplitude statistics. Local excursions in amplitude may often be called clutter spikes [1], but these are fully incorporated into the standard models for the PDF and Doppler spectrum.

There are additional large amplitude scattering events that are comparatively rare, compared with the continuum of Bragg scattering. These 'fast' scattering events can cause problems with false alarms in detectors and are more difficult to model statistically due to their transient nature. One important source of such clutter spikes are breaking waves or whitecaps. Breaking waves are an inherent aspect in the development of wind-driven waves. As the energy from the wind builds up, the wave grows until it can no longer be sustained. It then breaks, dispersing its energy into new waves with a longer wavelength. The breaking waves are seen as a local area of white water, which may also be whipped up by the wind. The whitecap radar returns have a large amplitude, observed as a spike that typically persists for 1 or 2 s with similar amplitude characteristics for both vertical and horizontal polarisations. Short-lived spikes are more commonly observed in the horizontal polarisation due to specular scattering from the crests of incipient breaking waves. These specular or 'burst' spikes may last up to about 200 ms and have a velocity consistent with the velocity of the wave top. These spikes are comparatively rare compared to whitecap spikes.

An example of a short-lived horizontally polarised spike is shown in Figure 2.15. This spike lasts for about 30 ms with a peak power about 16 dB above the surrounding mean level. Figure 2.16 then shows the PDF of the data with a fit to the K+noise distribution with $\nu = 2.8$ and a CNR of 16 dB. The effect of the spike on the tail of the distribution is clearly visible. Figure 2.16 also shows the PDF of the data with the spike removed where there is now a good fit to the K+noise distribution.

To better understand the behaviour of the sea-spike components, it is useful to isolate them from the main sea clutter return. There are three main methods of characterising the sea-spike component of the sea clutter. The first two methods fit relevant models to the PDF and the mean Doppler spectrum [33,37,62–64]. The third method does not assume any underlying statistical model but involves thresholding the magnitude of the data in order to distinguish between Bragg scattering and sea-spike events [65–69]. There are often secondary criteria imposed at this stage that also consider the spike width and/or the interval between spikes.

An example of the third method has been presented in [70] for the Ingara X-band sea clutter data. The approach taken is to threshold the magnitude of the raw backscatter data in the range/time domain at three standard deviations above the mean

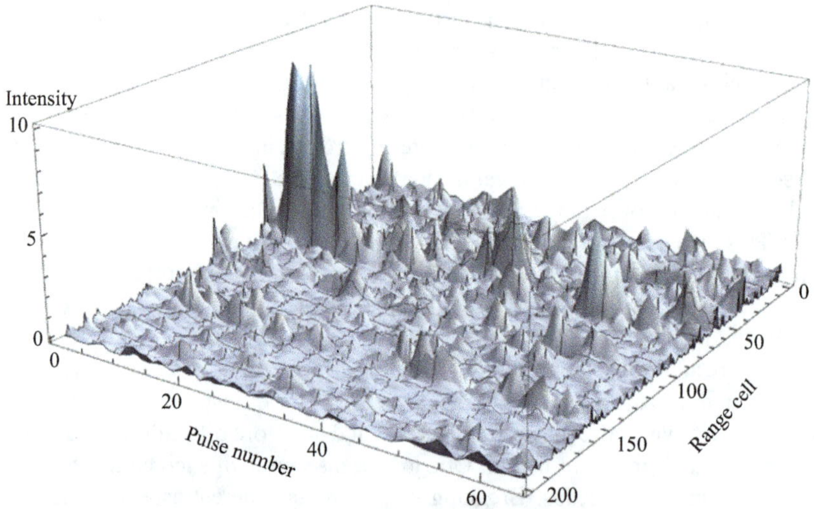

Figure 2.15 Example of a short-lived clutter spike from the Ingara X-band data set. The spike is 16 dB above the local mean

and then measure the percentage of sea spikes present in the data. The results show that the majority occur in the lower grazing angle region for the horizontal polarisation and are slightly higher in the cross-wind directions for the cross and vertical polarisations. An image processing algorithm can then be used to isolate the short- (discrete) and long-lived (persistent) scatterers. The persistent scatterers are attributed to whitecaps and last for at least 1 s. The results in Figures 2.17 and 2.18 show that the persistent whitecaps are spread quite evenly in grazing and azimuth angles for the horizontal polarisation, but with a clear preponderance in the cross-wind directions for the cross and vertical polarisations. Based on this information, the wave velocity, life time and spike decorrelation time of the whitecaps can be measured and are shown in Figure 2.19.

2.5 Doppler characteristics

Due to the action of the wind, currents and tide, the motion of the sea surface modulates the phase of the radar returns from successive pulses. In a coherent radar, this is characterised by the autocorrelation function, or equivalently the Doppler spectrum, estimated over a burst of pulses. The mean Doppler spectrum of sea clutter is often characterised in its simplest form as having a Gaussian shape. This will typically have a velocity spectrum width between 0 and 1 m/s, where 1 m/s is equivalent to a Doppler spectrum standard deviation of about 65 Hz at X-band [2,71]. The mean

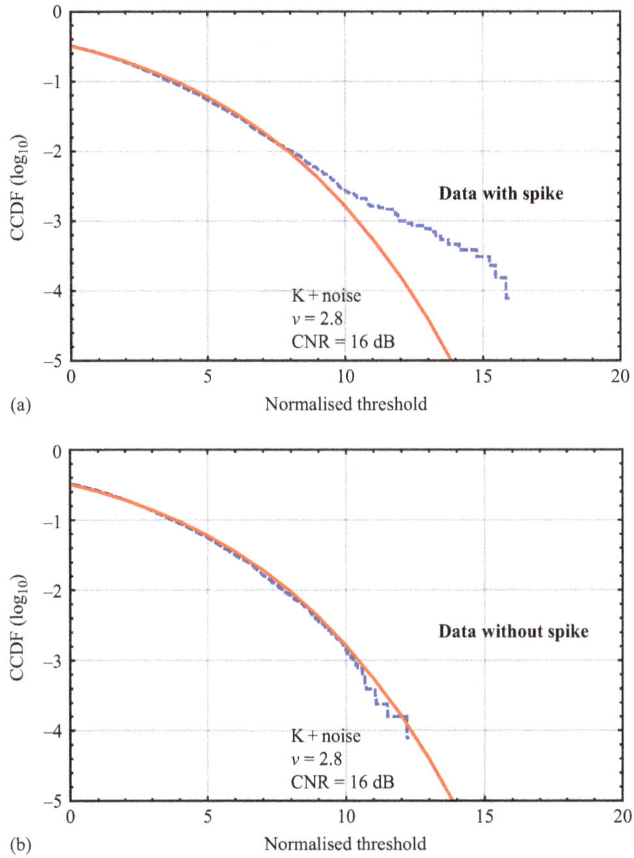

Figure 2.16 Data from Figure 2.15: (a) with and (b) without spike, showing fits to the K+noise distribution

Doppler shift is determined by the orbital velocity of the water particles in a wind-driven wave. When looking up-wind, it is typically between 1/4 and 1/6 of the wind velocity (see Chapter 4).

It is found that this simple model of the Doppler spectrum is more of a mathematical convenience than an accurate representation of observed behaviour. The radar returns from the sea surface will usually have time- and range-varying characteristics, with significant fluctuations relative to the mean spectrum. Further, these fluctuations result in the overall mean spectra being asymmetric and not Gaussian-shaped. An example from the CSIR X-band data set (see Chapter 1) is shown in Figure 2.20 and reveals the variation over time of the sea clutter Doppler spectrum. Here the power spectral density has been estimated over successive bursts of 64 pulses using a −55 dB Dolph–Chebyshev weighting prior to the fast Fourier transform (FFT).

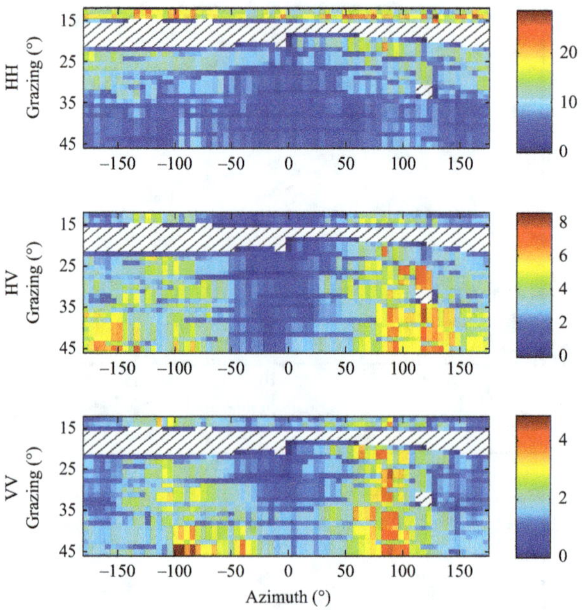

Figure 2.17 *Percentage of discrete sea-spike detections for the Ingara X-band data set [70]. © Commonwealth of Australia 2013*

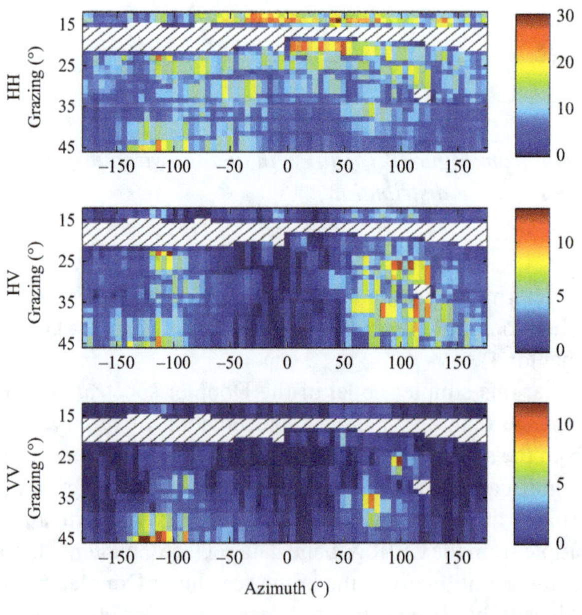

Figure 2.18 *Percentage of whitecap detections for the Ingara X-band data set [70]. © Commonwealth of Australia 2013*

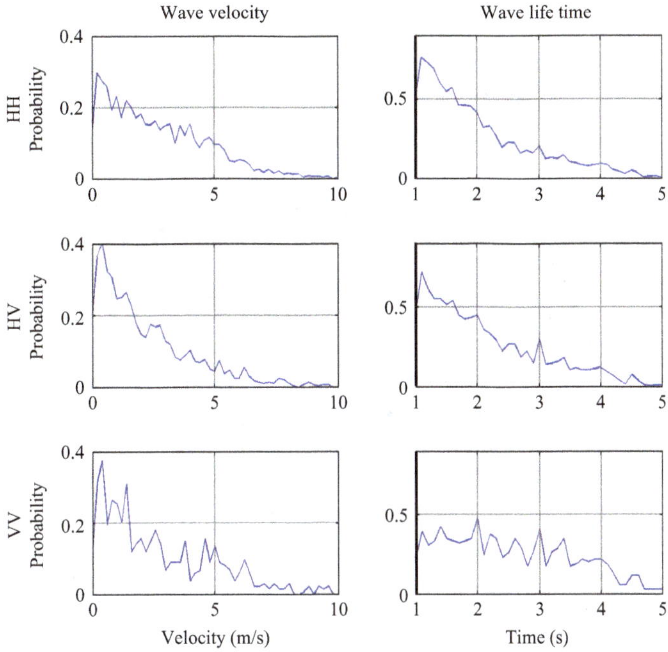

Figure 2.19 Sea-spike characteristic PDFs for the Ingara X-band data set: wave velocity and life time [70]. © Commonwealth of Australia 2013

This example was from a cliff-top looking at a range of 3 km with a grazing angle of about 1.3°. The modulation in intensity over time due to the sea wave pattern can be clearly seen, with a period of about 10 s along with variations of the shape and mean Doppler shift. The mean spectrum, averaged over the time period is then shown in Figure 2.21 as a function of frequency and is expressed as the mean CNR. The asymmetry shown here can be assessed in terms of contributions from the different clutter-spike components discussed in Section 2.4. The scattering component due to Bragg scattering gives rise to Doppler spectra where

- the average intensity of the vertical polarisation is stronger than the horizontal polarisation,
- the standard deviation fluctuates randomly about the mean level with values on the order of tens of Hz and is consistent with the expected phase velocity of capillary waves,
- the mean Doppler shift may not be zero and will vary with the wave and wind directions,
- the average spectrum width will be similar for both the vertical and horizontal polarisations.

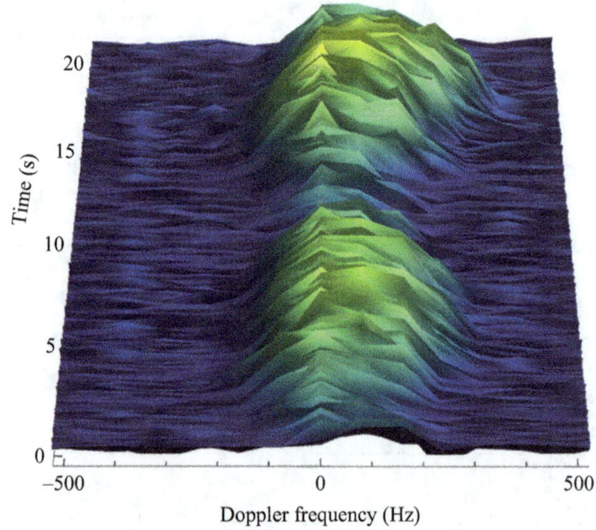

Figure 2.20 Successive power spectra from an X-band radar, vertical polarisation

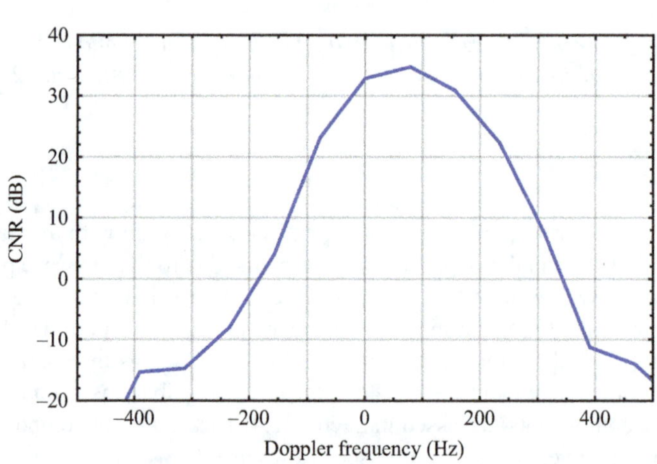

Figure 2.21 Average CNR against Doppler frequency for the spectra shown in Figure 2.20 (original time domain data had a CNR of 20 dB)

The Doppler spectra of whitecap spikes will be broader than the Bragg scattering component and the mean Doppler shifts may be higher, giving the potential for significant asymmetric broadening of spectra over the life time of a spike. Whitecap spikes have similar amplitudes and Doppler spectra for both polarisations. Finally specular or burst

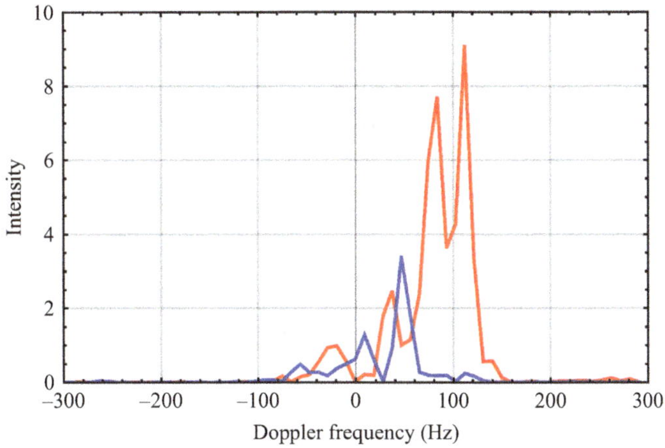

Figure 2.22 Two example power spectra for the horizontal polarisation and 37° grazing angle. The CNRs for the blue and red lines are 12 and 16 dB, respectively

spikes are more prominent in the horizontal polarisation and may have a high amplitude, a relatively narrow spectrum and a Doppler shift consistent with the velocity of the wave tops.

Figure 2.22 shows an example of two spectra from the Ingara X-band data set, recorded with horizontal polarisation at a grazing angle of about 37°, having CNRs of 12 and 16 dB, respectively. The spectra were formed at different ranges using a burst of 64 pulses. The lower intensity spectrum appears to be due to Bragg scattering while the larger spectrum has a higher mean Doppler shift and a larger width, due to both Bragg and fast scattering components.

2.5.1 Coherent amplitude statistics

The variability of the spectrum has a marked effect on the amplitude statistics observed over range or time at a single Doppler frequency (or more precisely in a single Doppler bin). The compound Gaussian models for the amplitude PDFs described in Section 2.3 can be used as the basis for analysis of coherent clutter. The coherent analogue to the speckle component in (2.6) is given by a circular complex normal distribution,

$$P_x(\mathbf{x}) = \frac{1}{(\pi)^N |\mathbf{M}|} \exp\left(-\mathbf{x}^H \mathbf{M}^{-1} \mathbf{x}\right) \tag{2.56}$$

where $\mathbf{x}$ is a complex single-look vector of length $N \times 1$ and $\mathbf{M} = \langle \mathbf{x}\mathbf{x}^H \rangle$ is its covariance. This complex distribution can now be used to represent the speckle in a coherent compound Gaussian distribution. If the normalised covariance matrix is assumed to be constant, the stochastic process is referred to as a spherically invariant random process (SIRP) and the overall clutter distribution is given by

$$P_y(\mathbf{y}) = \int_0^\infty \frac{1}{(\pi\tau)^N|\mathbf{G}|} \exp\left(-\frac{\mathbf{y}^H\mathbf{G}^{-1}\mathbf{y}}{\tau}\right) P_\tau(\tau)d\tau \tag{2.57}$$

where the texture modulates the speckle covariance, so that $\mathbf{M} = \tau\mathbf{G}$. Implicit in the definition of the SIRP is the assumption that the underlying power (or texture) remains constant across the coherent processing interval of the measurement vector. Furthermore, the assumption of a constant normalised covariance matrix, $\mathbf{G}$, implies that the power spectrum is fixed with local values scaled by a common power factor, τ. This means in the absence of receiver noise, all spectral components of the underlying clutter are scaled equally and the underlying power spectrum at different ranges will be identical up to a common scalar. In statistical terms, it means that the PDFs of clutter samples drawn from different frequency bins will have identical shape parameters and only differ in their scale values. The SIRP model therefore results in the clutter having the same PDF at each Doppler frequency as the time-domain data. In the presence of receiver noise, the combined PDF for the clutter-plus-noise will vary with Doppler frequency due to changes in the CNR, while the statistical properties of the underlying SIRP clutter component will remain constant across Doppler frequencies. This presumed stationarity of clutter statistics versus Doppler frequency is in contradiction with real-world observations. It has been known for many years [1] that the amplitude statistics of the clutter component vary significantly with Doppler frequency, often becoming very spiky at the edges of the spectra. This is because the Doppler spectra of sea clutter are often strongly time-varying and cannot be realistically modelled as an SIRP. Furthermore, the SIRP assumption, which implies that perfect correlation of texture across all Doppler frequencies is physically inconsistent with the existence of independent Bragg and fast scattering mechanisms.

This behaviour is illustrated here using the data plotted in Figure 2.20. When the returns over time are analysed as a function of Doppler frequency, f, it is found that they are distinctly non-Gaussian (spiky) and that this spikiness varies with Doppler frequency. Figure 2.23 shows as a function of frequency the mean intensity of the power spectrum of clutter-plus-noise, $\langle z'(f)\rangle$, and the normalised second intensity moments $\langle z'(f)^2\rangle / \langle z'(f)\rangle^2$. For Gaussian statistics, the normalised second intensity moments should have a value of 2, which can be seen in the noise-limited regions below −300 Hz. The most non-Gaussian behaviour is seen around 300 Hz. Referring to Figure 2.20 it can be seen that at this Doppler frequency the spectrum has a very low level, with occasional excursions of high intensity, probably due to wind and breaking waves. Another way of showing the non-Gaussian behaviour at different Doppler frequencies is to fit the data to a K+noise distribution. It may be noted that even if the time-domain data is a good match to a K+noise distribution, there is no theoretical reason why the data in a Doppler bin may also be modelled in this way. However, the analysis of real data has shown that in fact this is a reasonable approach to take. The average CNR of this data in each Doppler bin is shown in Figure 2.21. Using the first and second intensity moments from Figure 2.23, combined with a knowledge of the thermal noise level, the data can be fitted to the K+noise distribution. The resulting values of the K+noise distribution shape parameter at each frequency, $\nu'(f)$, are shown in Figure 2.24 where it can be seen that the clutter becomes extremely spiky (i.e. $\nu'(f) < 0.1$) at the edges of the spectrum. In regions where the CNR becomes very low, this has little effect on the overall statistics of the clutter-plus-noise, but as seen in Figure 2.23 the peak effect will be around a frequency of 300 Hz, where the clutter shape parameter $\nu'(f) \approx 0.05$ and the CNR is still significant, with $C'(f) \approx 10$ dB.

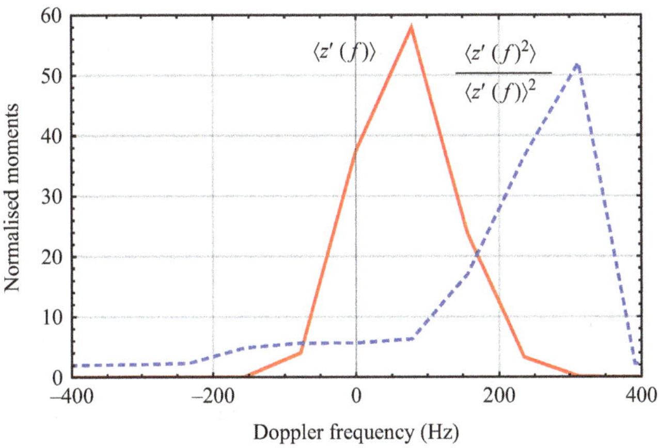

Figure 2.23 Moments of spectrum intensity against Doppler frequency

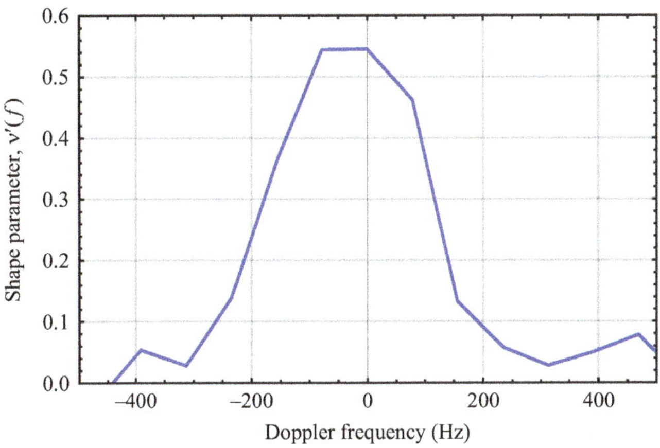

Figure 2.24 K+noise distribution shape parameter against Doppler frequency for the data in Figure 2.20 (original time domain data had a CNR of 20 dB and $\nu = 0.4$)

Conversely, at a frequency of -300 Hz, where the clutter component is again very spiky, the CNR is small, with $C'(f) \approx -15$ dB, and the overall statistics in Figure 2.23 appear noise-like.

2.5.2 Mean Doppler spectra

The different spectral widths and Doppler shifts of the different scattering components, combined with their transient or time-varying nature, result in average spectra (averaged over range or time) that are asymmetric. Various approaches have been made to represent the mean Doppler spectra of sea clutter. For low grazing angles, Lee *et al.* [72] proposed a two-component model, with a Gaussian shape for the Bragg scattered component and either a Lorentzian or a Voigtian shape for the fast scattering component. Figure 2.25 shows an example of the mean power spectral density for vertically polarised measurements made in a wave tank with the two components clearly identified.

Lamont-Smith [73] made wave tank and clifftop measurements at different grazing angles. He represented the vertically polarised spectra by a single Gaussian component suggesting that the Bragg scattering component was dominant, while the horizontally polarised spectra were modelled by two Gaussian components representing the Bragg and fast scattering components. Similar observations were made by Melville *et al.* [74] who also noted that specular scattering was virtually absent in their vertically polarised data.

As discussed, the intensity and spectra of the fast scattering component due to breaking waves are similar for both horizontal and vertical polarisations. The work by Jessup *et al.* [65] reported that the horizontal to vertical polarisation ratio which

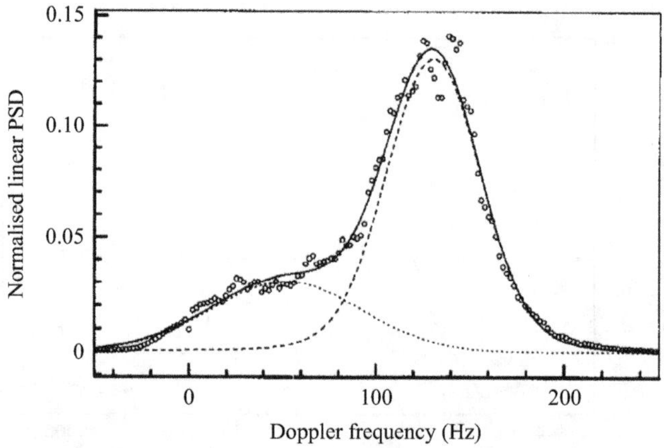

Figure 2.25 Mean Doppler spectrum example from a wave tank. Data is vertically polarised data and collected with a 7° grazing angle: (o) data, (- -) Voigtian component, (· · ·) Gaussian component and (—) combined mean spectrum [72]

is usually less than 1, approached unity as the grazing angle increased. Based on wave tank measurements, Walker [62] developed a model using three Gaussian components, associated with Bragg scattering, breaking waves and specular reflections, respectively.

At medium grazing angles, Stacy *et al.* [75] and subsequently Rosenberg *et al.* [63,64] reported on Doppler spectra measured from the Ingara X-band data set. They fitted the observed spectra to the Walker model with good agreement. However, further investigations [70] found that the Walker model did not totally describe the scattering, and a modification was required. Consequently, a new two-component Doppler spectrum model was introduced in [76] using the same Gaussian building blocks that both Walker [62] and Lamont-Smith [73] used.

2.5.3 Time and range varying spectra

From the examples in Figures 2.20 and 2.22 and the discussion earlier, it is clear that the mean spectrum does not describe the dynamic nature of the Doppler spectra. Characterising the possible variations over range and time is important when assessing the performance of coherent detection methods, such as described in Chapter 7. For example, the optimum detector for a coherent burst of Gaussian clutter at a given range requires knowledge of the covariance matrix (or spectrum) of the clutter. Further, estimating the local covariance matrix by averaging over the covariance matrices in adjacent bins may also fail to accurately represent the cell under test.

When observing spectra over range and time, it is found that the instantaneous power spectrum can often be represented by a fluctuation around an underlying Gaussian shape, as proposed for the various models of the mean spectra. However, both the normalised spectrum width, given by the standard deviation of the Gaussian power spectrum, and the mean Doppler shift are found to vary over range and time. Various methods for modelling this range- and time-varying behaviour have been proposed. Greco *et al.* described an autoregressive model [77], while Davidson proposed a method for recreating measured spectra [78]. A generic model for range and time evolution of spectra was later proposed by Watts [79] and is discussed here. The exploitation of this model and possible alternative approaches, such as that of McDonald and Cerutti-Maori, are described in Chapter 5.

The development of the evolving Doppler model by Watts [79] used data from the CSIR X-band data set (see Chapter 1). Examples of two spectra from the same range gate, but separated by a few seconds in time, are shown in Figure 2.26. These have been fitted to Gaussian shapes, characterised by a mean Doppler, m_f, and a standard deviation, w. It may be noted that although the values of w are similar, the value of m_f is higher for the higher intensity spectrum. Also, suitably scaled Gaussian shapes are a consistently good fit to the spectra for this data. Subsequent analyses of other data sets have confirmed this choice. In some cases when high intensity Bragg and fast scattering components are both present, there may be additional spreading of the spectrum. A bimodal variation of the model incorporating two Gaussian shapes can be used to accommodate this behaviour.

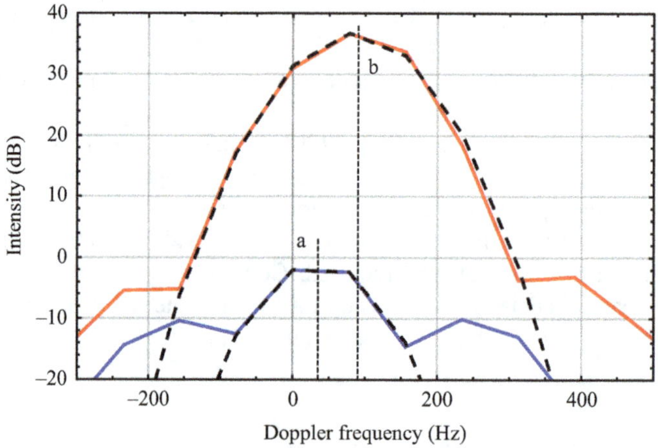

Figure 2.26 Two examples of clutter spectra from a single range gate, spaced by several seconds and fitted with a Gaussian model (dashed lines). a: $m_f = 36$ Hz, $w = 49$ Hz; b: $m_f = 91$ Hz, $w = 55$ Hz

Having established a normalised spectrum shape, the spectrum powers, their mean Doppler shifts (centre of gravity of the power spectra) and their standard deviations were measured for successive bursts in a single range gate, over total periods of 30–60 s. Bursts of 64 pulses were used and the spectra were estimated using an FFT with −55 dB Dolph–Chebyshev weighting. Figure 2.27 shows an example, with successive estimates averaged over groups of 10 bursts (a total time of 128 ms), a range of 3 km and a grazing angle of about 1.3°. It can be clearly seen that the mean Doppler is correlated in some manner with the intensity and that larger mean Doppler shifts are associated with larger intensities. A lag of a few seconds in response is also apparent, but investigation of the autocorrelation function between the mean Doppler and intensity shows that this only occurs at the highest values of intensity. The spectrum width fluctuates over time but appears to be uncorrelated with the local intensity. The peak in the estimated spectrum width at around 29 s is due to the very low CNR at this point with thermal noise corrupting the estimate. This highlights one of the difficulties of estimating the mean and standard deviation of individual spectra. Possible methods for mitigating the effect of thermal noise on these parameter estimates are discussed in [80]. The relationship between the mean Doppler shift and local intensity is further illustrated in Figure 2.28. Although the relationship is noisy, it can be seen that a linear fit between m_f and the normalised mean speckle power, $\tau / \langle \tau \rangle$, describes the general trend. The noise in this correlation may be partly explained by errors in estimating m_f, however, the relationship is likely to be more complicated than a simple linear correlation. It has been found that a model based on a simple linear relationship can reproduce many of the characteristics of real clutter that affect the performance of target detection algorithms and is useful for simulating

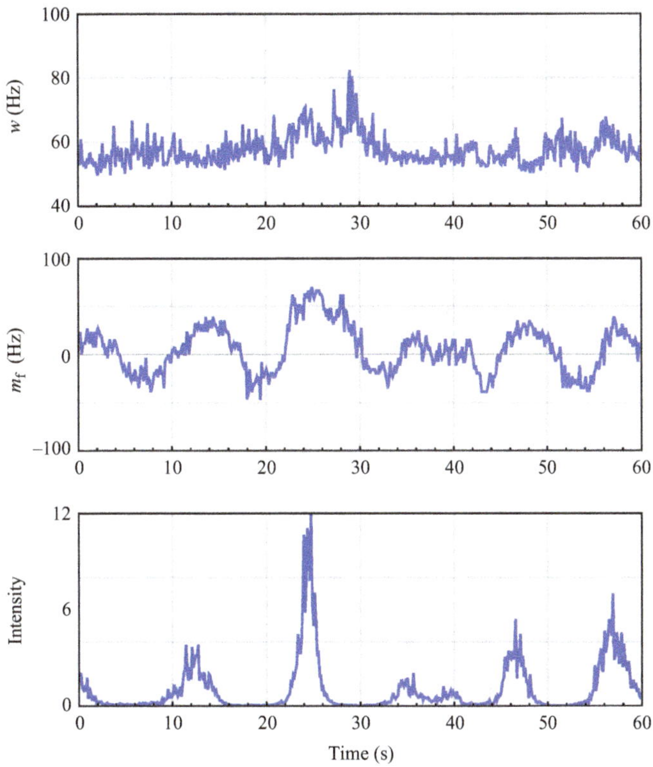

Figure 2.27 *Evolution of spectrum intensity from a single range gate over time showing the relationship with mean Doppler, m_f, and standard deviation, w*

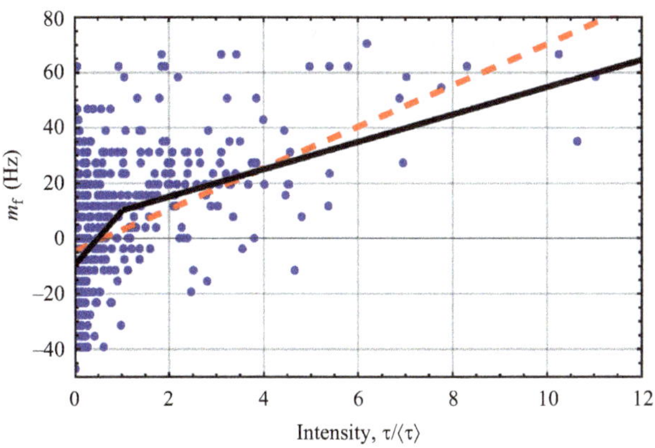

Figure 2.28 *Mean doppler, m_f, as a function of the normalised spectrum intensity: (•) spectrum mean estimates, (- -) linear fit, (—) bilinear fit*

data and undertaking performance prediction, as discussed in Chapters 5 and 6, respectively.

The model for evolving Doppler spectra takes the following form [79]:

$$G(f) = \frac{\tau}{\sqrt{2\pi}w} \exp\left(-\frac{(f - m_{\mathrm{f}}(\tau))^2}{2w^2}\right) \tag{2.58}$$

where $G(f)$ is the mean power spectral density for Doppler frequency, f, and τ is the local clutter intensity or texture, having a gamma distributed PDF if the clutter has a K-distribution. The mean Doppler frequency is related to the normalised mean speckle power, $\bar{\tau} = \tau / \langle\tau\rangle$ with

$$m_{\mathrm{f}}(\tau) = A_{\mathrm{c}} + B_{\mathrm{c}}\bar{\tau} + f_{\mathrm{lin}} \tag{2.59}$$

where the term f_{lin} was added to the model in [81] and captures the fluctuation around the model fit. It is described by a Gaussian random variable with zero mean and standard deviation, σ_{lin}. The final term in (2.58) is the spectrum width, w (standard deviation of the Gaussian-shaped spectrum), which has a PDF, $P_w(w)$, with mean m_{w} and standard deviation σ_{w}. As seen in Figure 2.27, the spectrum intensity may also be strongly correlated over range and time. The representation of this correlation is discussed in Section 2.6.

There has been extensive work to validate and extend this model. This includes the analysis of spectra as they evolve over range, which provides equivalent results to the evolution over time in a single range gate. These studies include further assessment of the monostatic low grazing angle CSIR data set [82], the low grazing angle bistatic NetRAD data set [83] and the medium grazing angle Ingara X-band data set [80]. This model has been extended for multiple phase centres in [84] and for a multiple phase centre scanning radar in [85,86].

As mentioned before, the mixture of Bragg and fast scattering components can give rise to a bimodal behaviour when the mean Doppler increases less rapidly with intensity at a certain level. To accommodate this behaviour, a bimodal extension to the model was presented in [81] as a mixture of two Gaussian-shaped spectra designed to model the Bragg and fast components,

$$\begin{aligned} G_{\mathrm{bi}}(f) = {} & \frac{\alpha\tau}{\sqrt{2\pi}w_1} \exp\left(-\frac{(f - m_{\mathrm{f}_1}(\tau))^2}{2w_1^2}\right) \\ & + \frac{(1-\alpha)\tau}{\sqrt{2\pi}w_2} \exp\left(-\frac{(f - m_{\mathrm{f}_2}(\tau))^2}{2w_2^2}\right) \end{aligned} \tag{2.60}$$

where α is a weighting factor and w_1 and w_2 are the spectrum widths, both modelled as random variables. The centre frequency is then given by

$$m_{f_1}(\tau) = \begin{cases} A_c + B_c\bar{\tau} + f_{bi}, & \bar{\tau} \leq \tau_{bi} \\ A_c + B_c\tau_{bi} + f_{bi}, & \bar{\tau} > \tau_{bi} \end{cases} \tag{2.61}$$

$$m_{f_2}(\tau) = A_c + B_c\bar{\tau} + f_{bi} \tag{2.62}$$

where f_{bi} is a zero-mean Gaussian random variable with standard deviation, σ_{bi}, which represents the noisy nature of the fit between the mean Doppler and local intensity. Dependent on the value of α, this model can have the effect of reducing the rate of increase of the mean Doppler shift with intensity and broadening the spectrum if $\bar{\tau}$ exceeds a threshold τ_{bi}. Also, to ensure a single Gaussian model, when $\bar{\tau} \leq \tau_{bi}$, the weighting factor, $\alpha = 0$. An example of the bimodal fit to the Doppler mean is shown by the bilinear fit in Figure 2.28. The mean and variance of the total spectrum can be expressed as

$$\begin{aligned} m_f(\tau) &= \alpha m_{f_1}(\tau) + (1-\alpha)m_{f_2}(\tau), \\ w^2(\tau) &= \alpha w_1^2 + (1-\alpha)w_2^2 \\ &\quad + \alpha(1-\alpha)\left(m_{f_1}^2(\tau) + m_{f_2}^2(\tau) - 2m_{f_1}(\tau)m_{f_2}(\tau)\right). \end{aligned} \tag{2.63}$$

If $\alpha = 0$ the model reverts to the single-mode case and when $\alpha > 0$, the effects of an additional fast scattering component are represented (see Figure 2.22). Values for the model parameters observed under different conditions are discussed in Chapter 4.

To estimate the parameters in (2.63), the first two moments of the Doppler spectrum, $H(n,k)$ can be used where n and k are the Doppler and range bin, respectively [87],

$$\begin{aligned} \widehat{m}_f(k) &= \frac{1}{h_n}\sum_{n=1}^{N}\left(\frac{nf_r}{N} - \frac{f_r}{2}\right)H(n,k), \\ \widehat{w}_{sp}^2(k) &= \frac{1}{h_n}\left(\sum_{n=1}^{N}\left(\frac{nf_r}{N} - \frac{f_r}{2} - \widehat{m}_f(k)\right)^2 H(n,k) - H_{noise}\frac{f_r^3}{12}\right) \end{aligned} \tag{2.64}$$

where w_{sp} is the width of the total Doppler spectrum, N is the CPI length, f_r is the PRF, H_{noise} is the noise power spectral density, $h_n = \left(\sum_{n=1}^{N} H(n,k) - H_{noise}f_r\right)$ is the normalising factor and the last term in (2.64) is a correction for the thermal noise contribution to the spectral width. To get an accurate measurement of underlying clutter spectra width, it is important to account for broadening due to the motion of the aircraft and the effect of the CPI length. If the broadening factor is modelled as a Gaussian function, then its standard deviation is given by

$$\sigma_{plat} = \frac{v_p\theta_{3dB}|\sin\theta_{plat}|}{\sqrt{2\ln 2}\lambda} \tag{2.65}$$

where θ_{plat} is the radar look direction relative to the platform heading, v_p is the platform velocity, λ is the radar wavelength and $\theta_{3\text{dB}}$ is the two-way azimuth 3 dB beamwidth. If is a window function applied in the time domain, then $Q_w(\cdot)$ is defined as the power spectral density of the window function. The effect of the CPI length can then be determined by calculating the normalised second-order moment,

$$\sigma_{\text{CPI}} = \left[\frac{\int_{-f_r/2}^{f_r/2} Q_w(f) f^2 df}{\int_{-f_r/2}^{f_r/2} Q_w(f) df} \right]^{1/2} \tag{2.66}$$

and hence the underlying spectral width of the clutter is given by

$$\widehat{w}^2(k) = \widehat{w}_{\text{sp}}^2(k) - \sigma_{\text{plat}}^2 - \sigma_{\text{CPI}}^2. \tag{2.67}$$

2.6 Texture correlation

The correlation of the texture can be characterised by two distinct components. The first is a correlation over a time period on the order of seconds, while the second component is along the range direction and is commonly referred to as spatial correlation. These are typically considered independently, as the characterisation of the two-dimensional texture is difficult and may not be necessary for short time periods or short-range extents. It is also difficult for airborne platforms to characterise the temporal component unless the radar is ground stabilised. The spatial correlation for the nth pulse and kth range bin can be determined by

$$\rho_{\text{spat}}(n,k) = \sum_{l=1}^{K-k-1} (Z(n,l+k) - \langle \mathbf{Z}(n) \rangle)(Z^*(n,l) - \langle \mathbf{Z}^*(n) \rangle) \tag{2.68}$$

where Z is the multilook intensity and K is the number of range bins. Similarly, for the temporal correlation,

$$\rho_{\text{temp}}(n,k) = \sum_{m=1}^{N-n-1} (Z(n+m,k) - \langle \mathbf{Z}(k) \rangle)(Z^*(m,k) - \langle \mathbf{Z}^*(k) \rangle) \tag{2.69}$$

where N is the number of pulses.

An example horizontally polarised range/time intensity plot from the Ingara X-band data set is shown in Figure 2.29 for a range extent of 150 m and a period of 0.8 s. The data intensities have been averaged over groups of 32 pulses to smooth the speckle component. This has revealed the periodic nature of the returns along range, although it may be noted that the sea wavelength is not uniform. Over time a small variation in range (range walk) is visible due to the motion of the waves, although this may also have been affected by uncompensated aircraft motion. The spatial correlation of the underlying clutter intensity is shown in Figure 2.30 with the correlation scaled

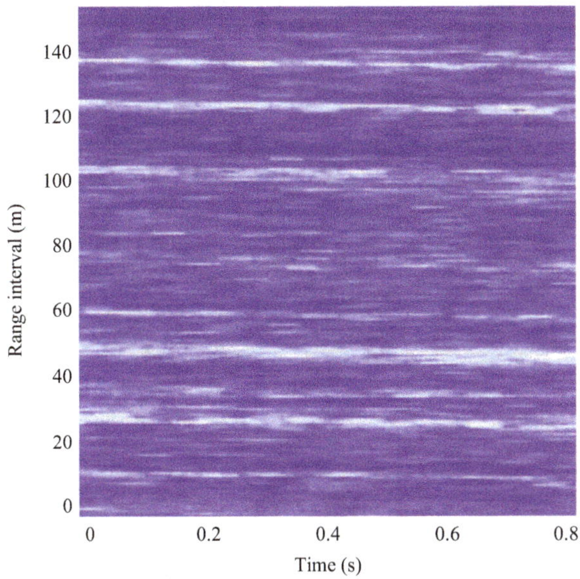

Figure 2.29 Horizontally polarised range/time intensity of the Ingara X-band data set, averaged over groups of 32 pulses

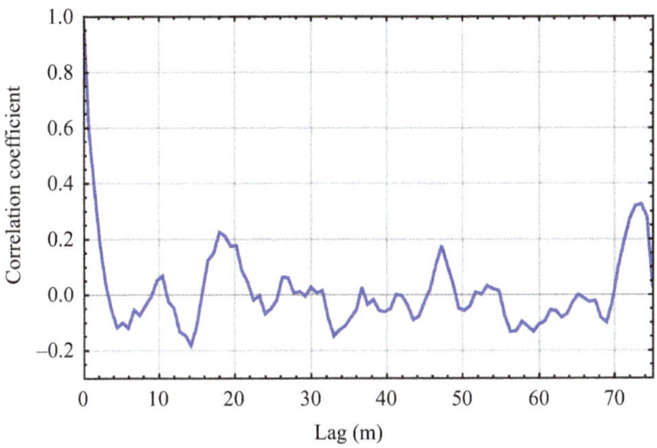

Figure 2.30 Average correlation coefficient of clutter intensity over range using the data from Figure 2.29

to have a maximum of 1. It may be noted that the correlation coefficient does not reduce to zero over a lag of 75 m due to the periodic and non-stationary nature of the returns.

As a further example, Figure 2.31 shows the two-sided spatial correlation for the vertical polarisation of the IPIX ground-based radar [88]. The data has been

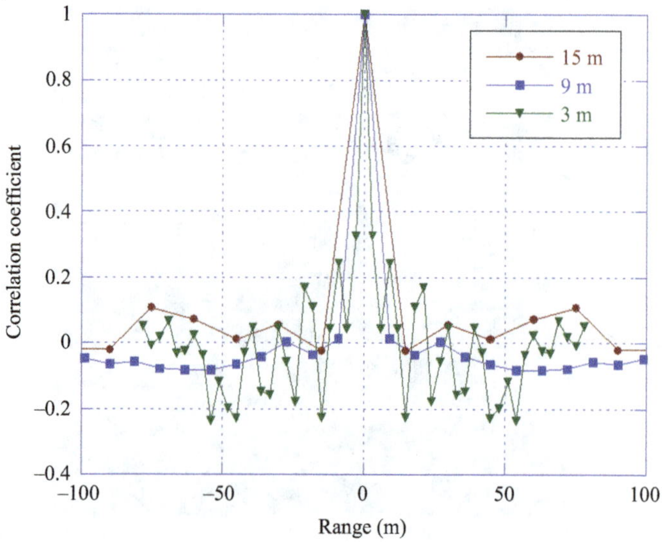

Figure 2.31 Spatial correlation for the IPIX data with vertical polarisation and different spatial resolutions [88]

averaged over 128 pulses to remove the speckle and processed for three different range resolutions, 3, 9 and 15 m. This result highlights that it is possible to resolve different periodicities by processing the data at finer resolutions.

References

[1] K. D. Ward, R. J. A. Tough, and S. Watts, *Sea clutter: Scattering, the K-distribution and radar performance*, 2nd ed. The Institute of Engineering Technology, London, 2013.

[2] F. E. Nathanson, J. P. Reilly, and M. N. Cohen, *Radar design principles*, 2nd ed. McGraw-Hill, New York, 1991.

[3] V. Gregers-Hansen and R. Mittal, "An improved empirical model for radar sea clutter reflectivity," *IEEE Transactions on Aerospace and Electronic Systems*, vol. 48, no. 4, pp. 3512–3524, 2012.

[4] G. Soriano and C. Guerin, "A cutoff invariant two-scale model in electromagnetic scattering from sea surfaces," *IEEE Geoscience and Remote Sensing Letters*, vol. 5, no. 2, pp. 199–203, 2008.

[5] P. Spiga, "Diffraction des ondes électromagnétiques par des surfaces rugueuses en incidence rasante," Ph.D. dissertation, 2008.

[6] S. Angelliaume, V. Fabbro, G. Soriano, and C. A. Guerin, "The GO-SSA extended model for all-incidence sea clutter modelling," *IEEE Geoscience and Remote Sensing Symposium*, 2014, pp. 5017–5020.

[7] M. I. Skolnik, *Introduction to radar systems*, 3rd ed. McGraw-Hill, New York, 2001.
[8] M. I. Skolnik, *Radar handbook*, 3rd ed. McGraw-Hill, New York, 2008.
[9] J. N. Briggs, *Target detection by marine radar*. The Institution of Electrical Engineers, London, 2004.
[10] L. Rogers, C. P. Hattan, and J. L. Krolik, "Using radar sea echo to estimate surface layer refractivity profiles," *International Geoscience and Remote Sensing Symposium*, vol. 1, pp. 658–662, 1999.
[11] S. Vasudevan, R. Anderson, S. Kraut, P. Gerstoft, L. Rogers, and J. Krolik, "Recursive Bayesian electromagnetic refractivity estimation from radar sea clutter," *Radio Science*, vol. 42, 2007.
[12] W. L. Patterson, "Advanced refractive effects prediction system (AREPS)," Space and Naval Warfare Systems Center, US Navy, Tech. Rep., 2021.
[13] T. Elfouhaily and C. A. Guérin, "A critical survey of approximate scattering wave theories from random rough surfaces," *Waves in Random and Complex Media*, vol. 14, no. 4, pp. 1–40, 2004.
[14] V. Kudryavtsev, D. Hauser, G. Caudal, and B. Chapron, "A semi-empirical model of the normalized radar cross-section of the sea surface, part I: The background model," *Journal of Geophysical Research*, vol. 108, pp. 2-1–2-24, 2003.
[15] V. Kudryavtsev, D. Hauser, G. Caudal, and B. Chapron, "A semi-empirical model of the normalized radar cross-section of the sea surface, part II: Radar modulation transfer function," *Journal of Geophysical Research*, vol. 108, pp. 3-1–3-16, 2003.
[16] D. E. Barrick, "Grazing behaviour of scatter and propagation above any rough surface," *IEEE Transactions on Antennas and Propagation*, vol. 46, pp. 73–83, 1998.
[17] V. I. Tatarskii and M. I. Charnotskii, "On the universal behaviour of scattering from a rough surface for small grazing angles," *IEEE Transactions on Antennas and Propagation*, vol. 46, pp. 67–72, 1998.
[18] C. Bourlier, "Azimuthal harmonic coefficients of the microwave backscattering from a non-Gaussian ocean surface with the first-order SSA model," *IEEE Transactions on Geoscience and Remote Sensing*, vol. 42, no. 11, pp. 2600–2611, 2004.
[19] Z. Guerraou, S. Angelliaume, L. Rosenberg, and C. Gurin, "Investigation of azimuthal variations from X-band medium-grazing-angle sea clutter," *IEEE Transactions on Geoscience and Remote Sensing*, vol. 54, no. 10, pp. 6110–6118, 10 2016.
[20] B. Spaulding, D. Horton, and H. Pham, "Wind aspect factor in sea clutter modeling," *International Radar Conference*, 2005, pp. 89–92.
[21] T. Johnsen, "Characterization of X-band radar sea-clutter in a limited fetch condition from low to high grazing angles," *IEEE Radar Conference*, 2015, pp. 109–114.
[22] X. Deng, C. López-Martínez, J. Chen, and P. Han, "Statistical modeling of polarimetric SAR data: A survey and challenges," *Remote Sensing*, vol. 9, pp. 1–34, 2017.

[23] G. V. Trunk and S. F. George, "Detection of targets in non-Gaussian sea clutter," *IEEE Transactions*, vol. AES-6, pp. 620–628, 1978.

[24] F. A. Fay, J. Clarke, and R. S. Peters, "Weibull distribution applications to sea clutter," *IEEE Conference Publication*, no. 155, 1977, p. 1012.

[25] E. Jakeman and P. N. Pusey, "A model for non-Rayleigh sea echo," *IEEE Transactions on Antennas and Propagation*, vol. AP-24, no. 6, pp. 806–814, 1976.

[26] K. D. Ward, "Compound representation of high resolution sea clutter," *Electronic Letters*, vol. 17, no. 16, pp. 561–563, 1981.

[27] P. R. Krishnaiah and L. Lin, "Complex elliptically symmetric distributions," *Communications in Statistics – Theory and Methods*, vol. 15, no. 12, pp. 3693–3718, 1986.

[28] E. Ollila, D. E. Tyler, V. Koivunen, and H. V. Poor, "Complex elliptically symmetric distributions: Survey, new results and applications," *IEEE Transactions on Signal Processing*, vol. 60, no. 11, pp. 5597–5625, 2012.

[29] K. D. Ward, R. J. A. Tough, and P. W. Shepherd, "Modelling sea clutter: correlation, resolution and non-Gaussian statistics," *IEEE International Radar Conference*, 1997, pp. 95–99.

[30] S. Watts, "Radar detection prediction in K-distributed sea clutter and thermal noise," *IEEE Transactions on Aerospace and Electronic Systems*, vol. AES-23, no. 1, pp. 40–45, 1987.

[31] S. Bocquet, "Parameter estimation for Pareto and K distributed clutter with noise," *IET Radar Sonar and Navigation*, vol. 9, no. 1, pp. 104–113, 2015.

[32] D. Blacknell and R. J. A. Tough, "Parameter estimation for the K-distribution based on [z log(z)]," *IEE Proceedings of Radar, Sonar and Navigation*, vol. 148, no. 6, pp. 309–312, 2001.

[33] L. Rosenberg, D. J. Crisp, and N. J. Stacy, "Analysis of the KK-distribution with medium grazing angle sea-clutter," *IET Proceedings of Radar Sonar and Navigation*, vol. 4, no. 2, pp. 209–222, 2010.

[34] M. Sekine and Y. Mao, *Weibull radar clutter*. IEE Radar, Sonar, Navigation and Avionics Series, Peter Peregrinus, London, 1990.

[35] L. Rosenberg, S. Watts, and S. Bocquet, "Application of the K+Rayleigh distribution to high grazing angle sea-clutter," *International Radar Conference*, 2014, pp. 1–6.

[36] A. Balleri, A. Nehorai, and J. Wang, "Maximum likelihood estimation for compound-Gaussian clutter with inverse gamma texture," *IEEE Transactions on Aerospace and Electronic Systems*, vol. 43, no. 2, pp. 775–779, 2007.

[37] Y. Dong, "Distribution of X-band high resolution and high grazing angle sea clutter," Defence Science Technology Organisation, Research Report DSTO-RR-0316, 2006.

[38] K. D. Ward and R. J. A. Tough, "Radar detection performance in sea clutter with discrete spikes," *International Radar Conference*, 2002, pp. 15–17.

[39] C. H. Gierull and I. C. Sikaneta, "A compound-plus-noise model for improved vessel detection in non-Gaussian SAR imagery," *IEEE Transactions on Geoscience and Remote Sensing*, pp. 1444–1453, 2017.

[40] J. Carretero-Moya, J. Gismero-Menoyo, Y. Blanco-del-Campo, and A. Asensio-Lopez, "Statistical analysis of a high-resolution sea-clutter database," *IEEE Transactions on Geoscience and Remote Sensing*, vol. 48, no. 4, pp. 2024–2037, 2010.

[41] E. Ollila, D. E. Tyler, V. Koivunen, and H. V. Poor, "Compound-Gaussian clutter modeling with an inverse Gaussian texture distribution," *IEEE Signal Processing Letters*, vol. 19, pp. 876–879, 2012.

[42] A. Mezache, F. Soltani, M. Sahed, and I. Chalabi, "Model for non-Rayleigh clutter amplitudes using compound inverse Gaussian distribution: An experimental analysis," *IEEE Transactions on Aerospace and Electronic Systems*, vol. 51, no. 1, pp. 142–153, 2015.

[43] T. Lamont-Smith, "Power transforms for the Weibull and the K-distribution," *IET Radar Sonar and Navigation*, vol. 13, pp. 522–529, 2019.

[44] K. D. Ward, C. J. Baker, and S. Watts, "Maritime surveillance radar. I. Radar scattering from the ocean surface," *Radar and Signal Processing, IEE Proceedings F*, vol. 137, no. 2, pp. 51–62, 1990.

[45] J. A. Northrop and A. Papandreou-Suppappola, "Computationally efficient estimation of compound K-distribution sea clutter in thermal noise and its application to sea echo reflectivity observations," *IEEE Transactions on Aerospace and Electronic Systems*, vol. 56, no. 3, pp. 2340–2350, 2020.

[46] M. A. Sletten, "Multipath scattering in ultrawide-band radar sea spikes," *IEEE Transactions on Antennas and Propagation*, vol. 46, no. 1, pp. 45–56, 1998.

[47] T. Lamont-Smith, "Translation to the normal distribution for radar clutter," *IEE Proceedings of Radar, Sonar and Navigation*, vol. 147, no. 1, pp. 17–22, 2000.

[48] P. Fayard and T. R. Field, "Optimal inference of the inverse gamma texture for a compound-Gaussian clutter," *IEEE International Conference on Acoustics, Speech and Signal Processing*, 2009, pp. 2969–2972.

[49] M. Farshchian and F. L. Posner, "The Pareto distribution for low grazing angle and high resolution X-band sea clutter," *IEEE Radar Conference*, 2010, pp. 789–793.

[50] G. V. Weinberg, "Assessing Pareto fit to high-resolution high-grazing-angle sea clutter," *IET Electronic Letters*, vol. 47, no. 8, pp. 516–517, 2011.

[51] L. Rosenberg and S. Bocquet, "Application of the Pareto plus noise distribution to medium grazing angle sea-clutter," *IEEE Journal of Selected Topics in Applied Earth Observations and Remote Sensing*, vol. 8, no. 1, pp. 255–261, 2015.

[52] D. Middleton, "New physical-statistical methods and model for clutter and reverberation: The KA-distribution and related probability structures," *IEEE Journal of Oceanic Engineering*, vol. 24, no. 3, pp. 261–284, 1999.

[53] S. Watts, K. D. Ward, and R. J. A. Tough, "The physics and modelling of discrete spikes in radar sea clutter," *International Radar Conference*, 2005, pp. 72–77.

[54] D. W. J. Stein, "A robust exponential mixture detector applied to radar," *IEEE Transactions on Aerospace and Electronic Systems*, vol. 35, no. 2, pp. 519–532, 1999.
[55] C. H. Gierull and I. C. Sikaneta, "Improved SAR vessel detection based on discrete texture," *European SAR Conference*, 2016, pp. 523–526.
[56] S. Angelliaume, L. Rosenberg, and M. Ritchie, "Modelling the amplitude distribution of radar sea clutter," *Remote Sensing*, vol. 11, no. 319, 2019.
[57] S. Bocquet, L. Rosenberg, and C. H. Gierull, "Parameter estimation for the trimodal discrete radar clutter model," *IEEE Transactions of Geoscience and Remote Sensing*, vol. 58, no. 10, pp. 7062–7073, 2020.
[58] L. Rosenberg and S. Angelliaume, "Characterisation of the tri-modal discrete sea clutter model," *International Radar Conference*, 2018, pp. 1–6.
[59] L. Rosenberg and V. Duk, "Land clutter statistics from an airborne passive bistatic radar," *IEEE Transactions on Geoscience and Remote Sensing*, 2021.
[60] F. Gini, M. S. Greco, A. Farina, and P. Lombardo, "Optimum and mismatched detection against K-distributed plus Gaussian clutter," *IEEE Transactions on Aerospace and Electronic Systems*, vol. 34, no. 3, pp. 860–876, 1998.
[61] P. Lombardo, C. J. Oliver, and R. J. A. Tough, "Effect of noise on order parameter estimation for K-distributed clutter," *IEE Proceedings of Radar, Sonar and Navigation*, vol. 142, no. 1, pp. 33–40, 1995.
[62] D. Walker, "Doppler modelling of radar sea clutter," *IEE Proceedings of Radar, Sonar and Navigation*, vol. 148, no. 2, pp. 73–80, 2001.
[63] L. Rosenberg and N. J. Stacy, "Analysis of medium angle X-band sea-clutter Doppler spectra," *IEEE Radar Conference*, 2008, pp. 1898–1903.
[64] L. Rosenberg, D. J. Crisp, and N. J. Stacy, "Characterisation of low-PRF X-band sea-clutter Doppler spectra," *International Radar Conference*, 2008, pp. 100–105.
[65] A. T. Jessup, W. K. Melville, and W. C. Keller, "Breaking waves affecting microwave backscatter, 1. detection and verification," *Journal of Geophysical Research*, vol. 96, no. C11, pp. 20547–20559, 1991.
[66] L. Liu and S. J. Frasier, "Measurement and classification of low-grazing-angle radar sea spikes," *IEEE Transactions on Antennas and Propagation*, vol. 46, no. 1, pp. 27–40, 1998.
[67] H. W. Melief, H. Greidanus, P. van Genderen, and P. Hoogeboom, "Analysis of sea spikes in radar sea clutter data," *IEEE Transactions on Geoscience and Remote Sensing*, vol. 44, no. 4, pp. 985–993, 2006.
[68] D. Walker, "Model and characterisation of radar sea clutter," Ph.D. dissertation, 2001.
[69] M. Greco, P. Stinco, and F. Gini, "Identification and analysis of sea radar clutter spikes," *IET Journal of Radar, Sonar and Navigation*, vol. 4, no. 2, pp. 239–250, 2010.
[70] L. Rosenberg, "Sea-spike detection in high grazing angle X-band sea-clutter," *IEEE Transactions on Geoscience and Remote Sensing*, vol. 51, no. 8, pp. 4556–4562, 2013.

[71] L. B. Wetzel, *Radar handbook*, 3rd ed. McGraw Hill, New York, 2008, ch. Sea Clutter.
[72] P. H. Y. Lee, J. D. Barter, B. M. Lake, and H. R. Thompson, "Lineshape analysis of breaking-wave Doppler spectra," *IEE Proceedings of Radar, Sonar and Navigation*, vol. 145, no. 2, pp. 135–139, 1998.
[73] T. Lamont-Smith, "Investigation of the variability of Doppler spectra with radar frequency and grazing angle," *IEE Proceedings of Radar, Sonar and Navigation*, vol. 151, no. 5, pp. 291–298, 2004.
[74] W. K. Melville, A. D. Rozenberg, and D. C. Quigley, "Laboratory study of polarized microwave scattering at grazing incidence," *IEEE International Geoscience and Remote Sensing Symposium*, 1995, pp. 951–953.
[75] N. J. S. Stacy, M. Preiss, and D. Crisp, "Polarimetric characteristics of X-band SAR sea clutter," *IEEE International Conference on Geoscience and Remote Sensing*, 2006, pp. 4017–4020.
[76] L. Rosenberg, "Characterisation of high grazing angle X-band sea-clutter Doppler spectra," *IEEE Transactions on Aerospace and Electronic Systems*, vol. 50, no. 1, pp. 406–417, 1 2014.
[77] M. Greco, F. Bordoni, and F. Gini, "X-band sea-clutter nonstationary: Influence of long waves," *IEEE Journal of Oceanic Engineering*, vol. 29, no. 2, pp. 269–293, 2004.
[78] G. Davidson, "Simulation of coherent sea clutter," *IET Radar Sonar and Navigation*, vol. 4, no. 2, pp. 168–177, 2010.
[79] S. Watts, "Modeling and simulation of coherent sea clutter," *IEEE Transactions on Aerospace and Electronic Systems*, vol. 48, no. 4, pp. 3303–3317, 2012.
[80] S. Watts, L. Rosenberg, S. Bocquet, and M. Ritchie, "The Doppler spectra of medium grazing angle sea clutter; part 1: Characterisation," *IET Radar Sonar and Navigation*, vol. 10, no. 1, pp. 24–31, 2016.
[81] S. Watts, L. Rosenberg, S. Bocquet, and M. Ritchie, "The Doppler spectra of medium grazing angle sea clutter; part 2: Exploiting the models," *IET Radar Sonar and Navigation*, vol. 10, no. 1, pp. 32–42, 2016.
[82] M. Ritchie, A. G. Stove, S. Watts, K. Woodbridge, and H. D. Griffiths, "Application of a new sea clutter Doppler model," *IEEE International Radar Conference*, 2013, pp. 560–565.
[83] M. Ritchie, A. Stove, A. Woodbridge, and H. Griffiths, "NetRAD: Monostatic and bistatic sea clutter texture and Doppler spectra characterization at S-band," *IEEE Transactions on Geoscience and Remote Sensing*, vol. 54, no. 9, pp. 5533–5543, 9 2016.
[84] S. Kemkemian, L. Lupinski, J. Degurse, V. Corretja, R. Cottron, and S. Watts, "Performance assessment of multi-channel radars using simulated sea clutter," *International Radar Conference, Radar 2015*, 2015, pp. 1015–1020.
[85] L. Rosenberg and S. Watts, "Coherent simulation of sea-clutter for a scanning radar," *NATO SET-239 Workshop on Maritime Radar Surveillance from Medium and High Grazing Angle Platforms*, 2016, pp. 1–10.

[86] L. Rosenberg, S. Watts, and S. Bocquet, "Scanning radar simulation in the maritime environment," *IEEE Radar Conference*, 2020, pp. 1–6.
[87] S. Watts, L. Rosenberg, and M. Ritchie, "Characterising the Doppler spectra of high grazing angle sea clutter," *International Radar Conference*, 2014, pp. 1–6.
[88] M. Greco, F. Gini, and M. Rangaswamy, "Statistical analysis of measured polarimetric clutter data at different range resolutions," *IET Proceedings on Radar, Sonar and Navigation*, vol. 153, no. 6, pp. 473–481, 2006.

Chapter 3
Bistatic sea clutter

3.1 Introduction

Bistatic radar operation is defined by the Institute of Electrical and Electronic Engineers (IEEE) [1] as 'a radar using antennas for transmission and reception at sufficiently different locations that the angles or ranges to the targets are significantly different'. Where multiple monostatic and bistatic radar nodes are used together, the operation may be described as multistatic. Bistatic radars and their operation are described in detail in [2,3]. One of the attractions for using two or more receiving positions is the potential gain from exploiting different observation directions (angular diversity) and a potential increase in detection performance at microwave frequencies against stealthy targets. In military applications, bistatic operation also provides some protection for the receiver location, which is not radiating a radar signal. Passive bistatic radar, using illuminators of opportunity, also provides the potential for covert operations and operation in spectrally congested environments. In general, the potential benefits of active bistatic operation must be offset against the difficulty in synchronising the transmitter and receiver. This is more complicated in some geometries, as the area coverage may also involve 'pulse chasing' [2], which adds further system complexity. However, the trade-off between complexity and performance can only be made when the performance of bistatic systems is better understood.

For an assessment of bistatic radar detection performance in sea clutter, it is necessary to have a good understanding of the bistatic characteristics of both targets and clutter. At present, relatively little sea clutter data has been collected with bistatic radars when compared with monostatic radars. Consequently, there are still very few models available for bistatic sea clutter. This chapter looks at the bistatic characteristics of sea clutter. In Section 3.2, the bistatic geometry is described, which includes the definition of the bistatic range resolution and bistatic Doppler shift. In Section 3.3, a number of active bistatic measurement campaigns are described that have been reported in the open literature, together with the observed sea clutter characteristics. Section 3.4 then describes two passive bistatic radar experiments and presents an analysis of the sea clutter.

3.2 Bistatic geometry

The viewing geometry for a bistatic radar illuminating a surface clutter patch is shown in Figure 3.1 [4]. The grazing angles for the transmitter and receiver at the centre of the clutter patch are ϕ_1 and ϕ_2, the bistatic angle is β and the azimuth scattering angle is ψ_B. When $\psi_{B,deg} = 0°$ or $180°$, the scattering geometry is described as 'in-plane'; otherwise, it is known as 'out-of-plane'. For a monostatic radar, $\psi_{B,deg} = 180°$ and $\phi_1 = \phi_2$. For forward scattering, $\psi_{B,deg} = 0°$, with the case of $\phi_1 = \phi_2$ corresponding to specular scattering. The bistatic angle is given by

$$\beta = \cos^{-1}(\sin\phi_1 \sin\phi_2 - \cos\phi_1 \cos\phi_2 \cos\psi_B). \tag{3.1}$$

The bistatic range is $R_{Tx} + R_{Rx}$ and the minimum resolvable range difference for a point target in the slant plane is ΔR_B. The effective bistatic range resolution is illustrated in Figure 3.2(a). Assume two equal-sized point targets are situated at points A and B and lie along a line at an angle ψ to the bisector of the bistatic angle. These two targets will be resolved at the bistatic receiver if they are spaced by

$$\Delta R_B \approx \frac{c_0}{2B\cos(\beta/2)\cos\psi} \tag{3.2}$$

where B is the pulse bandwidth and c_0 is the speed of light. It is common to define the bistatic range resolution for $\psi = 0$ with the targets to be resolved lying along the bisector of the bistatic angle, as illustrated in Figure 3.2(b) with

$$\Delta R_B \approx \frac{c_0}{2B\cos(\beta/2)}. \tag{3.3}$$

Another common definition is when targets lie along a radial line from the transmitter or receiver, where $\psi = \beta/2$ and

$$\Delta R_B \approx \frac{c_0}{2B\cos^2(\beta/2)}. \tag{3.4}$$

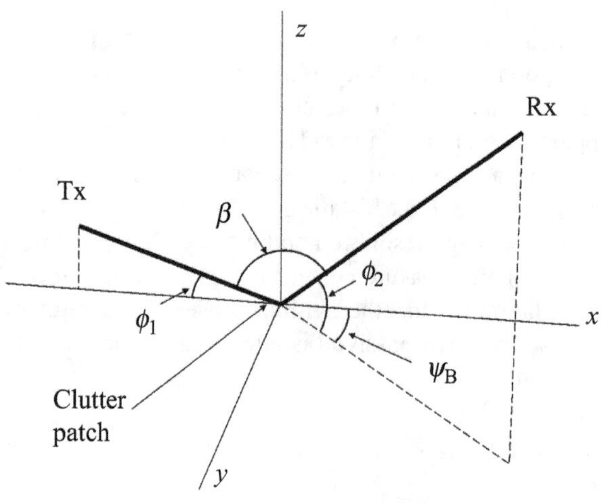

Figure 3.1 Bistatic geometry

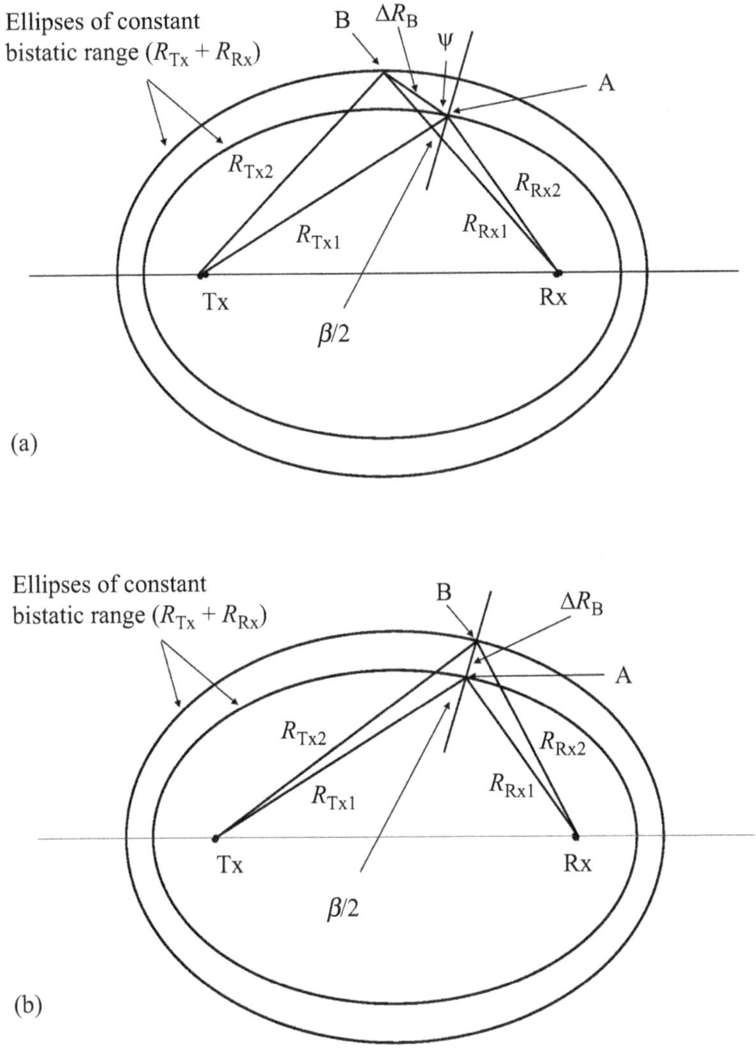

Figure 3.2 Bistatic range resolution: (a) targets A and B on a line at an angle ψ to the bisector of the bistatic angle, (b) targets A and B lying along the bisector of the bistatic angle

When $\beta = 0$ and $\psi = 0$, the standard result for a monostatic radar is obtained, $\Delta R_B = c_0/2B$. If the targets lie on the ground or sea surface, their resolvable range as measured along the surface must take into account the local grazing angles. In general, the bistatic range resolution projected onto the ground for targets lying along the bisector of the bistatic angle can be written as [5]

$$\Delta R_{BG} = \frac{c_0}{B\sqrt{\cos^2\phi_1 + \cos^2\phi_2 + 2\cos\beta\cos\phi_1\cos\phi_2}} \quad (3.5)$$

where ϕ_1 and ϕ_2 are the transmit and receive grazing angles, as shown in Figure 3.1. If $\phi_1 = \phi_2 = \phi$, then $\Delta R_{BG} \approx c_0/(2B\cos(\beta/2)\cos\phi)$ and if $\beta = 0$ the standard monostatic result is obtained with $\Delta R_{BG} = c_0/(2B\cos\phi)$.

The standard radar range equation can be readily adapted to the bistatic case with the received power proportional to $1/\left(R_{Tx}^2 R_{Rx}^2\right)$ and an equivalent monostatic range sometimes given as $\sqrt{R_{Tx}R_{Rx}}$. As shown in Figure 3.2, lines of constant bistatic range are ellipses and lines of constant bistatic signal power with constant $R_{Tx}R_{Rx}$ are known as ovals of Cassini [2].

When assessing clutter returns, the size of the clutter patch is determined by the pulse bandwidth and the intersection of the transmit and receive antenna patterns. The size of the illuminated clutter patch is shown in Figure 3.3 for idealised square antenna patterns. The size of the cell may be limited by the pulse length and the smaller of the two beamwidths (range-limited cell) or by the two beamwidths (beam-limited cell). The effective area of the cell is best evaluated by numerical integration, using knowledge of the antenna beamshapes, range resolution and viewing geometry. A simple approximation for the range-limited case at low grazing angles is given by [3]

$$A_{cl} \approx R_{Tx}\theta_{3dB,Tx}\frac{c_0}{2B}\sec^2(\beta/2), \qquad \theta_{3dB,Tx} \le \frac{R_{Rx}\theta_{3dB,Rx}}{R_{Tx}} \ll 2\text{ rad} \tag{3.6}$$

or

$$A_{cl} \approx R_{Rx}\theta_{3dB,Rx}\frac{c_0}{2B}\sec^2(\beta/2), \qquad \theta_{3dB,Rx} \le \frac{R_{Tx}\theta_{3dB,Tx}}{R_{Rx}} \ll 2\text{ rad} \tag{3.7}$$

where $\theta_{3dB,Tx}$ and $\theta_{3dB,Rx}$ are the transmit and receive 3 dB beamwidths, respectively.

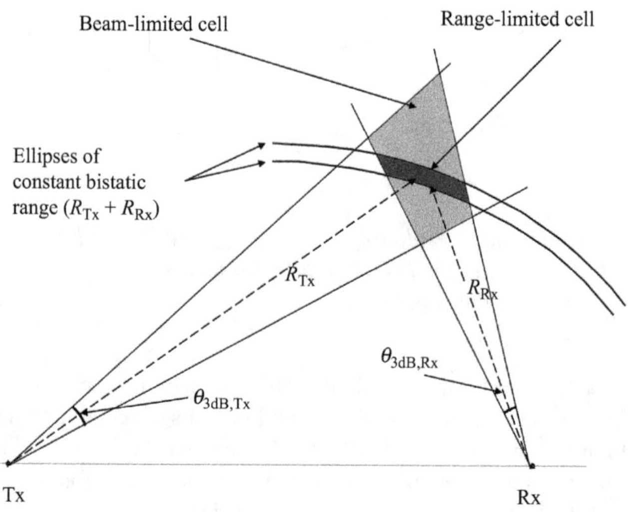

Figure 3.3 Clutter patch size for a bistatic geometry

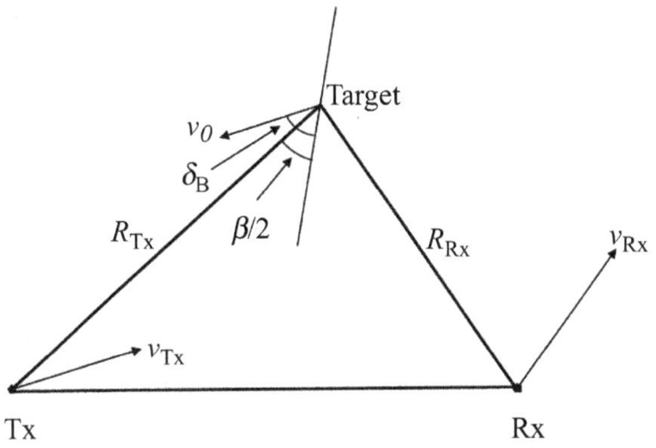

Figure 3.4 Bistatic geometry with a moving target, transmitter and receiver

The bistatic Doppler shift due to target and radar platform motion is also of interest in the study of bistatic clutter [2]. Figure 3.4 shows a target with bistatic range $R_{Tx} + R_{Rx}$ moving with velocity v_0 at some angle δ_B to the bisector of the bistatic angle. If the transmitter and receiver are stationary, then the Doppler shift at the receiver due to target motion is given by

$$f_B = \frac{1}{\lambda}\frac{d}{dt}(R_{Tx} + R_{Rx}) \tag{3.8}$$

Now dR_{Tx}/dt and dR_{Rx}/dt are determined by the resolved component of the target velocity along the lines of sight to the transmitter and receiver, respectively, so that

$$f_B = \frac{v_0}{\lambda}\left(\cos\left(\delta_B - \frac{\beta}{2}\right) + \cos\left(\delta_B + \frac{\beta}{2}\right)\right) = \frac{2v_0}{\lambda}\cos\delta_B\cos\left(\frac{\beta}{2}\right). \tag{3.9}$$

If the transmitter and receiver are also moving with velocities v_{Tx} and v_{Rx}, respectively, the resolved combined velocities along the lines of sight of the transmitter and receiver to the target must be used. The geometry for this case is shown in Figure 3.4, where the effect of platform motion of the transmitter and/or receiver is more complex than the monostatic case and the Doppler spread across a clutter patch must be determined from the difference between the minimum and maximum Doppler shifts at different regions across the cell [2]. The solution for a specific set of conditions will depend on the range and bearing of the target, the range and angular resolutions of the bistatic radar and the velocity vectors of the transmitter and receiver platforms.

3.3 Active bistatic radar

In this section, a summary of the bistatic measurement campaigns reported in the open literature is described, together with the key sea clutter characteristics that have been observed. These include the in-plane and out-of-plane reflectivity, amplitude statistics, correlation, Doppler spectra and a study of sea spikes. Further detail on modelling the bistatic reflectivity is given in Chapter 4.

3.3.1 Sea clutter measurements

Compared to monostatic radar, relatively few bistatic measurement campaigns have been conducted. A comprehensive list up to the year 2007 is given in [3]. Since then, a significant number of measurements have been made with the NetRAD radar system. Table 3.1 describes the major measurement campaigns. Note that HH refers to horizontal transmit and receive, while VV is vertical transmit and vertical receive.

Johns Hopkins Applied Physics Laboratory

In 1965, Pidgeon [6] used a land-based, vertically polarised, C-band continuous wave (CW) radar to illuminate the sea. The receiver was in an aircraft and had both horizontal and vertically polarised receive channels. Measurements were made in-plane, with the aircraft flying over the transmitter from the land out to sea. The receive antenna was pointed in the direction of aircraft flight and had a broad elevation coverage. An area of interest on the sea surface was defined by the selection of two Doppler frequencies within the received clutter spectrum and was dependent on the aircraft height and speed as well as the intersection of the transmit and receive beam patterns. In a further set of trials, Pidgeon [7] made measurements at X-band, using a land-based pulsed transmitter having 40 kW peak power, a pulse length of 0.5 μs and horizontal polarisation. The receiver was in an aircraft that flew at 1,000 ft (304 m) towards and away from the transmitter in the direct line of the transmit illumination. The transmitter depression angle varied from 1.2° to 8° and the receiver depression angle varied from 12° to 55°.

Table 3.1 Bistatic sea clutter measurement campaigns

Organisation	Freq. band	Pol.	Sea state	$\psi_{B,deg}$	$\phi_{1,deg}$	$\phi_{2,deg}$	β_{deg}
Johns Hopkins University (1966–67)	C X	VV, VH HH	1–3, 5	180 180	0.2–3 1.2–8	10–90 12–55	10–87 11–37
GEC Stanmore (1967)	X	VV, HH	5	0, 165, 180	0–90	0–90	0–180
Georgia Institute of Technology (1982–84)	X	VV, HH	3–4	95–157	< 0.1	< 0.1	23–85
MIT Lincoln Labs (1991)	X	VV, VH	1	180	0.3	5–40	4.7–39.7
NetRAD (UCL/UCT) (2010–18)	S	VV, VH, HH, HV	3–4	180, 165, 150, 120, 90	0.7–4.0	0.7–4.0	0, 15, 30, 60, 90

General Electric Company (GEC) Stanmore

Domville [8–10] used an airborne CW X-band transmitter with receivers mounted in the same aircraft for monostatic measurements and in a second aircraft for bistatic measurements. Most of the measurements of sea clutter were obtained with the second aircraft flying astern with horizontal and vertical polarised antennas pointing along the line of flight, measuring in-plane scattering. A limited number of out-of-plane measurements were made for backscatter azimuth angles ranging from 180° (in-plane) to about 170°.

Georgia Institute of Technology

Ewell and Zehner [11] made simultaneous monostatic and out-of-plane bistatic measurements at X-band for bistatic angles covering 23°–85° and both vertical and horizontal polarisations. The transmitter was at a height of 22.9 m above the sea surface, the receiver was at a height of 8.2 m and the bistatic baseline was 3.5 km. The radar used a pulse repetition frequency of 1 kHz and a pulse length of 0.2 μs. Measurements were made with the transmitter pointing at either 90° or 150° relative to the baseline.

Massachusetts Institute of Technology (MIT) Lincoln Labs

Kochanski *et al.* [12] used a similar collection geometry to Pidgeon [6]. An X-band CW illuminator was situated at a height of 4.5 m with the centre of the illuminated patch sub-tending an angle of 0.3° to the transmitter. The receiver was in an aircraft that flew directly in-line with the transmitter and clutter patch, collecting data with depression angles varying from 7° to 40°.

NetRAD experiments

University College London (UCL), in conjunction with the University of Cape Town (UCT), developed a fully coherent bistatic radar systems operating at 2.4 GHz [13,14]. The netted radar or NetRAD system had one node with a transmitter and receiver and two further receive nodes that could be deployed over long baselines with synchronisation achieved using GPS and wireless links. Each antenna could be configured for operation with horizontal or vertical polarisations and the transmitted waveforms were generated digitally. The main radar parameters are given in Table 3.2, with a full description of the systems and the method used to synchronise the nodes given in [15].

Table 3.2 NetRAD radar parameters [15]

Parameter	Value
Frequency	2.4 GHz
Bandwidth	50 MHz
Monostatic range resolution	3 m
Noise figure	~5 dB
Pulse length	0.1–10 μs
Transmit power	500 W
Antenna gain	23 dBi
Antenna one-way 3 dB beamwidths	9° (azimuth)/11° (elevation)
Polarisation	Vertical/horizontal

Table 3.3 NetRAD clutter measurement campaigns in South Africa [15]

Year	Month	Site	Pol.	Baseline (m)	β_{deg}	Tx, Rx height (m)	ϕ_{deg}
2010	5 Oct	Cape Peninsular	HH, VV, HV	416	15, 30, 60, 90	72, 65	
2010	10 Oct	Cape Peninsular	HH, VV	1827	60, 90, 120	80, 65	
2011	June	Misty Cliffs	HH, VH, VV	1830	65, 75, 85, 90, 95, 105, 120		0.7–1.12

Early measurements were made at Peacehaven in Southern England, and subsequently in South Africa, which provided most of the sea clutter measurements reported in the literature. The key parameters for the trials in South Africa are given in Table 3.3. In all cases, the transmitter and receiver were at similar heights and the antennas were pointed to achieve the required bistatic angle with the clutter patch at the vertex of an isosceles triangle relative to the baseline. As a result, $R_{\text{Tx}} = R_{\text{Rx}}$ and $\phi = \phi_1 = \phi_2$, with their values determined by the baseline and the value of β selected. During these collections, the radar was located on a coastal bay facing the Atlantic Ocean, and the wind and wave directions were approximately from the North West.

3.3.2 In-plane reflectivity

In-plane reflectivity measurements are slightly easier to interpret as the azimuth angle $\psi_{\text{B,deg}} = 0°$ or $180°$. The values of reflectivity reported by Pidgeon [6] at C-band and sea states 1–3 are shown in Figure 3.5. These were not fully developed seas and the wind speeds measured for sea states 1, 2 and 3 were 2.6 m/s (5 kts) crossing, 10.3–15.4 m/s (20–30 kts) crossing and 5.2 m/s (10 kts) down-wind, respectively. The receiver grazing angle values were in the region $10° \leq \phi_{2,\text{deg}} \leq 90°$ and it was found that the reflectivity was determined by the transmitter grazing angle, ϕ_1. The VV polarisation values were consistently higher than VH, but with smaller differences at higher grazing angles. Pidgeon noted that under these conditions, the bistatic reflectivity was the same order of magnitude as the equivalent monostatic reflectivity over the same grazing angles as the transmitter.

In a further set of trials [7], Pidgeon made measurements at X-band for wind speeds corresponding to a Douglas sea state between 4 and 5. He obtained reflectivity values of about -35 dBm^2/m^2, independent of the transmitter grazing angle, $1.2° \leq \phi_{1,\text{deg}} \leq 5°$ or the receiver grazing angle, $12° \leq \phi_{2,\text{deg}} \leq 55°$. Kochanski [12] made measurements at X-band using a similar CW arrangement to Pidgeon. The sea was calm (sea state 1) and values of bistatic reflectivity were observed in the range $-65 \leq \sigma^0_{\text{B,dB}} \leq -35$ dBm^2/m^2, apparently independent of the transmit and receive grazing angles. An average value of -45 dBm^2/m^2 was reported.

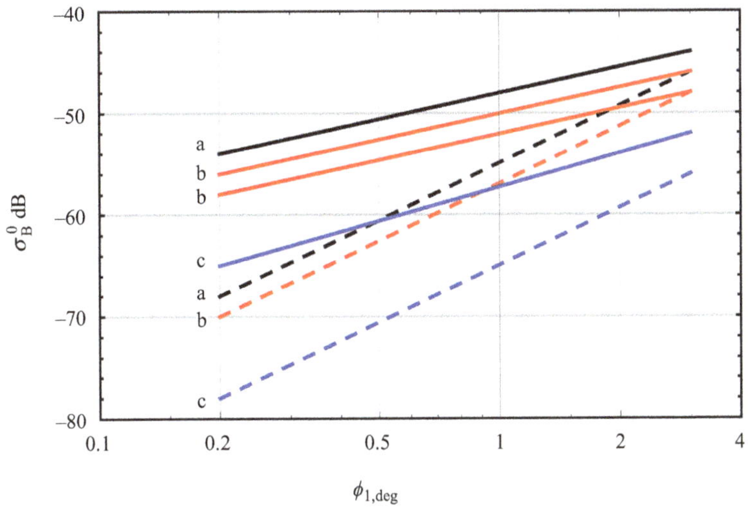

Figure 3.5 Bistatic C-band reflectivity reported by Pidgeon with $10° \leq \phi_{2,deg} \leq 90°$. a: Sea state 3, b: sea state 2, c: sea state 1. Solid lines: VV, dashed lines: VH [6]

Domville [8–10] made in-plane and out-of-plane measurements at X-band for both backscatter and forward scatter with wind speeds in the range 4–10 m/s. It was observed for backscatter geometries with very low transmit grazing angles that the values of reflectivity were almost independent of the receive grazing angle. Kochanski [12] reported that his measurements agreed with those of Domville. Domville produced an empirical model for in-plane bistatic reflectivity that is discussed in Chapter 4.

3.3.3 *Out-of-plane reflectivity*

Domville made a small number of out-of-plane measurements and reported that small changes of $\psi_{B,deg}$ away from 180° or 0° resulted in a reduction in reflectivity. For backscatter with VV polarisation, this was 1 dB per degree over the region $170° \leq \psi_{B,deg} \leq 180°$. For forward scatter over the region $0° \leq \psi_{B,deg} \leq 30°$, reflectivity with the HH polarisation reduced by 0.39 dB per degree and with the VV polarisation, by 0.33 dB per degree. The results are symmetrical with equivalent behaviour for $\psi_{B,deg} \geq 180°$ or $\psi_{B,deg} \leq 0°$.

Ewell and Zehner [11] reported three sets of simultaneous measurements of monostatic and out-of-plane bistatic reflectivity with transmit and receive grazing angles that were very low in all cases. Table 3.4 gives some examples of the bistatic geometry for the case when the transmitter was pointed at 90° to the baseline. On days 1 and 2, the visually estimated wave heights were 1.2–1.8 m and on day 3 they were 0.9 m. The seas on day 2 were described as 'somewhat disorganised' and the transmitter was either pointed at 90° or 150° to the bistatic baseline with the

Table 3.4 Bistatic geometry for trials reported by Ewell and Zehner [11]

β_{deg}	R_{Tx} **(km)**	R_{Rx} **(km)**	$\phi_{1,deg}$	$\phi_{2,deg}$
20	9.6	10.2	0.1	0.01
60	2.0	4.0	0.64	0.09

Table 3.5 σ_B^0/σ^0 in dB measured by Ewell and Zehner for different bistatic angles, β [11]

	Day	**VV**					**HH**				
β_{deg}		22	32	36	45	60	22	32	36	45	60
$\sigma^0_{B,dB} - \sigma^0_{dB}$, Tx. 90°	1	–5	–5	–14.5	–23	–26	–6	–5	–11	–17.5	–19.5
$\sigma^0_{B,dB} - \sigma^0_{dB}$, Tx. 90°	3	0	–4.5	–7	–9	–9	–4	–6.5	–7	–10	–12
β_{deg}		24	34	51	84		24	34	51	84	
$\sigma^0_{B,dB} - \sigma^0_{dB}$, Tx. 150°	2	–1.5	–5	–3	–6		–8	–8	–11	–11	

receiver pointing direction adjusted to achieve the required bistatic angle. The results summarised in Table 3.5 show the ratio of median bistatic to monostatic reflectivities as a function of bistatic angle. The overall trend was that the bistatic reflectivity was always less than the monostatic reflectivity and the bistatic reflectivity reduced as the bistatic angle increased.

In 2010, two significant calibrated sets of bistatic reflectivity measurements were made with the NetRAD system in sea states 2 and 4, respectively, [16–18]. Table 3.6 shows the bistatic geometry and reflectivity values obtained in sea state 2, with a bistatic baseline of 416 m, while Table 3.7 shows the results obtained in sea state 4, over a baseline of 1,827 m. Only those results which had sufficient clutter-to-noise ratio (CNR) to estimate reflectivity values have been included. For the sea state 2 measurements in Table 3.6, $\sigma^0_{B,dB} - \sigma^0_{dB}$ was effectively 0 dB for $\beta_{deg} \leq 60°$, while for $\beta_{deg} = 90°$, $\sigma^0_{B,dB} - \sigma^0_{dB} \approx -10$ dB for the VV polarisation and -16 dB for HH. Another interesting result is that for the HV polarisation, $\sigma^0_B > \sigma^0$ for $\beta_{deg} = 90°$ which is probably due to the high polarisation 'rotation' obtained at bistatic angles around 90°. The same phenomenon may also account for the much lower values of σ^0_B observed for both VV and HH polarisations with $\beta_{deg} = 90°$. For sea state 4, Table 3.7 shows measurements for $\beta_{deg} \geq 60°$. It may be noted that the values of σ^0_B were always less than σ^0 but there was little discernible variation with bistatic angle, β. For both of these sets of measurements, the time of data collection is given. It may be noted for the sea state 2 measurements with $\beta_{deg} = 90°$, the sea clutter reflectivity changed significantly over the period of 1 h.

Table 3.6 Reflectivity measurements with NetRAD on 5 October 2010, sea state 2 [17]

Time	β_{deg}	Pol.	R_{Tx} (m)	$\phi_{1,deg}$, $\phi_{2,deg}$	$\sigma^0_{B,dB}$	σ^0_{dB}
14:46	30		805	1.42	−59	−59
14:55	60	VV	418	2.75	−47.9	−47.1
15:09	90		296	3.88	−55	−44.5
16:08	90		296	3.88	−47.8	−37.6
16:26	90	HV	296	3.00	−55	−59.8
17:16	90	HH	296	3.88	−61	−44.8

Table 3.7 Reflectivity measurements with NetRAD on 10 October 2010, sea state 4 [17]

Time	β_{deg}	Pol.	R_{Tx} (m)	$\phi_{1,deg}$, $\phi_{2,deg}$	$\sigma^0_{B,dB}$	σ^0_{dB}
15:51	60		1,826	0.63	−51.5	−38.4
16:03	90	VV	1,292	0.89	−52.6	−38.6
16:17	120		1,055	1.09	−51	−38.8
12:20	60		1,826	0.63	−58.1	−51.2
12:33	60	HH	1,826	0.63	−58.2	−49.5
12:44	90		1,292	0.89	−57.1	−44.3
12:53	120		1,055	1.09	−58.9	−49.8

3.3.4 Amplitude statistics

Ewell and Zehner [11] reported that the amplitude distributions of their data were nearly log–normal in shape. They provided a scatter diagram of the log–normal standard deviation (LSD) for simultaneous monostatic and bistatic measurements. Typical values ranged from 4 to 18 dB for bistatic data and 5 to 9 dB for monostatic data. There was a tendency for the monostatic returns to have a larger LSD than the bistatic returns (i.e. they had a longer tail or were more spiky). The HH-polarised returns had a larger LSD than VV.

The amplitude statistics of the measurements made with NetRAD in 2010 (see Table 3.3) have been extensively analysed by Al-Ashwal [14,18], who generated fits to log–normal, Weibull, Pareto, K+noise and KA-distributions. The best fit in all cases was achieved with the KA-distribution. However, it is easier to make comparisons with other results by considering the fits to the K+noise distribution, which was still quite good in most cases. Tables 3.8 and 3.9 show results for the simultaneous monostatic shape parameter, ν and bistatic shape parameter, ν_B for the data sets obtained in sea states 2 and 4, respectively.

Fioranelli *et al.* [19] analysed a further data set collected in 2011, which included HH, VV, HV and VH monostatic and bistatic data. Figure 3.6 shows the fits to the

Table 3.8 Fits to the K+noise distribution for NetRAD measurements on 5 October 2010, sea state 2 [18]

Time	β_{deg}	**Pol.**	R_{Tx} **(m)**	$\phi_{1,deg}, \phi_{2,deg}$	ν_B	ν
14:46	30		805	1.42	3.0	1.2
14:55	60	VV	418	2.75	0.3	0.3
15:09	90		296	3.88	0.8	0.3
16:08	90		296	3.88	0.4	0.2
16:26	90	HV	296	3.00	0.5	0.1
17:16	90	HH	296	3.88	0.1	0.04

Table 3.9 Fits to the K+noise distribution for NetRAD measurements on 10 October 2010, sea state 4 [18]

Time	β_{deg}	**Pol.**	R_{Tx} **(m)**	$\phi_{1,deg}, \phi_{2,deg}$	ν_B	ν
15:51	60		1,826	0.63	0.5	0.6
16:03	90	VV	1,292	0.89	0.9	1.6
16:17	120		1,055	1.09	2.2	4.6
12:20	60		1,826	0.63	0.1	0.1
12:33	60	HH	1,826	0.63	0.2	0.2
12:44	90		1,292	0.89	0.3	0.1
12:53	120		1,055	1.09	1.2	0.5

K+noise shape parameter as a function of the bistatic angle. In these plots, the two bistatic nodes are N1 and N2, with N3 being the monostatic node. The dependency on bistatic angle, β, does not seem very strong except around $\beta_{deg} = 90°$ when the HV and VV bistatic signals all have larger values of shape parameter (i.e. they are less spiky) than at other values of β. The HH monostatic and bistatic returns were very spiky with $\nu \approx 0.1$. For the VV and VH polarisations, ν_B was generally greater (less spiky) than ν.

3.3.5 Correlation

The correlation of the texture can be characterised by two distinct components. The first is a correlation over a time period on the order of seconds, while the second component is along the range direction and is commonly referred to as spatial correlation. Calculation of the correlation is the same as for monostatic clutter and was described in Chapter 2.

For the NetRAD data shown in Figure 3.7, the spatial correlation has been analysed by Fioranelli *et al.* [19]. For this particular data set (sea state 4 with $\beta_{deg} = 85°$), there is clear evidence of a periodic structure in range that varies over time. It may

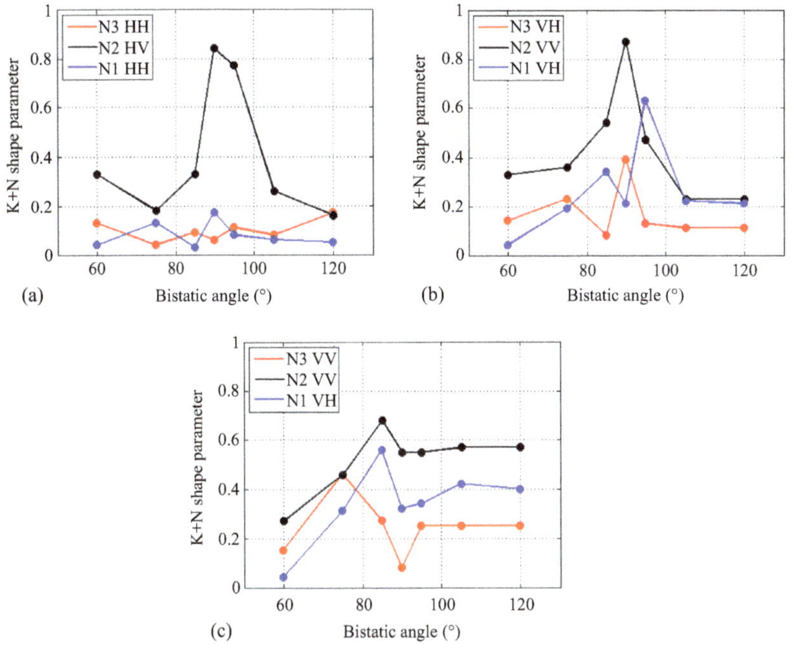

Figure 3.6 NetRAD K+ *noise shape parameter as a function of bistatic angle where N1 and N2 are bistatic nodes and N3 is the monostatic node: (a) data sets 1–7, (b) data sets 8–14, (c) data sets 15–21 [19]*

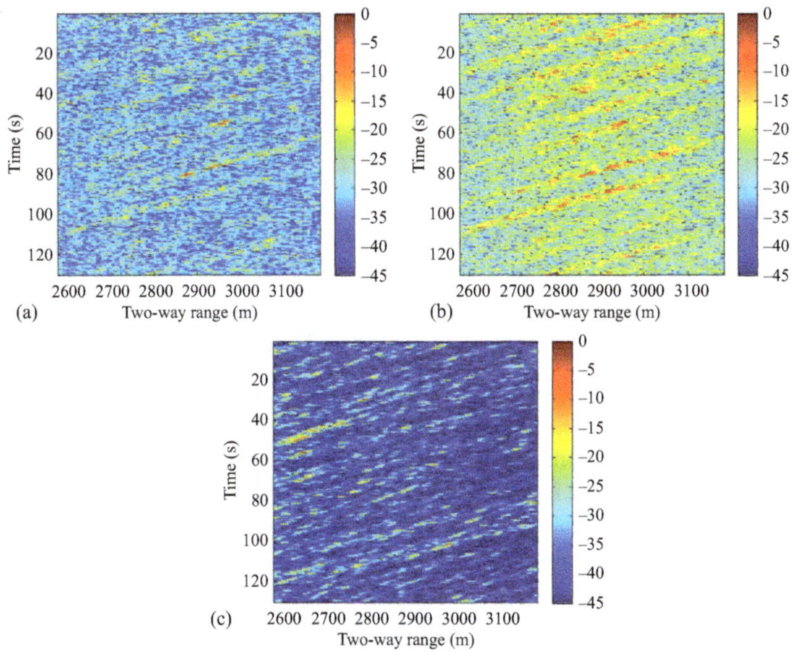

Figure 3.7 NetRAD examples of simultaneously collected range/time intensity plots for $\beta_{\mathrm{deg}} = 85°$: (a) bistatic HH, (b) bistatic HV, (c) monostatic HH [19]

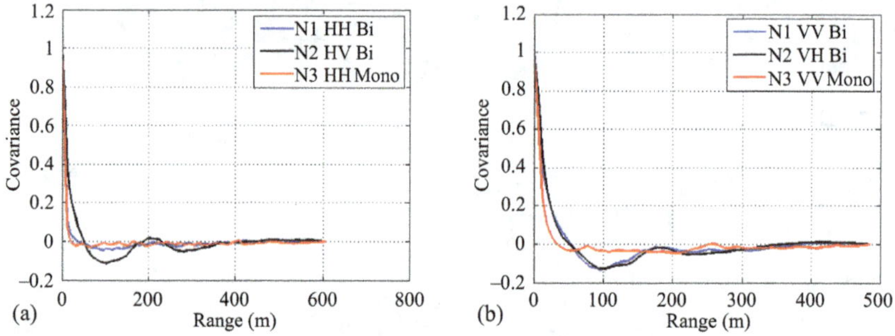

Figure 3.8 NetRAD correlation coefficients for co-polarised monostatic and co- and cross-polarised bistatic clutter intensity for $\beta_{\text{deg}} = 85°$: (a) horizontal transmit polarisation, (b) vertical transmit polarisation [19]

be noted that for this particular set of sea conditions and viewing geometry, the periodic structure of the monostatic data has a significantly higher spatial frequency than the bistatic data. The HH and HV bistatic plots show similar features, but with different relative intensities. This is clarified in Figure 3.8 which shows plots of the spatial correlation of the underlying texture. When averaged over range, the monostatic data shows little periodicity in the correlation coefficient. However, the bistatic data with VV, VH and HV polarisations do show the same strong periodic component. Palamà [20] studied further data sets and noted that while some showed very marked spatial structure, the correlation from the same area a few days later was entirely devoid of apparent structure.

3.3.6 Doppler spectra

The data collected by NetRAD has been extensively analysed to assess the characteristics of bistatic Doppler spectra and to compare them with monostatic characteristics [21,22]. In these publications, Ritchie *et al.* showed that bistatic Doppler spectra have many of the same characteristics as monostatic Doppler spectra, such as the time- and range variations described in [23] and Chapter 2. They showed that the clutter amplitude statistics varied with Doppler frequency, appearing to be very spiky at the edges of the spectra. This is evidence that clutter spectra cannot be modelled as a spherically invariant random process. It was also noted that the monostatic and bistatic spectra exhibited different mean Doppler centroids, depending on the wind direction and bistatic geometry. This could be exploited in multistatic systems by selecting the bistatic combination with the best clutter conditions for detecting a particular target.

More detailed analysis of the Doppler spectra in sea state 4 was undertaken in [24]. In this work, the Doppler spectra were fitted to the evolving Doppler

Table 3.10 NetRAD Doppler spectrum model fits reported in [24]

β_{deg}	Pol.	Bistatic			Monostatic		
		m_w (Hz)	A_c (Hz)	B_c (Hz)	m_w (Hz)	A_c (Hz)	B_c (Hz)
60		13.5	1	6	14.0	–21	0.5
90	HH	12.7	2	4	14.0	–21	–1.5
120		10.9	–2.5	3.5	13.9	–24	–4.5
60		12.3	–0.5	6	12.5	–12	2
90	VV	15.6	–0.5	3	11.7	–13	1
120		10.9	0.5	2	11.9	–14	–2

spectrum model described in Chapter 2. The model comprises a Gaussian-shaped spectrum,

$$G(f) = \frac{\tau}{\sqrt{2\pi}w}\exp\left(-\frac{(f - m_f(\tau))^2}{2w^2}\right) \tag{3.10}$$

where f is the frequency and the mean Doppler frequency is related to the normalised mean speckle power, $\bar{\tau} = \tau / \langle\tau\rangle$ with

$$m_f(x) = A_c + B_c\bar{\tau}. \tag{3.11}$$

The spectral width (standard deviation of the Gaussian shape) is given by w, which itself fluctuates randomly having a mean value, m_w, and standard deviation σ_w.

For the NetRAD data set, the average values of the parameters m_w, A_c and B_c are shown in Table 3.10. In these results, the mean spectrum widths are very similar apart from the bistatic VV polarisation for $\beta_{deg} = 90°$, where the data may have been corrupted [24]. For these particular combinations of viewing geometries and wind directions, the mean Doppler centroids, $A_c + B_c$, of the monostatic data had significant negative values, while the bistatic spectra had small positive mean Doppler centroids. The actual Doppler centroids will depend on the wind speed and direction and viewing geometry, but the differences between monostatic and bistatic values highlight the potential for exploitation in a target detection scheme.

Further examples of bistatic spectra have been produced by Fioranelli *et al.* [19]. Figure 3.9 shows spectrograms for a bistatic angle of $\beta_{deg} = 105°$. The bistatic data with HH and HV polarisations show similar Doppler centroids but have distinctly different temporal variations. The HH monostatic data has a different Doppler centroid, as expected and also a different temporal structure.

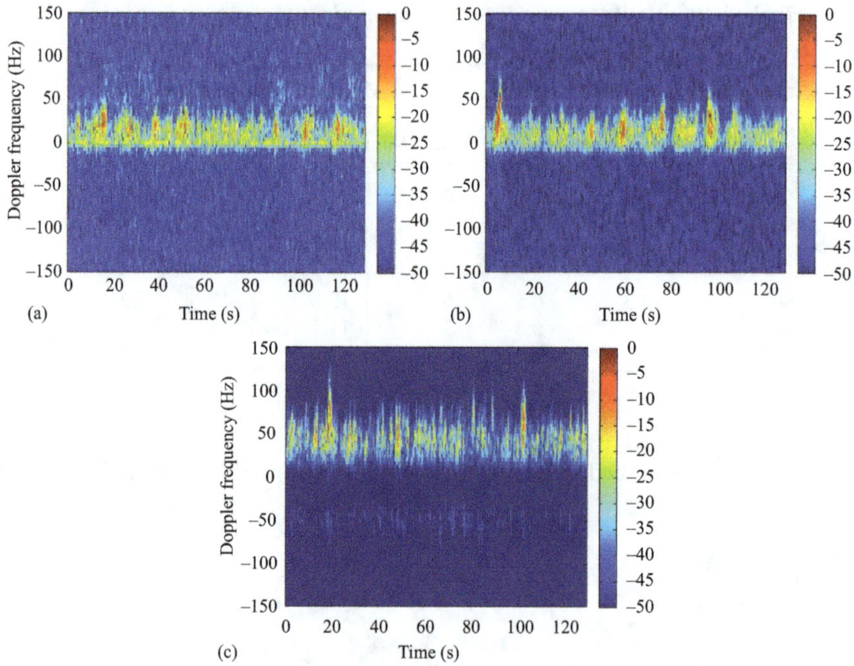

Figure 3.9 NetRAD examples of simultaneously collected Doppler spectrograms for $\beta_{deg} = 105°$: (a) bistatic node 1 HH, (b) bistatic node 2 HV, (c) monostatic node 3 HH [19]

3.3.7 Sea spikes

Palamà *et al.* [25,26] have analysed the NetRAD data to assess the behaviour of sea spikes. The presence of discrete spikes might explain the good fit to the KA-distribution achieved by Al-Ashwal [18]. In this work, sea spikes were defined in terms of their power relative to the mean background, their duration and their separation in time. The power threshold was set at six times the mean power of the returns and the minimum spike lifetime was 80 ms. Returns that exceeded these thresholds were declared as spikes with those separated by less than the minimum spike interval of 150 ms merged into a single spike. Figure 3.10 shows an example of simultaneously gathered bistatic and monostatic returns for $\beta_{deg} = 30°$ and VV polarisation. The events identified as spikes are shown in red and it can be seen that spikes are evident in both monostatic and bistatic returns with some qualitative evidence that the bistatic spikes may be lasting longer with a smaller amplitude. There does not appear to be any strong correlation between spikes events in the monostatic and bistatic data.

The sea spikes in the data were also analysed to assess the percentage of spikes with respect to the total number of samples, P_{sp}, the mean spike duration, d_{sp}, and the mean interval between spikes, T_{sp}. Data was analysed for $\beta_{deg} = 30°$, 50°, 60°

and 90° with the results for $\beta_{deg} = 30°$ and 50° collected on a different day. These collections had a similar sea state but a different wind direction compared to the results for $\beta_{deg} = 60°$ and 90°. The average spike characteristics for $\beta_{deg} = 30°$ and 90° are shown in Table 3.11. In these results, the bistatic data has longer spikes than the monostatic VV data and most of the HH data. The mean interval between spikes was also generally higher for the bistatic data.

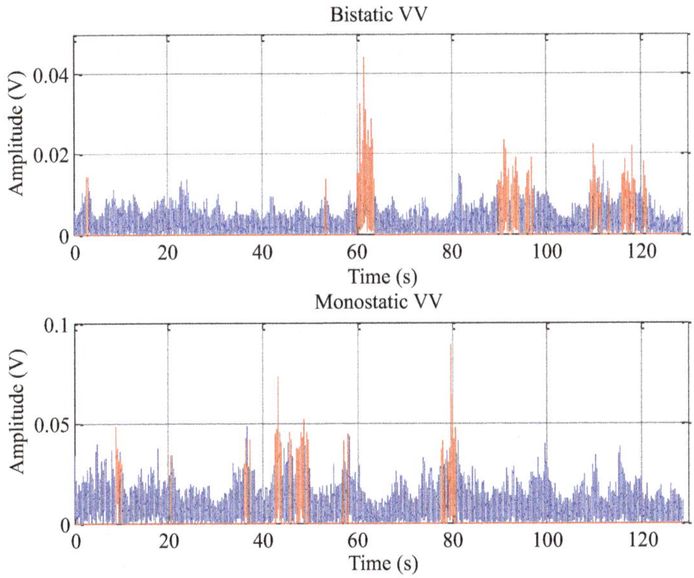

Figure 3.10 NetRAD time history of amplitude clutter samples for the VV polarisation, $\beta_{deg} = 30°$ with events identified as spikes shown in red. The data has a sea state 4 and grazing angle of 3° [26]. © IEEE 2015

Table 3.11 NetRAD sea-spike characteristics in sea state 4 and grazing angles of 0.9° ($\beta_{deg} = 90°$) and 3.0° ($\beta_{deg} = 30°$) [26]

	β_{deg}	**Bistatic HH**	**Monostatic HH**	**Bistatic VV**	**Monostatic VV**
P_{sp} (%)	30	8.4	13.3	11.9	12.2
	90	5.0	5.3	7.1	3.4
d_{sp} (s)	30	0.94	0.75	1.21	1.00
	90	0.46	0.37	0.44	0.32
T_{sp} (s)	30	12.2	6.5	14.1	8.62
	90	8.3	6.1	6.1	9.4

3.4 Passive bistatic radar

Over the past decade, there has been significant research into the exploitation of passive bistatic radar systems for imaging and the detection and tracking of ground, maritime and airborne targets [27]. The attraction of passive radar is its covert operation, wide coverage area and reduced cost. In this section, the statistics of sea clutter using passive illumination from the Sirius-XM S-band and the Optus-D3 Ku-band broadcast satellites are presented. By characterising the sea clutter statistics, target detection schemes can be developed, expected radar performance can be modelled and realistic simulation of bistatic radar sea clutter in a passive radar context can be created.

In the maritime domain, passive radar operation has been proposed using a number of illuminators of opportunity. For operation near the coast, potential illuminators include WiMAX, [28], FM radio [29], Digital Video Broadcast [30] and Sirius-XM digital radio [31,32]. However, to be effective beyond the shoreline, illuminators must be spaceborne and provide persistent coverage. Some potential systems include Global Navigation Satellite System [33] and communication satellites such as Iridium, INMARSAT and Optus [34,35].

Target detection in a passive radar is typically performed in the delay/Doppler domain where the extent of the clutter dominated delay bins depends on the system setup and characteristics. The rest of the display is dominated by the ambiguity noise floor [29]. However, to fully characterise the passive return, the sea clutter statistics must also be understood. A comprehensive study of the clutter statistics from the Sirius-XM and Optus D3 satellite illuminators has been given in [35] and includes the amplitude statistics, temporal correlation and Doppler spectra. A summary of this work is presented in this section.

Two passive bistatic sea clutter data sets are first described in Section 3.4.1, before Section 3.4.2 describes the data processing. Section 3.4.3 then presents an analysis of the amplitude statistics for the clutter and noise-dominated regions in the delay/Doppler domain. Details on the time/Doppler amplitude statistics, temporal correlation and the time evolution of the Doppler spectrum are then given in Section 3.4.4.

3.4.1 Sea clutter measurements

In this section, two different satellite-based passive radar experiments are described. The first was conducted by the US Naval Research Laboratory at the tail end of Hurricane Maria in September 2017 using a Sirius-XM illuminator, while the 'Passive Seas trial' was conducted in April 2017 by the Australian Defence Science and Technology Group off the coast of Fremantle, Western Australia using the Optus-D3 communication satellite.

3.4.1.1 Sirius-XM data set

The Sirius-XM constellation consists of multiple satellites positioned in geosynchronous and geostationary orbits over the United States. During this experiment, the XM3 geostationary satellite, located at 85° West, was selected as the transmitter of opportunity with an S-band frequency centred at 2,333.47 MHz. It transmits with a left-hand circular polarisation and the transmitted data is quadrature phase-shift key modulated with each side-band having a half-power bandwidth of 1.84 MHz. On the ground, the saturating flux density is approximately −94 dBW/m^2, making it very suitable as an illuminator of opportunity. During the collection time, the sea conditions were measured by a nearby wavebuoy and the wind speed and direction by an anemometer. Both the wind and swell directions were from the North–North-East direction, in which the significant wave height was 4.5 m and the average wind speed was 16 m/s, resembling a Douglas sea state 6. This extreme sea state is useful to determine the worst case impact of sea clutter for a passive radar using this illuminator.

In an effort to avoid shoaling waves and shore reflections, the experimental equipment was installed at the end of the 550 m long research pier at the Army Corps of Engineers field research facility (FRF) in Duck, NC. An aerial photograph of the pier is shown in Figure 3.11 with the experimental set-up shown in Figure 3.12. The bistatic, coherent, unobtrusive imager and tracker receiver was used for data recording and is described in [31]. The direct-path signal received by the reference antenna was left-hand circularly polarised to appropriately match the transmitted polarisation of the satellite. For the surveillance channel, both horizontal and vertically polarised wire-grid antennas were used with half power beamwidths of 9° and 11°, respectively. The wire-grid antennas have an approximate gain of 24 dBi and were chosen for their cross-polarised isolation with inherent low physical cross section to reduce wind loading. As shown in Figure 3.13, the antennas were positioned at a height of 10 m from the mean sea level, with the centre of the beam pointing at a grazing or scattering angle of $\phi_{2,\text{deg}} = 10°$. Also to avoid an instrumentation tower located on the Southern

Figure 3.11 Field research facility pier, Duck NC (Photo courtesy of the US Army Corps of Engineers) [31]. © IEEE 2018

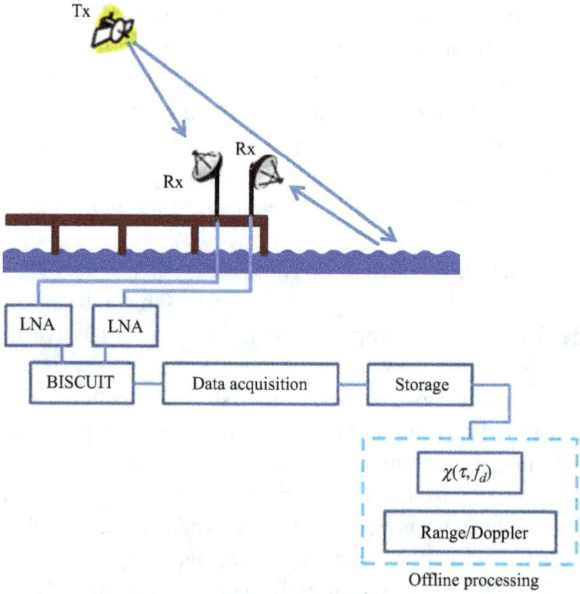

Figure 3.12 A block diagram of the experimental setup showing the hardware and software configuration [31]. © IEEE 2018

Figure 3.13 Grid antennas used for receiving the horizontal (left) and vertical (right) polarisations. The super-imposed photo on the top right shows the left-hand circularly polarised dish antenna used for sampling the reference signal [31]. © IEEE 2018

corner of the pier and a nearby WiMAX transmitter, the antennas had to point away from the shore. As a result, four azimuthal angles were selected for data collection as shown in Figure 3.14. To reduce the impact of the WiMAX transmission, the direct path antenna was positioned behind a large science Conex-type container and in the hardware, a filter was added and the mixing scheme altered to push the WiMAX interference as far out of band as possible. In this figure, a yellow circle is shown at 36 m from the end of the pier and marks the location where the one-way 3 dB elevation point of the radar beam intersects the sea surface. Table 3.12 shows the calculated bistatic angles and range resolution for each look direction, with the incidence angle of the satellite, $\phi_{1,\mathrm{deg}} = 46.9°$ and the scattered elevation angle at a grazing angle of 5° corresponding to the 3 dB beam edge where the sea clutter is dominant. Across the elevation beam, there will be a reduction in the bistatic range resolution as the

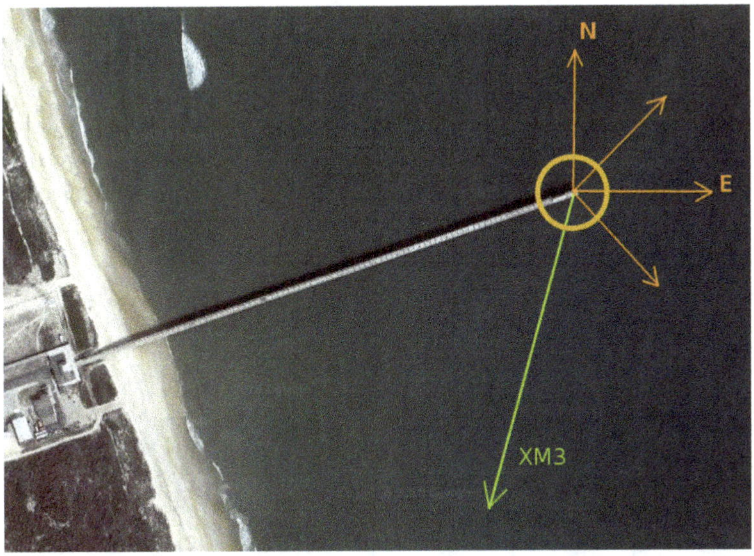

Figure 3.14 Azimuthal look angles relative to the FRF pier. The Southern direction was limited by an instrument tower that restricted the view. The XM3 satellite is shown as oriented to the pier and the yellow circle at 36 m shows the intersection of the radar beam with the sea surface [31]. © IEEE 2018

Table 3.12 Sirius-XM bistatic collection geometry

Direction	**N**	**NE**	**E**	**SE**
Bistatic angle, β_{deg}	44.0	49.0	75.7	105.7
Azimuth (out of plane) angle, $\psi_{\mathrm{B,deg}}$	164.4	150.6	105.6	60.6
Bistatic range resolution, ΔR_{B}	104.4	106.3	121.7	156.1

bistatic angle changes. However, this change is expected to be minor over the clutter dominated region as there are only 4–5 dominant clutter bins.

3.4.1.2 Optus-D3 Passive Seas data set

For the Passive Seas trial, the geosynchronous Ku-band Optus-D3 communication satellite was used as an illuminator of opportunity. This is located at 156° East and simultaneously transmits orthogonal waveforms for both horizontal and vertical polarisations [36]. The transmitted content of the Optus-D3 satellite includes subscription television, free-to-air television and commercial radio. The host platform was the research vessel Whale Song, shown in Figure 3.15 and the passive radar system is an evolution of the one described in [37]. The surveillance antenna was a 65 cm dish located at 3.24 m above the water level. For the receiver system, there were two 85 cm direct path dish antennas mounted on top of the boat to capture the line of sight signal for up to two satellites. For the data sets studied here, the horizontal polarisation was used with a saturating flux density of approximately −91 dBW/m^2 and a 500 MHz bandwidth. A snapshot of the direct path signal is shown in Figure 3.16 and comprises 12 × 41.5 MHz channels covering 11.7–12.2 GHz, each having a distinct windowed shape. Within each channel, there is 36 MHz of data with a 5.5 MHz gap between channels to avoid cross-channel interference. This means that the maximum coherent spectrum for all 12 channels is 432 MHz. The impact of the window is to reduce sidelobes in the time domain, while the spectral gaps can cause artefacts in the final delay/Doppler image. This has been studied in the previous work [38,39] with some potential methods for 'gap-filling'.

Using an anemometer on the boat, the wind direction was measured as 170° with respect to North and the speed was 4.6 m/s corresponding to a Douglas sea state 3. Sea

Figure 3.15 Whale Song research ship used for the data collection. The two direct path antennas are located within white radomes on top of the ship. The surveillance antenna is a black dish to the left of these antennas [35]. © *Commonwealth of Australia 2020*

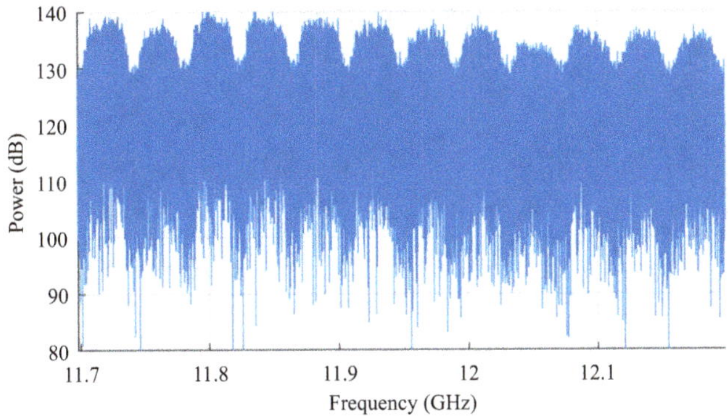

Figure 3.16 Received data spectrum from the direct path. The spectrum comprises 12 × 36 MHz channels with distinctive windowed shapes [35].

Table 3.13 Optus-D3 bistatic collection geometry

Direction	**Up-wind**	**Cross-wind**	**Down-wind**
Bistatic angle, β_{deg}	143.8	103.2	34.5
Mean receiver grazing angle, $\phi_{2,\mathrm{deg}}$	2.0	0.3	0.08
Azimuth (out of plane) angle, $\psi_{\mathrm{B,deg}}$	350	74	169
Bistatic range res. (1 channel) (m)	14.1	7.3	4.7
Bistatic range res. (12 channels) (m)	1.2	0.6	0.4

clutter was collected for up-, down-wind and both cross-wind directions. The up-wind geometry is very close to being in-plane, while the other three have an out-of-plane bistatic geometry. As the antenna system was not stabilised, it fluctuated within $\pm 2.5°$ over the 9.8 s collection period. The average measured elevation angles are given in Table 3.13 along with the bistatic angles and range resolutions with the grazing angle from the satellite, $\phi_{1,\mathrm{deg}} = 33°$.

3.4.2 Data processing

To process the bistatic return, the desired spectral region of the data is first isolated and a Weiner–Hopf filter is used to remove the zero-Doppler clutter and any direct path interference [40]. The delay/Doppler return is determined by a cross-correlation between the direct path channel, $h_{\mathrm{d}}(t)$, and the surveillance channel, $h_{\mathrm{r}}(t)$, with a -55 dB Dolph–Chebyshev window, $w_{\mathrm{D}}(t)$, used to reduce sidelobes in the frequency domain. Note that this will also modulate the data in the time domain, but the impact on the clutter amplitude statistics is expected to be minor. If t_D represents the delay and f is the Doppler frequency, then the processed signal is

$$\chi(t_D, f) = \int_0^{T_I} w_D(t) h_d(t) h_r^*(t + t_D) e^{j2\pi f t} dt \tag{3.12}$$

where T_I is the coherent integration time. In terms of implementation, the cross-correlation can be calculated in the frequency domain to improve the computational time. Note that (3.12) assumes a narrowband assumption for the data, which is reasonable as the bandwidth to centre frequency ratios are 0.08% and 4.2% for the Sirius-XM and Optus-D3 waveforms, respectively. The noise region is referred to as 'ambiguity noise' since the power levels are typically higher than the thermal noise floor in the receiving radar system.

Figure 3.17 shows the delay/Doppler map of the Sirius-XM North-East data set with an integration time of $T_I = 0.5$ s. The observed sea clutter is visible for the first 4–5 delay bins (425.2–531.5 m) and both polarisations look similar. This is due to the coarse bistatic range resolution, which has averaged out any detail in the sea clutter along the range direction. Figure 3.18 then shows the delay/Doppler map for the Optus-D3 up- and cross-wind directions with the maximum bandwidth. Due to the finer bistatic range resolution, there are now periodic variations observed over bistatic range in the up-wind direction covering both positive and negative Doppler. This is not present in cross-wind where there is a less distinctive spread of energy with both positive and negative Doppler components. For this example, the 50 delay bins correspond to 60 and 30 m for the up- and cross-wind directions, respectively.

For the Optus-D3 data, significant oscillation is observed. To clarify whether this is due to a wind-driven wave pattern or the radar platform, the Doppler shift due to

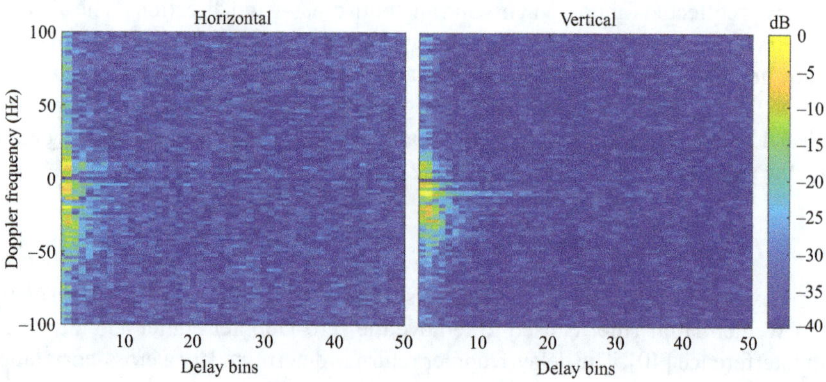

Figure 3.17 Sirius-XM delay/Doppler image for the North-East direction. The 50 delay bins correspond to 5,315 m [35]. © Commonwealth of Australia 2020

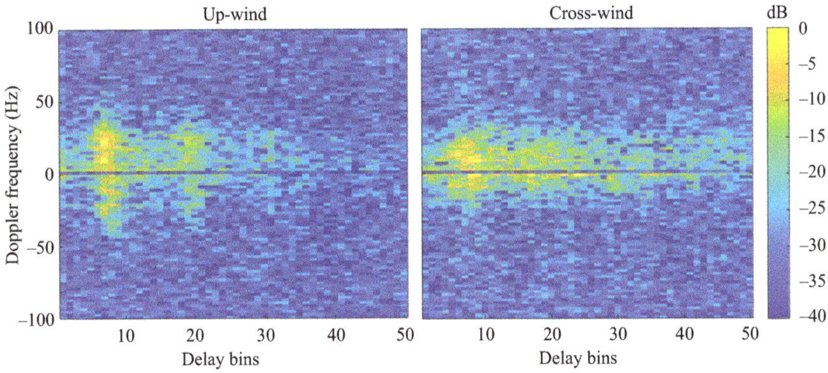

Figure 3.18 Optus-D3 delay/Doppler images using the finest bistatic range resolution. The 50 delay bins correspond to 60 m for up-wind and 30 m for cross-wind [35]. © Commonwealth of Australia 2020

the boat motion can be determined from both the reference and surveillance channels and then removed from the processed data. This is achieved using the measured roll, pitch and yaw of the boat as it varies over time [41]. The analysis in [35] showed that the maximum absolute Doppler shifts due to the boat motion were relatively small compared to the Doppler shifts shown in Figure 3.18.

3.4.3 Delay/Doppler amplitude statistics

Target detection in passive bistatic radar typically occurs in the delay/Doppler domain, and hence it is useful to characterise the amplitude statistics in this domain. Using an integration time of $T_I = 0.5$ s, two data blocks have been extracted from the delay/Doppler maps of each data set corresponding to the clutter-plus-noise and noise-only regions. To increase the number of samples, the data is pooled using a number of data blocks over the collection time. This gives sample sizes on the order of 2×10^3 for the clutter-plus-noise regions and 2×10^5 for the noise-only regions. A K+Rayleigh distribution (see Chapter 2) has then been fitted to both data regions where the shape and Rayleigh mean power has been estimated using the $z \log z$ method [42]. This distribution is described by a shape, ν_r and Rayleigh power p_r. If the noise mean power, p_n, is known or estimated from the data, then the influence of the extra Rayleigh component can be measured by the ratio of the mean Rayleigh component to the clutter mean power, p_c by $k_r = p_r/p_c$.

Figure 3.19 shows the complementary cumulative distribution functions (CCDF) for the Sirius-XM data in the North-East direction. The horizontal polarisation has a slightly longer tail with shape values of 0.4 and 12.4 for the clutter-plus-noise and noise-only regions, while the vertical polarisation has shape values of 0.7 and 34.8. For the noise component, the shape value of 12.4 indicates that the received data cannot be represented by an exponential distribution. The corresponding Rayleigh ratios

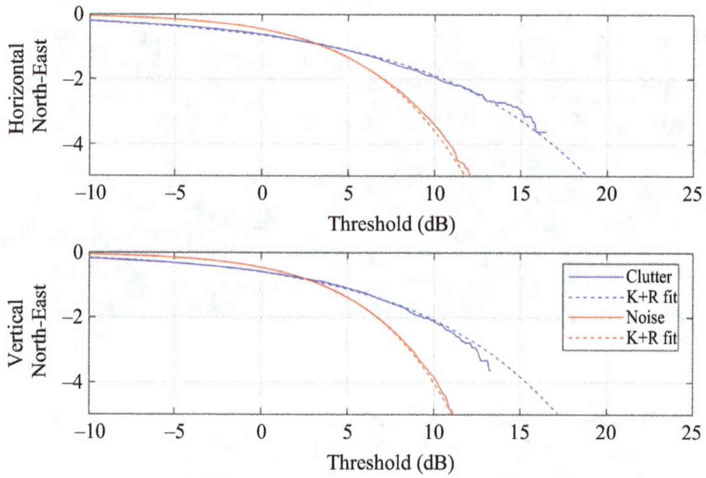

Figure 3.19 Sirius-XM amplitude distributions in the delay/Doppler domain for the North-East direction [35]. © Commonwealth of Australia 2020

are 0.04 and 0 for the horizontal polarisation and 0 for both regions in the vertical polarisation indicating that the dominant Rayleigh component is from the ambiguity noise. The results are very similar for the other look directions and are detailed in [35]. The most likely explanation for the low shape values is interference from a nearby WiMAX transmission. It is likely transmitting with a vertical polarisation as these results are the most impacted by the interference. As a result, the clutter-plus-noise estimates are also going to be similarly impacted by the interference.

For the Optus-D3 data, the same analysis is performed with different clutter-plus-noise and noise-only regions determined for each data block. The shape values for the clutter-plus-noise regions are 0.4, 0.8 and 0.3 with Rayleigh ratios of 0.07, 0 and 0.2. For the noise-only regions, the data is effectively exponential with shape values above 50. To see how the statistics vary with bistatic range resolution, 1, 3, 6, 9 and 12 spectral blocks of the spectrum have been processed corresponding to 36, 108, 216, 324 and 432 MHz of bandwidth. The result in Figure 3.20 shows a clear increase in the shape value with increasing bistatic range resolution, which is expected as the clutter cell size is increasing and the data is becoming less spiky. Interestingly, the lowest shape values are observed in the down-wind direction, although all the shape values are quite similar.

The next results investigate the impact of the integration time on the clutter statistics. A longer integration time enables greater CNR for the processed signal. However, due to the larger computational effort, the cross-correlation processing takes longer and any motion of the target or clutter will be averaged out and may span multiple range and/or frequency bins. Figure 3.21 shows the variation with the Sirius-XM data, where many of the estimates are similar across polarisations. This is

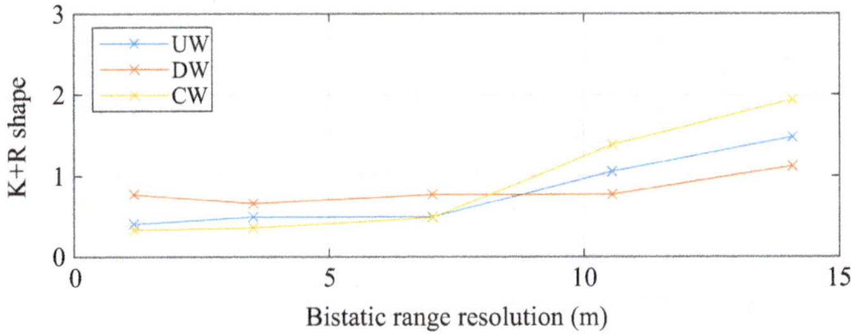

Figure 3.20 Optus-D3 K+ *Rayleigh shape values in the delay/Doppler domain as a function of bistatic range resolution [35]. © Commonwealth of Australia 2020*

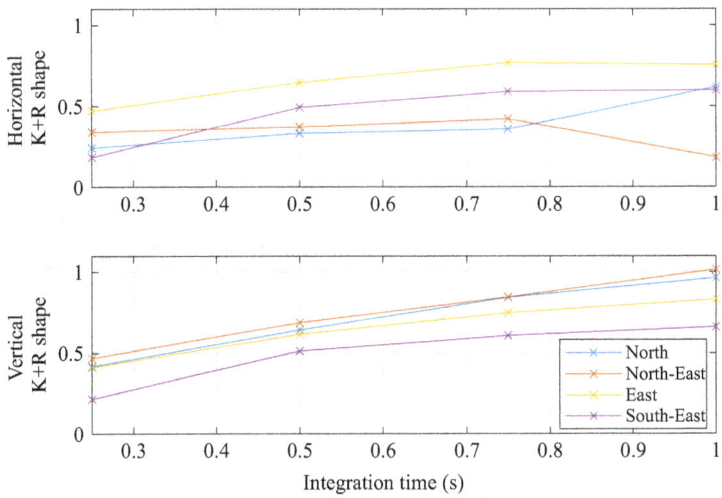

Figure 3.21 Sirius-XM shape variation with integration time in the delay/Doppler domain [35]. © Commonwealth of Australia 2020

an interesting result which highlights the differences between monostatic and bistatic scattering. For the remaining results, an integration time of 0.5 s is used as it achieves a good CNR in a reasonable processing time.

3.4.4 Time domain amplitude statistics and correlation

To study the evolution of the sea clutter over time, spectrograms are formed by extracting a range slice for each delay/Doppler map and then shifting the integration window in 0.1 s increments through each data set. At the centre of the spectrum, the zero-frequency signal component has been replaced with values that are interpolated from

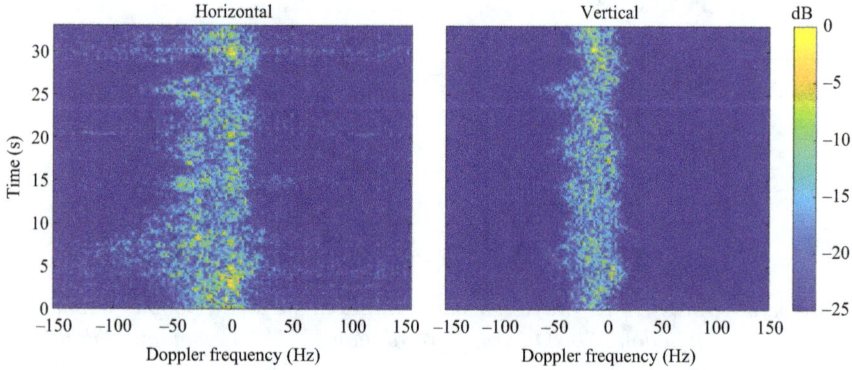

Figure 3.22 Sirius-XM spectrograms for the North-East direction [35]. © Commonwealth of Australia 2020

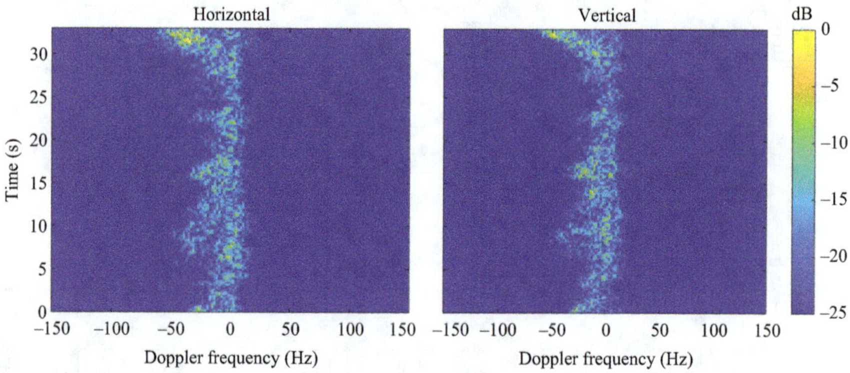

Figure 3.23 Sirius-XM spectrograms for the South-East direction [35]. © Commonwealth of Australia 2020

the surrounding frequency bins. This ensures that the clutter characteristics can be calculated correctly. Figures 3.22 and 3.23 show the time-varying nature of the Doppler spectrum for the second delay bin of the Sirius-XM data in the North-East and South-East directions, respectively. It is interesting to observe that the horizontal polarisation is much broader than the vertical one, and the overall width in the South-East data is significantly less than the North-East. This is due to the scattering mechanisms which vary with the wind direction. For the North-East direction, the look direction is close to up-wind, while the South-East direction is close to cross-wind.

For the Passive Seas data, the Doppler shifts due to the platform motion are also removed by shifting the spectrum by the sum of the surveillance and direct path Doppler values. Spectrograms for the up- and cross-wind directions are given

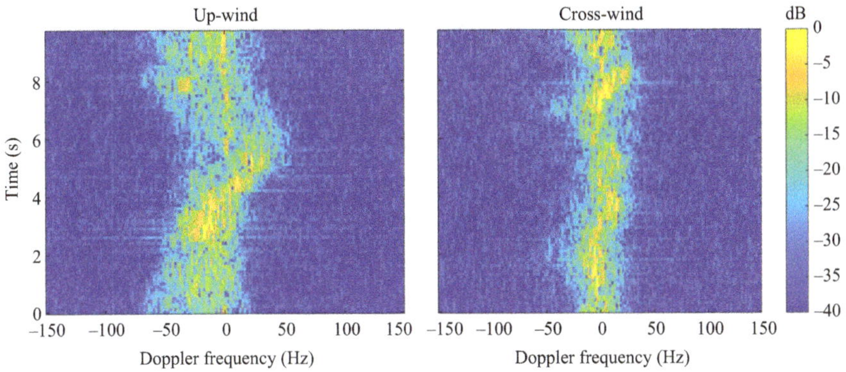

Figure 3.24 Optus-D3 spectrograms [35]. © Commonwealth of Australia 2020

in Figure 3.24 for the delay bin with the greatest energy. The up-wind result has the broadest spectrum with Doppler components between −70 and 50 Hz. There is also a clear sinusoidal variation over time revealing a wind-driven wave pattern moving through the scene. Interestingly, when compared to the cross- and down-wind directions (not shown), the period of this oscillation is smallest in the up-wind direction. To compare the different bistatic resolutions in the following analysis, the statistical results are compared with smaller blocks of the frequency spectrum.

To study the time-domain amplitude distribution, the noise mean power is first estimated from noisy regions of the spectrum and then an inverse Fourier transform is taken along the frequency dimension. The data is then histogrammed and a K+Rayleigh distribution is estimated from the resulting 4×10^4 time samples. The results for the Sirius-XM North-East data set are shown in Figure 3.25 and show a good match for the model with estimated shape values of 0.3 and 0.2 for the horizontal and vertical polarisations, respectively, indicating quite spiky sea clutter. The associated Rayleigh ratio values are 0.5 and 0.7 indicating the presence of significant Rayleigh scattering. For the Optus-D3 data sets, the model also matches quite accurately with a Rayleigh ratio of 0 observed for all look directions. The top plot in Figure 3.26 shows the shape values for each look direction as a function of the bistatic range resolution. As the bistatic range resolution increases, the shape value increases with the up-wind direction having the lowest shape value and hence the spikiest sea clutter. This is followed by cross-wind and then down-wind with the latter two becoming very high at the coarsest resolution, implying that the sea clutter intensity is becoming exponentially distributed.

Due to the small number of delay bins in the data set, only the temporal correlation of the texture was characterised in [35]. Figure 3.27 shows the temporal correlation with very little periodicity shown as is commonly observed in non-stationary sea clutter. The decorrelation time is measured when the correlation decays by $1/e \approx 0.37$ and for the horizontal and vertical polarisations, these values are 0.4 and 0.3 s. A

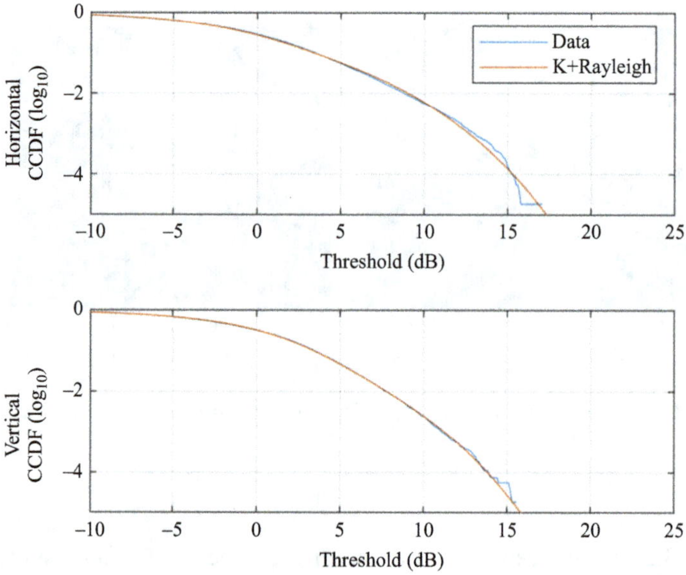

Figure 3.25 Sirius-XM CCDF plots for the North-East direction [35]. © Commonwealth of Australia 2020

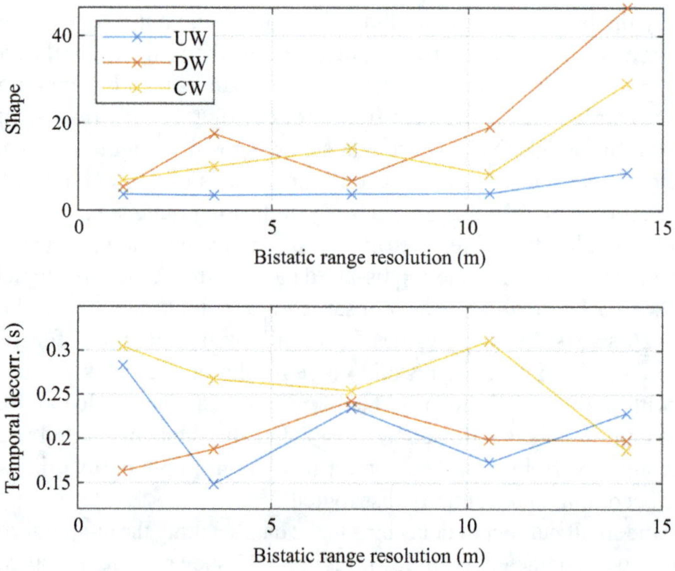

Figure 3.26 Optus-D3 shape and temporal correlation as a function of bistatic range resolution [35]. © Commonwealth of Australia 2020

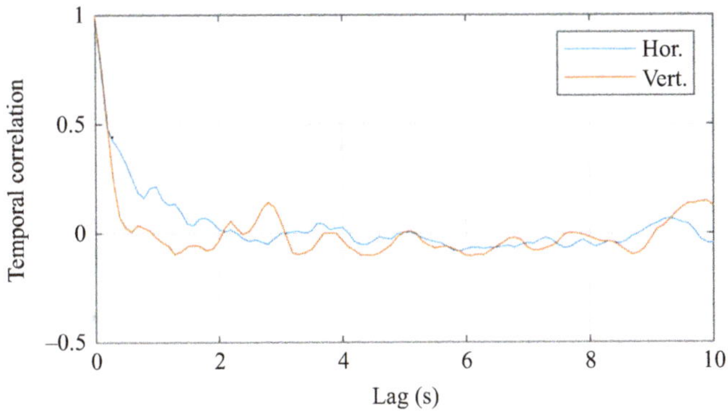

Figure 3.27 Sirius-XM temporal correlation for the North-East direction [35]. © Commonwealth of Australia 2020

Table 3.14 Summary of the Sirius XM model parameters for the second delay bin

	Direction	N	NE	E	SE
	K+Rayleigh distribution shape, ν_r	0.2	0.3	0.3	0.2
	K+Rayleigh distribution ratio, k_r	0.6	0.7	0.5	0.5
	Temporal decorrelation (s)	0.3	0.4	0.2	0.2
H	Doppler centre point, A_c (Hz)	–6.7	–14.0	–5.1	2.4
	Doppler centre point, B_c (Hz)	4.8	3.3	–14.3	–8.2
	Doppler spread mean, m_w (Hz)	12.0	15.3	12.7	8.6
	Doppler spread std. dev., σ_w (Hz)	7.0	11.5	8.2	5.8
	K+Rayleigh distribution shape, ν_r	0.2	0.2	0.4	0.2
	K+Rayleigh distribution ratio, k_r	0.6	0.7	0.5	0.5
	Temporal decorrelation (s)	0.3	0.3	0.2	0.2
V	Doppler centre point, A_c (Hz)	–18.2	–14.1	–7.9	2.8
	Doppler centre point, B_c (Hz)	11.0	3.4	0.06	–9.3
	Doppler spread mean, m_w (Hz)	8.3	10.2	10.5	7.9
	Doppler spread std. dev., σ_w (Hz)	5.2	6.4	6.6	4.9

similar result is observed for the Optus-D3 data set, with the results for all three look directions shown in the bottom plot of Figure 3.26. These results show little change with the bistatic range resolution. The shortest decorrelation time is in the up-wind direction, followed by down- and cross-wind.

For the Sirius-XM data set, results for the shape and decorrelation times for the other look directions are summarised in Table 3.14. From these results, it can be observed that the shape parameter is always less than 1 indicating spiky sea clutter and there is extra Rayleigh scattering present in each look direction. There does not appear to be a lot of difference in the statistics between the two polarisations.

3.4.5 Doppler spectra

The mean Doppler spectrum is often characterised by a single Gaussian component with an offset and spread [43]. This is in contrast to the asymmetrical spectra which is often observed due to the interplay of both Bragg and fast scattering mechanisms [44, ch. 2]. The Bragg scattering response dominates for vertically polarised radar systems, while the fast scattering is dominant for low grazing angle data and horizontally polarised systems. While these models are useful to capture the mean of the Doppler spectrum, realistic characterisation of sea clutter requires models that capture the evolution over both range and time. The evolving Doppler model was developed for this purpose and is used in this analysis [23].

To analyse the parameters of the evolving Doppler model, the estimation methods outlined in Chapter 2 are used. The first parameter is the centre frequency which is plotted for the Sirius-XM North-East direction against the normalised intensity in Figure 3.28. The measured parameters, A_c and B_c for the linear fit in (3.11) are -14 and 3.3 for the horizontal polarisation and -14.1 and 3.4 for the vertical. This indicates that there is a reasonable relationship between the intensity and the centre frequency of the spectrum. However, for the vertical polarisation, the model over-fits beyond intensity values of 1.6 as there are insufficient data points. Figure 3.29 then shows the PDF of the width with a gamma model fit. The mean and standard deviation, m_w and σ_w, are 15.3 and 11.5 for the horizontal polarisation and 10.2 and 6.4 for the vertical.

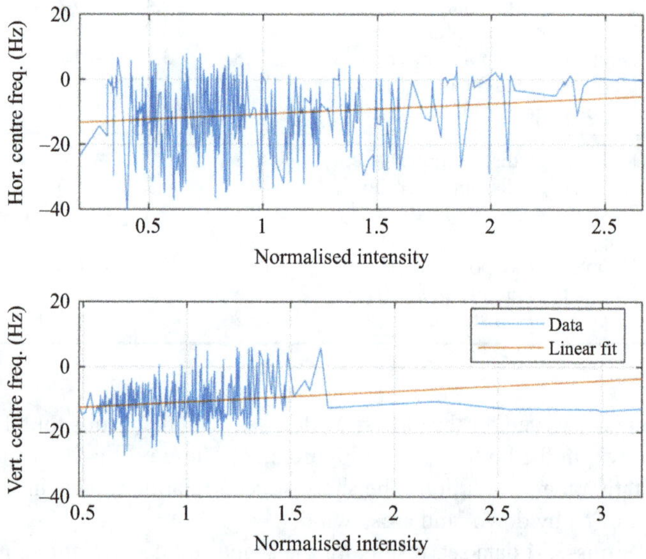

Figure 3.28 Sirius-XM Doppler centre point for the North-East direction [35]. © Commonwealth of Australia 2020

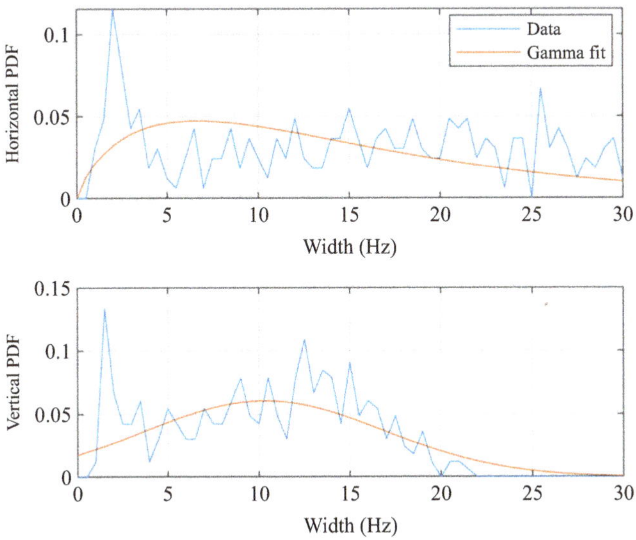

Figure 3.29 Sirius-XM spectral width probability for the North-East direction [35]. © Commonwealth of Australia 2020

This model only shows a reasonable fit to the vertical data but does not capture the peak at 2 Hz which is common to both polarisations.

Also important in understanding the Doppler characteristics are the CNR and the K+noise distribution shape across frequency. Figure 3.30 shows the CNR on the top plot and the shape values for the endo-clutter region on the bottom. These results reveal that the horizontal polarisation has a much larger spread and the shape values are fairly constant across the spectrum with occasional peaks where the clutter is exponentially distributed. The mean shape values are 1.3 for the horizontal polarisation and 2.9 for vertical. Table 3.14 then shows the model parameters for each look direction. For the Doppler centre point, the linear relationship with intensity is largest for the horizontal data set in the East direction. For the Doppler spread, the mean and standard deviation are very similar for each look direction.

The last results consider the evolving Doppler spectrum model for the Optus-D3 data set. Figures 3.31 and 3.32 show how the Doppler model parameters for each look direction vary with bistatic range resolution. The first figure shows the mean Doppler parameters A_c and B_c, which are fairly constant for each value of bistatic range resolution. There is a greater fluctuation in the up-wind spectrum as shown by the smaller values of B_c. The second result in Figure 3.32 then shows how the mean and spread of the Doppler width vary. There is little variation with bistatic range resolution for most results with the exception of the down-wind direction which shows both parameters decreasing with increasing resolution. The up-wind result has the largest mean and spread, followed by the cross- and down-wind.

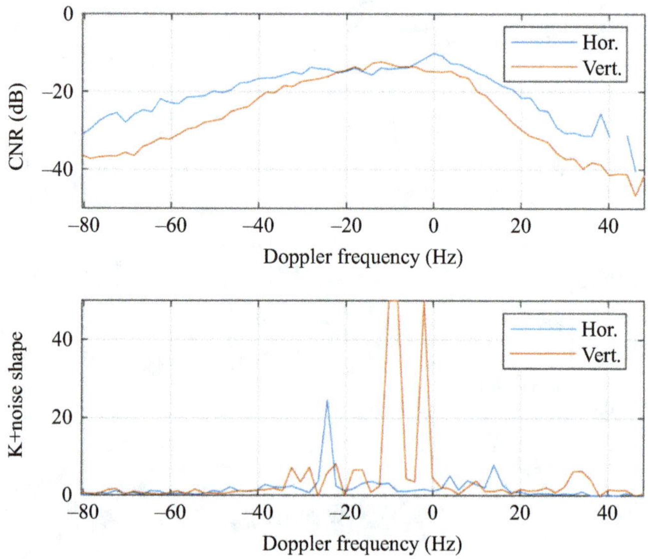

Figure 3.30 Sirius-XM CNR and K+noise distribution shape across frequency for the North-East direction [35]. © Commonwealth of Australia 2020

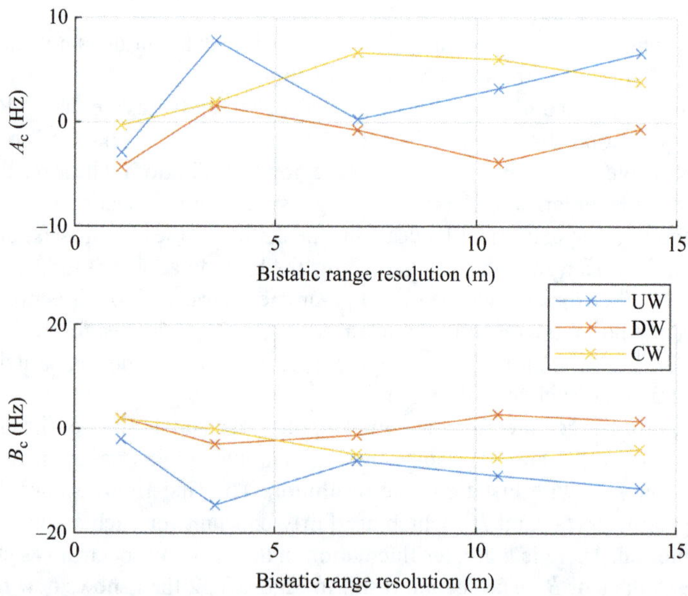

Figure 3.31 Optus-D3 Doppler mean linear model parameters as a function of bistatic range resolution [35]. © Commonwealth of Australia 2020

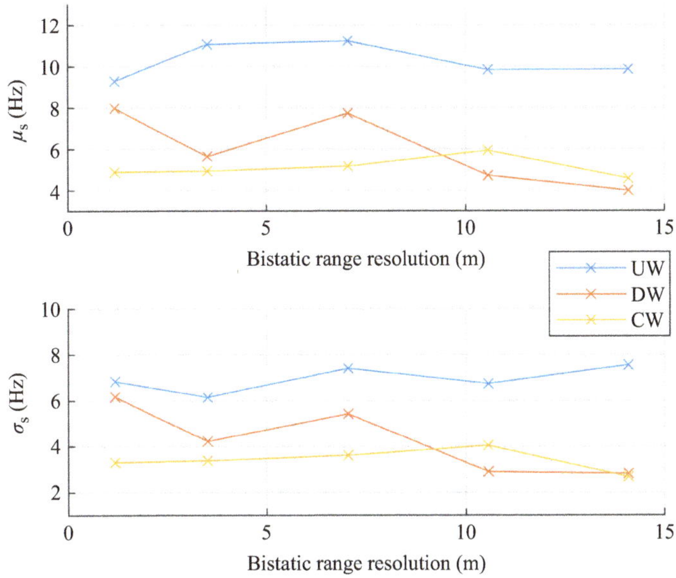

Figure 3.32 Optus-D3 Doppler width mean and spread as a function of bistatic range resolution [35]. © Commonwealth of Australia 2020

3.4.6 Summary

In this section, the characteristics of bistatic sea clutter collected by two satellite illuminators of opportunity were presented. These were processed using a cross-correlation between the direct and surveillance channels. Analysis of the sea clutter included a study of the delay/Doppler and time/Doppler-domain amplitude statistics, temporal correlation and the time evolution of the Doppler spectrum. In the delay/Doppler domain, there was not a lot of differences in the amplitude statistics between the two datasets, with the clutter-plus-noise data showing spiky statistics and the noise data being nearly exponential. When looking at the variation in integration time, there was an increase observed in the shape values with the largest observed in the cross-wind direction of the Optus-D3 data. To study the time variation, spectrograms were formed from a single delay bin. These revealed the time-varying nature of the Doppler spectrum and reflected the wave pattern moving through the scene. For the Sirius-XM data set, there was a high sea state and a very spiky amplitude distribution observed with little difference in the statistics between the two polarisations. Analysis of the evolving Doppler model found a greater variation in Doppler spread for the East horizontal data set. For the Optus-D3 data, the up-wind data set had the broadest Doppler spread and spikiest statistics. There was also a clear sinusoidal variation over time for each look direction. As the bistatic range resolution became coarser, the shape value increased for each look direction and the clutter became less spiky. These results indicate that target detection in the clutter-dominated region will be more difficult in the up-wind direction due to the broader Doppler spectrum and longer tails of the amplitude distribution.

To further characterise the sea clutter statistics, more data needs to be studied to distinguish the impact of wind direction versus the bistatic angle. Also, if the noise dominates the sea clutter at less than 500 m bistatic range, the impact of sea clutter may not be perceived as a problem in a practical system. However, this could be quite different if the passive receiver was on an airborne platform with a larger receive footprint.

3.5 Discussion on bistatic sea clutter

This chapter has reviewed the available measurements of bistatic sea clutter and shown how their statistical characteristics can be represented in models for performance prediction and radar design purposes. This knowledge can now contribute to the wider task of designing bistatic radar systems.

It has been previously noted that there may be operational advantages in employing active or passive bistatic radars for maritime surveillance. However, any such advantages must be weighed against the increased complexity of bistatic systems. At the present time there are few, if any, operational bistatic radars being used for maritime surveillance and hence it is difficult to compare the relative detection performance of monostatic and bistatic systems. Sea clutter is a significant factor in such performance and again, relatively few measurement campaigns have been undertaken to collect bistatic sea clutter data. From what is available, it appears that the main characteristics of reflectivity, amplitude statistics and Doppler spectra are comparable with monostatic operation. It is not obvious that either has a distinct advantage in terms of detection performance, when considered over all operational conditions. Perhaps the most important point is that they are different, and this may give an advantage in certain conditions and viewing geometries. It is reasonable to assume that the diversity in the viewing geometry achieved in multistatic operation could bring significant detection gains. However, this would be at the expense of considerable complexity compared to monostatic operation, especially in a rapidly changing airborne surveillance scenario.

References

[1] *IEEE Standard for Radar Definitions*, Std., 2017. [Online]. Available: https://ieeexplore.ieee.org/document/8048479/

[2] N. J. Willis, *Bistatic radar*. SciTech Publishing Inc., London, 1995.

[3] N. Willis and H. D. Griffiths, *Advances in bistatic radar*. SciTech Publishing, London, 2007.

[4] M. C. Jackson, "The geometry of bistatic radar systems," *Proceedings of the IEE, Part F*, vol. 133, no. 7, pp. 604–612, 1986.

[5] W. Dower and M. Yeary, "Bistatic SAR: Forecasting spatial resolution," *IEEE Transactions on Aerospace and Electronic Systems*, vol. 4, no. 55, pp. 1584–1595, 2019.

[6] V. Pidgeon, "Bistatic cross section of the sea," *IEEE Transactions on Antennas and Propagation*, vol. 14, no. 3, pp. 405–406, 1966.

[7] V. W. Pidgeon, "Bistatic cross-section of the sea for Beaufort 5 sea," *Science and Technology*, vol. 17, pp. 447–448, 1968, use of Space Systems for Planetary Geology and Geophysics.

[8] A. R. Domville, "The bistatic reflection from land and sea of X-band radio waves, part I," GEC (Electronics) Ltd, Stanmore, England, Tech. Rep., Jul. 1967, memorandum SLM1802.

[9] A. R. Domville, "The bistatic reflection from land and sea of X-band radio waves, part II," GEC (Electronics) Ltd, Stanmore, England, Tech. Rep., Jul. 1968, memorandum SLM2116.

[10] A. R. Domville, "The bistatic reflection from land and sea of X-band radio waves, part II – Supplement," GEC (Electronics) Ltd, Stanmore, England, Tech. Rep., Jul. 1969, memorandum SLM2116.

[11] G. W. Ewell and S. P. Zehner, "Bistatic sea clutter return near grazing incidence," *IEE Conference*, no. 216, pp. 188–192, 1982.

[12] T. P. Kochanski, M. J. Vanderhill, J. V. Zolotarevsky, and T. Fariss, "Low illumination angle bistatic sea clutter measurements at x-band," *OCEANS 92 – Proceedings of Mastering the Oceans Through Technology*, vol. 1, Oct. 1992, pp. 518–523.

[13] M. Inggs, G. Inggs, S. Sandenbergh, W. Al-Ashwal, K. Woodbridge, and H. Griffiths, "Multistatic networked radar for sea clutter measurements," *IEEE International Geoscience and Remote Sensing Symposium*, 2011, pp. 4449–4452.

[14] W. A. Al-Ashwal, "Measurement and modelling of bistatic sea clutter," Ph.D. dissertation, Department of Electronic and Electrical Engineering, 2011.

[15] M. Inggs, A. Balleri, W. A. Al-Ashwal, *et al.*, "NetRAD multistatic sea clutter database," *IEEE International Geoscience and Remote Sensing Symposium*, Jul. 2012, pp. 2937–2940.

[16] W. A. Al-Ashwal, C. J. Baker, A. Balleri, *et al.*, "Statistical analysis of simultaneous monostatic and bistatic sea clutter at low grazing angles," *Electronics Letters*, vol. 47, no. 10, pp. 621 –622, 2011.

[17] W. A. Al-Ashwal, K. Woodbridge, and H. D. Griffiths, "Analysis of bistatic sea clutter – part I: Average reflectivity," *IEEE Transactions on Aerospace and Electronic Systems*, vol. 50, no. 2, pp. 1283–1292, 2014.

[18] W. A. Al-Ashwal, K. Woodbridge, and H. D. Griffiths, "Analysis of bistatic sea clutter – part II: Amplitude statistics," *IEEE Transactions on Aerospace and Electronic Systems*, vol. 50, no. 2, pp. 1293–1303, 2014.

[19] F. Fioranelli, M. Ritchie, H. Griffiths, S. Sandenbergh, and M. Inggs, "Analysis of polarimetric bistatic sea clutter using the NetRAD radar system," *IET Radar, Sonar and Navigation*, vol. 10, pp. 1356–1366, 2016.

[20] R. Palamà, M. Ritchie, H. Griffiths, *et al.*, "Correlation analysis of simultaneously collected bistatic and monostatic sea clutter," *IEEE Radar Conference*, 2017, pp. 1466–1471.

[21] M. A. Ritchie, W. A. Al-Ashwal, A. G. Stove, K. Woodbridge, and H. D. Griffiths, "Statistical analysis of monostatic and bistatic sea clutter Doppler spectrum," *CIE International Conference on Radar*, vol. 1, Oct. 2011, pp. 816–820.

[22] M. A. Ritchie, W. A. Al-Ashwal, A. G. Stove, K. Woodbridge, and H. D. Griffiths, "Coherent analysis of horizontally-polarized monostatic and bistatic sea clutter," *IET International Conference on Radar*, Oct . 2012, pp. 1–5.

[23] S. Watts, "Modeling and simulation of coherent sea clutter," *IEEE Transactions on Aerospace and Electronic Systems*, vol. 48, no. 4, pp. 3303–3317, 2012.

[24] M. Ritchie, A. Stove, A. Woodbridge, and H. Griffiths, "NetRAD: Monostatic and bistatic sea clutter texture and Doppler spectra characterization at S-band," *IEEE Transactions on Geoscience and Remote Sensing*, vol. 54, no. 9, pp. 5533–5543, 9 2016.

[25] R. Palamà, M. Greco, P. Stinco, and F. Gini, "Analysis of sea spikes in NetRAD clutter," *European Radar Conference*, Oct. 2014, pp. 109–112.

[26] R. Palamà, M. S. Greco, P. Stinco, and F. Gini, "Statistical analysis of bistatic and monostatic sea clutter," *IEEE Transactions on Aerospace and Electronic Systems*, vol. 51, no. 4, pp. 3036–3054, 2015.

[27] H. D. Griffiths and C. J. Baker, *An introduction to passive radar*. Artech House, London, 2017.

[28] K. Chetty, K. Woodbridge, H. Guo, and G. E. Smith, "Passive bistatic WiMAX radar for maritime surveillance," *IEEE Radar Conference*, 2010, pp. 188–193.

[29] M. Malanowski, R. Haugen, M. S. Greco, D. W. O'Hagan, R. Pišek, and A. Bernard, "Land and sea clutter from FM-based passive bistatic radars," *IET Radar Sonar and Navigation*, vol. 8, no. 2, pp. 160–166, 2014.

[30] J. Raout, "Sea target detection using passive DVB-T based radar," *International Radar Conference*, 2008, pp. 695–700.

[31] J. D. Ouellette and D. J. Dowgiallo, "Sea surface scattering of hurricane Maria remnants using bistatic passive radar with S-band satellite illumination," *IEEE Radar Conference*, 2018, pp. 462–466.

[32] L. Rosenberg, J. D. Ouelette, and D. J. Dowgiallo, "Analysis of S-band passive bistatic sea clutter," *IEEE Radar Conference*, 2019, pp. 1–6.

[33] F. Pieralice, D. Pastina, F. Santi, and M. Bucciarelli, "Multi-transmitter ship target detection technique with GNSS based passive radar," *International Radar Conference*, 2017, pp. 1–6.

[34] L. Daniel, S. Hristov, X. Lyu, A. G. Stove, M. Cherniakov, and M. Gashinova, "Design and validation of a passive radar concept for ship detection using communication satellite signals," *IEEE Transactions on Aerospace and Electronic Systems*, vol. 53, no. 6, pp. 3115–3134, 2017.

[35] L. Rosenberg, J. D. Ouellette, and D. J. Dowgiallo, "Passive bistatic sea clutter statistics from spaceborne illuminators," *IEEE Transactions on Aerospace and Electronic Systems*, vol. 56, no. 5, pp. 7062–7073, 2020.

[36] Optus, "Optus D series satellite payload information, technical note," 2013.
[37] J. Palmer, S. Palumbo, A. Summers, D. Merrett, and S. Howard, "DSTO's experimental geosynchronous satellite based PBR," *International Radar Conference*, 2009, pp. 1–6.
[38] W. Feng, J. M. Friet, G. Cherniak, and M. Sata, "Passive radar imaging by filling gaps between ISDB digital TV channels," *Journal of Selected Topics in Applied Earth Observation and Remote Sensing*, vol. 12, no. 7, pp. 2055–2068, 2019.
[39] W. Qiu, E. Giusti, A. Bacci, *et al.*, "Compressive sensing based algorithm for passive bistatic ISAR with DVB-T signals," *IEEE Transactions on Aerospace and Electronic Systems*, vol. 51, no. 3, pp. 2166–2180, 2015.
[40] J. E. Palmer and S. J. Searle, "Evaluation of adaptive filter algorithms for clutter cancellation in passive bistatic radar," *IEEE Radar Conference*, 2012, pp. 493–498.
[41] V. Chen and R. Lipps, "ISAR imaging of small craft with roll, pitch and yaw analysis," *IEEE International Radar Conference*, 2000, pp. 493–498.
[42] S. Bocquet, "Parameter estimation for Pareto and K distributed clutter with noise," *IET Radar Sonar and Navigation*, vol. 9, no. 1, pp. 104–113, 2015.
[43] L. B. Wetzel, *Radar handbook*, 3rd ed. McGraw Hill, New York, 2008, ch. Sea Clutter.
[44] K. D. Ward, R. J. A. Tough, and S. Watts, *Sea clutter: Scattering, the K-distribution and radar performance*, 2nd ed. The Institute of Engineering Technology, London, 2013.

Chapter 4
Parametric modelling

4.1 Introduction

This chapter presents a number of parametric models that relate to the statistical models presented in Chapters 2 and 3. The practical use of these statistical models requires a way to relate the model parameters to the sea conditions, collection geometry, polarisation and radar parameters, including spatial resolution and frequency. This can be done either empirically, based on measured data, or with electromagnetic modelling using physical models of the sea surface [1–3]. However, electromagnetic modelling approaches are usually too computationally intensive to be used for most applications.

In Section 4.2, a number of monostatic mean backscatter models are presented. These were developed over many years (1974 onwards) and collectively cover large regions of validity, i.e. frequency, grazing angle, wind speed and direction and sea state. A number of bistatic mean reflectivity models are then given in Section 4.3 for both in-plane and out-of-plane geometries. Section 4.4 describes a number of models for the K-distribution shape parameter, as it is the most widely used amplitude distribution model. These have been developed over the years to now cover both low- and medium grazing angle regions, all wind directions and a range of sea states. Parametric models for the Doppler spectrum are covered in Section 4.5 and include models for the mean Doppler spectrum, temporal decorrelation and a generic model for representing parameters of the evolving Doppler spectrum model. Finally, Section 4.6 presents two models for the spatial decorrelation length, which is important for modelling the range correlation of the texture.

4.2 Monostatic mean backscatter

Due to its importance in radar modelling, there have been many mean backscatter models for monostatic radar proposed over the years. These are typically empirical and depend on the sea state, collection geometry, wind speed and direction and radar frequency. Three early models designed for low grazing angles include the Royal Radar Establishment (RRE) [1], Sittrop [4] and the Georgia Institute of Technology (GIT) [5] models. There are also many empirical models based on fits to Nathanson's tables [6, Ch. 7] which comprise around 60 different experimental results. These cover a range of frequencies, geometric and environmental conditions, but are averaged

Table 4.1 Summary of mean backscatter models, with the published regions of validity

Model	Freq. (GHz)	Grazing (°)	Wind direction (°)	Sea state
RRE [1] (1974)	9–10	0.1–10	Averaged	1–6
Sittrop [4] (1977)	8–18	0.1–10	Up/cross-wind	–
GIT [5] (1978)	1–100	0.1–10	0–360	0–5
Hybrid [14] (1990)	0.5–35[a]	0.1–30	0–360	3–6
TSC [8] (1990)	0.5–35	0–90	0–360	0–5
Masuko [11] (1986)	10, 34	0–70	0–360	2–6
IRSG [12] (2008)	8–12	20–45	0–360	2–6
NRL [9] (2012)	0.5–35	0.1–60	Averaged	0–8
DSTG continuous [13] (2017)	8–12	0.1–45	0–360	2–6

[a]This was later extended to 70 GHz in the NAAWS model [15].

over wind direction. They include the hybrid [7], Technology Service Corporation (TSC) [8] and the Naval Research Laboratory (NRL) [9] models.

Another group of mean backscatter models are designed to fit the relationships described by Ulaby *et al.* [10]. These include the Masuko [11] and Ingara Imaging Radar Systems Group (IRSG) [12] models. Finally, the Defence Science and Technology Group (DSTG) continuous model [13] uses the GIT model at low grazing angle and a fit to the Ingara X-band data set at medium grazing angles.

Table 4.1 summarises the mean backscatter models presented in this chapter, with the sea state, S_{ss}, used to represent the sea condition. The sea state is defined in terms of wave height, represented by the root-mean square (RMS), H_{rms}, average, H_{avg}, or significant wave height, $H_{1/3}$. The significant wave height is the mean height of the highest third of the waves and is related to the RMS and average heights by $H_{1/3} = 4H_{rms} = 1.6H_{avg}$. An approximate relationship between wind speed and average wave height is given by

$$H_{avg} = 0.00452U^{2.5} \tag{4.1}$$

and to convert between wind speed U and sea state, the following relationship can be used, $U = 3.16S_{ss}^{0.8}$ [8]. The wind speed is typically defined at a height of 10 m, although it should be noted that these are approximate empirical relationships and other definitions may be found in the literature. The exact relationships will vary depending on the wind and sea conditions.

4.2.1 Royal Radar Establishment model

During the 1970s, the Royal Radar Establishment (RRE) in the United Kingdom produced an X-band mean backscatter model [1] using data from a number of sources [16,17]. It is valid for grazing angles <10°, sea states 1–6 and both horizontally transmit and receive (HH) and vertically transmit and receive (VV) polarisations.

Table 4.2 Royal Radar Establishment mean backscatter model coefficients

Pol.		Sea state					
		1	2	3	4	5	6
HH	a_{RRE}	–52	–46	–42	–39	–37	–35.5
	b_{RRE}	21	17.5	12.5	10.5	7	3.5
	c_{RRE}	1.015	3.39	2.03	1.35	2.03	2.37
VV	a_{RRE}	–51.5	–45.5	–41	–38.5	–36	–34.5
	b_{RRE}	15	12	11.5	11	9.5	8
	c_{RRE}	8.2	9.5	8	7.5	7	6.5

It is averaged over all wind directions, although an additional cosine term has been proposed to increase the mean backscatter by 3 dB in the up-wind direction and decrease it by 3 dB in the down-wind direction. The original model is given by

$$\sigma^0_{\mathrm{dB}} = \begin{cases} a_{\mathrm{RRE}} + b_{\mathrm{RRE}}\log_{10}\left(\phi_{\mathrm{deg}}\right), & \phi_{\mathrm{deg}} \le 1^\circ \\ a_{\mathrm{RRE}} + c_{\mathrm{RRE}}\log_{10}\left(\phi_{\mathrm{deg}}\right), & 1^\circ \le \phi_{\mathrm{deg}} < 10^\circ \end{cases} \tag{4.2}$$

where ϕ is the grazing angle and the model coefficients, a_{RRE}, b_{RRE} and c_{RRE} are given in Table 4.2 for different sea states and polarisations.

4.2.2 Sittrop's model

Sittrop's model [4] provides estimates of clutter reflectivity for X- and Ku-bands, horizontal and vertical polarisations and for up- and cross-wind directions. The range of grazing angles is not explicitly mentioned, but the data used in the original paper covered the range 0.1°–10°. The model is given by

$$\sigma^0_{\mathrm{dB}} = a_{\mathrm{Sit}} + b_{\mathrm{Sit}}\log_{10}\left(\frac{\phi}{\phi_0}\right) + \left(c_{\mathrm{Sit}}\log_{10}\left(\frac{\phi}{\phi_0}\right) + d_{\mathrm{Sit}}\right)\log_{10}\left(\frac{U}{U_0}\right) \tag{4.3}$$

where $\phi_{0,\mathrm{deg}} \equiv 0.5^\circ$ and $U_0 \equiv 5$ m/s are the reference grazing angle and wind velocity, respectively. The X- and Ku-band model parameters, a_{Sit}, b_{Sit}, b_{Sit} and d_{Sit} are given in Table 4.3.

4.2.3 Georgia Institute of Technology model

The GIT model is a piecewise continuous model for predicting the mean backscatter for both horizontal and vertical polarisations [5]. It covers a frequency range of 1–100 GHz, wave heights up to 3.5 m, all wind directions and grazing angles from 0.1° to 10°. Any data that had evidence of anomalous propagation was excised prior to model fitting and as a result, Nathanson's measured sea clutter data does not agree

Table 4.3 Sittrop's mean backscatter X- and Ku-band model coefficients

Pol.	Freq. band	Wind direction	a_{Sit}	b_{Sit}	c_{Sit}	d_{Sit}
HH	X	Up-wind	−50	12.6	−13.2	34
HH	X	Cross-wind	−53	6.5	0	34
VV	X	Up-wind	−49	17	−12.4	30
VV	X	Cross-wind	−58	19	−33	50
HH	Ku	Up-wind	−46	13.6	−10	26
HH	Ku	Cross-wind	−48	13	−6.6	21
VV	Ku	Up-wind	−46	17	−9	22
VV	Ku	Cross-wind	−47	13	−8.4	22

very well with the GIT model, particularly at low sea states. The model also has discontinuities at 10 GHz for the horizontal polarisation and at both 3 and 10 GHz for the vertical polarisation. Since publication of the original paper in 1978, a typographical error has been discovered [18] and is corrected in the description given here.

The model uses both average wave height, H_{avg}, and wind speed, U, to account for environmental conditions with (4.1) describing their relationship. This assumes a fully developed sea, where the wind and swell are aligned and the wind is consistent with a speed that is highly correlated with the wave height. The mean backscatter for the horizontal and vertical polarisations are given by

$$\sigma^0_{\text{HH,dB}} = \begin{cases} 10\log_{10}\left(\lambda\phi^{0.4}a_{\text{GIT}}b_{\text{GIT}}c_{\text{GIT}}\right) - 54.089, & 1 \le f_{\text{RF,GHz}} < 10 \\ 10\log_{10}\left(\phi^{0.547}a_{\text{GIT}}b_{\text{GIT}}c_{\text{GIT}}\right) - 54.381, & 10 \le f_{\text{RF,GHz}} < 100 \end{cases} \tag{4.4}$$

$$\sigma^0_{\text{VV,dB}} = \begin{cases} \sigma^0_{\text{HH,dB}} - 1.73\ln(H_{\text{avg}} + 0.015) + 3.76\ln\lambda \\ \quad + 2.46\ln(\phi + 0.0001) + 22.2, & 1 \le f_{\text{RF,GHz}} < 3 \\ \sigma^0_{\text{HH,dB}} - 1.05\ln(H_{\text{avg}} + 0.015) + 1.09\ln\lambda \\ \quad + 1.27\ln(\phi + 0.0001) + 9.7, & 3 \le f_{\text{RF,GHz}} < 10 \\ \sigma^0_{\text{HH,dB}} - 1.38\ln(H_{\text{avg}}) + 3.43\ln\lambda \\ \quad + 1.31\ln\phi + 18.55, & 10 \le f_{\text{RF,GHz}} < 100 \end{cases} \tag{4.5}$$

where λ and f_{RF} are the radar wavelength and carrier frequency, respectively. The multipath or interference factor is defined by

$$a_{\text{GIT}} = \frac{\sigma^4_{\text{GIT}}}{1 + \sigma^4_{\text{GIT}}} \tag{4.6}$$

and assuming a Gaussian distribution of surface wave heights, the multipath roughness parameter is given by

$$\sigma_{\text{GIT}} = \frac{(14.4\lambda + 5.5)\phi H_{\text{avg}}}{\lambda + 0.015}. \tag{4.7}$$

The GIT model assumes a fully developed sea where the wind and sea directions are aligned with the azimuth angle, θ. The sea direction factor is given by

$$b_{\text{GIT}} = \begin{cases} \exp\left(0.2\cos\theta(1-2.8\phi)(\lambda+0.015)^{-0.4}\right), & 1 \le f_{\text{RF,GHz}} < 10 \\ \exp\left(0.25\cos\theta(1-2.8\phi)(\lambda+0.015)^{-0.33}\right), & 10 \le f_{\text{RF,GHz}} < 100 \end{cases} \tag{4.8}$$

and the wind speed factor as

$$c_{\text{GIT}} = \begin{cases} \left(\dfrac{1.94U}{1+U/15.4}\right)^{1.1(\lambda+0.015)^{-0.4}}, & 1 \le f_{\text{RF,GHz}} < 10 \\ \left(\dfrac{1.94U}{1+U/15.4}\right)^{1.93(\lambda+0.015)^{-0.04}}, & 10 \le f_{\text{RF,GHz}} < 100. \end{cases} \tag{4.9}$$

4.2.4 The hybrid model

The hybrid mean backscatter model was presented by Reilly and Dockery [14] and combines features from Nathanson's data and the GIT model. It is valid for grazing angles 0.1°–30°, sea states 3–6, frequencies from 0.5 to 35 GHz, all wind directions and both horizontal and vertical polarisations. It was later published as the NATO anti-air warfare systems (NAAWS) model [15] with the stated frequency range extended to 70 GHz. Ducting is not explicitly included, although the model is described as 'based on published empirical fits to data that most likely already include some degree of ducting'. The model is based on a series of adjustments to a reference backscatter coefficient, σ_0, which corresponds to a sea state of 5, grazing angle of $\phi_{0,\text{deg}} = 0.1°$, azimuth of 0° (up-wind) and vertical polarisation. This reference level is then adjusted according to

$$\sigma^0_{\text{dB}} = \sigma_0 + a_{\text{hyb}} + b_{\text{hyb}} + c_{\text{hyb}} + d_{\text{hyb}} \tag{4.10}$$

where a_{hyb}, b_{hyb}, c_{hyb} and d_{hyb} are the grazing angle, sea state, polarisation and wind direction adjustments, respectively. The reference reflectivity is determined by

$$\sigma_0 = \begin{cases} 24.4\log_{10} f_{\text{RF,GHz}} - 65.2, & 0.5 \le f_{\text{RF,GHz}} < 12.5 \\ 3.25\log_{10} f_{\text{RF,GHz}} - 42, & 12.5 \le f_{\text{RF,GHz}} < 35. \end{cases} \tag{4.11}$$

For the grazing variation term, a_{hyb}, requires the definition of a transition angle

$$\phi_t = \operatorname{asin}(0.066\lambda/H_{\text{rms}}) \tag{4.12}$$

where the RMS wave height, $H_{\text{rms}} = 0.031S_{\text{ss}}^2$. Then, if $\phi_{\text{t}} \geq \phi_0$,

$$a_{\text{hyb}} = \begin{cases} 0, & \phi < \phi_0 \\ 20\log_{10}(\phi/\phi_0), & \phi_0 \leq \phi \leq \phi_{\text{t}} \\ 20\log_{10}(\phi_{\text{t}}/\phi_0) + 10\log_{10}(\phi/\phi_{\text{t}}), & \phi_{\text{t}} < \phi \leq \pi/6 \end{cases} \tag{4.13}$$

and if $\phi_{\text{t}} < \phi_0$,

$$a_{\text{hyb}} = \begin{cases} 0, & \phi \leq \phi_0 \\ 10\log_{10}(\phi/\phi_0), & \phi > \phi_0. \end{cases} \tag{4.14}$$

The sea state term is given by

$$b_{\text{hyb}} = 5(S_{\text{SS}} - 5) \tag{4.15}$$

and for the vertical polarisation, $c_{\text{hyb}} = 0$, while for the horizontal polarisation,

$$c_{\text{hyb}} = \begin{cases} 1.7\ln(H_{\text{av}} + 0.015) - 3.8\ln\lambda & \\ \quad - 2.5\ln(\phi + 0.0001) - 22.2, & 0.5 \leq f_{\text{RF,GHz}} < 12.5 \\ 1.1\ln(H_{\text{av}} + 0.015) - 1.1\ln\lambda & \\ \quad - 1.3\ln(\phi + 0.0001) - 9.7, & 3 \leq f_{\text{RF,GHz}} < 10 \\ 1.4\ln H_{\text{av}} - 3.4\ln\lambda - 1.3\ln\phi - 18.6, & 10 \leq f_{\text{RF,GHz}} < 35 \end{cases} \tag{4.16}$$

where $H_{\text{av}} = 0.08S_{\text{ss}}^2$. The final component accounts for the wind direction,

$$d_{\text{hyb}} = \left(2 + 1.7\log_{10}(0.1/\lambda)\right)(\cos\theta - 1). \tag{4.17}$$

4.2.5 The Technology Service Corporation model

The TSC model combines a low grazing angle component based on a model fit to Nathanson's tables [16] and a high grazing angle component based on the fitting of experimental data and theoretical considerations in [19]. Specifically, the model assumes that two different mechanisms dominate depending on the grazing angle: quasi-specular reflection at large grazing angles and diffuse scattering at low grazing angles, with the total backscatter being the sum of these two components. The model was first described in [8] and later modified in [20] to correct a predicted minimum in the down-wind direction. Also some of the data used at low grazing angles was collected under anomalous propagation conditions (ducting) and hence the mean backscatter does not reduce as rapidly at low grazing angles when compared to the GIT model. The model is specified as

$$\sigma^0_{\mathrm{HH,dB}} = 10\log_{10}\left(1.7\times10^{-5}\phi^{0.5}a_{\mathrm{TSC}}b_{\mathrm{TSC}}c_{\mathrm{TSC}}d_{\mathrm{TSC}}/(3.2808\lambda+0.05)^{1.8} + \sigma^0_{\mathrm{quasi}}\right), \tag{4.18}$$

$$\sigma^0_{\mathrm{VV,dB}} = \begin{cases} \sigma^0_{\mathrm{HH,dB}} - 1.73\ln(8.225\sigma_{\mathrm{h}}+0.05) + 3.76\ln\lambda \\ \quad + 2.46\ln(\sin\phi+0.0001) + 24.2672, & 0.5 \le f_{\mathrm{RF,GHz}} < 2 \\ \sigma^0_{\mathrm{HH,dB}} - 1.05\ln(8.225\sigma_{\mathrm{h}}+0.05) + 1.09\ln\lambda \\ \quad + 1.27\ln(\sin\phi+0.0001) + 10.945, & 2 \le f_{\mathrm{RF,GHz}} \le 35 \end{cases} \tag{4.19}$$

where the sea surface height standard deviation is given by $\sigma_{\mathrm{h}} = 0.03505S_{\mathrm{ss}}^{1.95}$. The low grazing angle factor is given by

$$a_{\mathrm{TSC}} = \sigma_{\mathrm{TSC}}^{1.5}/(1+\sigma_{\mathrm{TSC}}^{1.5}), \tag{4.20}$$

$$\sigma_{\mathrm{TSC}} = 4.5416\phi(3.2808\sigma_{\mathrm{h}}+0.25)/\lambda \tag{4.21}$$

and the wind speed factor,

$$b_{\mathrm{TSC}} = ((1.9438U+4)/15)^{A_{\mathrm{TSC}}} \tag{4.22}$$

where the wind speed is related to the sea state by $U = 3.189S_{\mathrm{ss}}^{0.8}$ and the constant A_{TSC} is given by

$$\begin{aligned} A_{\mathrm{TSC}} &= 2.63A_{\mathrm{TSC},1}/(A_{\mathrm{TSC},2}A_{\mathrm{TSC},3}A_{\mathrm{TSC},4}), \\ A_{\mathrm{TSC},1} &= (1+(\lambda/0.009144)^3)^{0.1}, \\ A_{\mathrm{TSC},2} &= (1+(\lambda/0.03048)^3)^{0.1}, \\ A_{\mathrm{TSC},3} &= (1+(\lambda/0.09144)^3)^{\phi^{0.6}/3}, \\ A_{\mathrm{TSC},4} &= 1+0.35\phi^{0.6}. \end{aligned} \tag{4.23}$$

The aspect factor is given by

$$c_{\mathrm{TSC}} = \begin{cases} 1, & \phi_{\mathrm{deg}} = 90^\circ \\ \exp\left(\dfrac{0.3\cos\theta\exp(-\phi/0.17)}{(10.7636\lambda^2+0.005)^{0.2}}\right), & \phi_{\mathrm{deg}} < 90^\circ \end{cases} \tag{4.24}$$

and the azimuth correction factor,

$$d_{\mathrm{TSC}} = 1 - 0.6\sin^2\theta. \tag{4.25}$$

Finally, the quasi-specular component of the reflectivity, σ^0_{quasi}, that becomes dominant with high grazing angles is defined by

$$\sigma^0_{\text{quasi}} = \mu_{\text{TSC}} \cot^2 \beta_{\text{TSC}} \exp\left(-\frac{\tan^2(\pi/2-\phi)}{\tan^2 \beta_{\text{TSC}}}\right) \tag{4.26}$$

where

$$\mu_{\text{TSC, dB}} = \begin{cases} -5, & \lambda < 0.05 \\ -5 + 12.5(\log_{10}\lambda - \log_{10} 0.05), & \lambda \geq 0.05 \end{cases} \tag{4.27}$$

and the angle β_{TSC} determines the width of the quasi-specular term as a function of sea state. It is given as

$$\beta_{\text{TSC,deg}} = \begin{cases} 10.1 + 1.65 S_{\text{ss}}, & S_{\text{ss}} \leq 2 \\ 13.4 + 0.7(S_{\text{ss}} - 2), & S_{\text{ss}} > 2. \end{cases} \tag{4.28}$$

4.2.6 The Masuko models

The work presented by Masuko *et al.* [11] covers a series of models developed for both X- and Ka-bands. These do not produce a complete model of the mean backscatter but rather individually model the variation with azimuth, grazing angle and wind speed. For the azimuth variation, a second-order cosine model is used and features maxima in the up- and down-wind directions with minima in between:

$$\sigma^0_{\text{Mas}}(\theta) = \sum_{n=0}^{2} a_{\text{Mas}}(n) \cos(n\theta) \tag{4.29}$$

where $a_{\text{Mas}}(n)$ are the coefficients to be fitted by regression and are not given in [11]. The variation with grazing angle is modelled as

$$\sigma^0_{\text{Mas}}(\phi) = \exp\left(-\sum_{n=0}^{N_{\text{Mas}}} b_{\text{Mas}}(n)\,(\pi/2-\phi)^n\right) \tag{4.30}$$

where N_{Mas} is the model order and $b_{\text{Mak}}(n)$ are the coefficients to be fitted by regression, also not provided in the original publication. Masuko *et al.* find that a model order of 4–5 is necessary to fit their data, which covers grazing angles from 0° to 70°. The variation with wind speed is modelled using a power law of the following form:

$$\sigma^0_{\text{Mas,dB}}(U_{19.5}) = 10\left(c_{\text{Mas}}(\theta,\phi) + d_{\text{Mas}}(\theta,\phi)\log_{10} U_{19.5}\right) \tag{4.31}$$

where c_{Mas} and d_{Mas} are the coefficients to be fitted by regression and $U_{19.5}$ is the wind speed measured at 19.5 m above the surface. This wind speed can be related back to U, which is assumed to be measured at a standard height of 10 m, using the relationships described in [11]. If atmospheric instability is ignored, then for $1 \leq U \leq 20$ m/s, $U_{19.5}$ can be approximated by

$$U_{19.5} = 0.03 + 1.04U + 0.0022U^2. \tag{4.32}$$

Table 4.4 The wind speed power law model coefficients estimated by Masuko et al. for both X- and Ka-band [11]

Pol.	Freq. band	Wind direction		Grazing angle 30°	40°	50°	60°
HH	X	Up-wind	c_{Mas}	–4.93	–4.55	–3.83	–2.83
			d_{Mas}	2.49	2.36	2.08	1.72
HH	X	Down-wind	c_{Mas}	–5.40	–5.18	–4.27	–3.00
			d_{Mas}	2.70	2.82	2.30	1.70
HH	X	Cross-wind	c_{Mas}	–5.64	–5.03	–4.15	–3.13
			d_{Mas}	2.63	2.37	1.86	1.63
VV	X	Up-wind	c_{Mas}	–4.28	–4.08	–3.53	–2.76
			d_{Mas}	2.30	2.24	1.98	1.70
VV	X	Down-wind	c_{Mas}	–4.36	–4.33	–3.81	–3.01
			d_{Mas}	2.31	2.49	2.10	1.79
VV	X	Cross-wind	c_{Mas}	–4.60	–4.46	–3.81	–3.07
			d_{Mas}	2.05	2.12	1.69	1.55
HH	Ka	Up-wind	c_{Mas}	–4.67	–4.80	–3.72	–2.36
			d_{Mas}	2.21	2.54	1.95	1.18
HH	Ka	Down-wind	c_{Mas}	–4.96	–5.48	–4.69	–2.52
			d_{Mas}	2.15	2.91	2.61	1.08
HH	Ka	Cross-wind	c_{Mas}	–5.10	–5.09	–5.00	–2.32
			d_{Mas}	2.11	2.30	2.58	0.68
VV	Ka	Up-wind	c_{Mas}	–3.96	–3.63	–2.80	–2.63
			d_{Mas}	1.88	1.79	1.29	1.49
VV	Ka	Down-wind	c_{Mas}	–4.16	–4.12	–3.40	–2.31
			d_{Mas}	1.81	2.11	1.72	1.00
VV	Ka	Cross-wind	c_{Mas}	–4.90	–5.17	–4.33	–3.32
			d_{Mas}	2.13	2.60	2.16	1.67

The model is valid for sea states 2–6 with the coefficients shown in Table 4.4 for both X- and Ka-band data, HH and VV polarisations and a range of azimuth and grazing angles.

4.2.7 The Imaging Radar Systems Group model

The IRSG model [12,21] was designed to fit a series of sea clutter measurements from the Ingara X-band data set (see Chapter 1). These were based on a large amount of data covering 360° in viewing direction, grazing angles from 20° to 45° and a range of wind speeds covering sea states 2–6. Two models were originally presented, with the linear model presented here, 'IRSG-lin', found to be simpler and more robust.

The model is based on work by Ulaby *et al.* [10] and contains three parts. The first uses three parameters, a_{IRSG}, b_{IRSG} and c_{IRSG} to capture the variation in azimuth with peaks at both up- and down-wind and minima in the cross-wind directions:

$$\sigma^0 = a_{\mathrm{IRSG}} + b_{\mathrm{IRSG}} \cos\theta + c_{\mathrm{IRSG}} \cos(2\theta). \tag{4.33}$$

These three parameters are then determined by the following relationships:

$$\begin{aligned} a_{\mathrm{IRSG}} &= (\sigma_{\mathrm{u}}^0 + \sigma_{\mathrm{d}}^0 + 2\sigma_{\mathrm{c}}^0)/4, \\ b_{\mathrm{IRSG}} &= (\sigma_{\mathrm{u}}^0 - \sigma_{\mathrm{d}}^0)/2, \\ c_{\mathrm{IRSG}} &= (\sigma_{\mathrm{u}}^0 + \sigma_{\mathrm{d}}^0 - 2\sigma_{\mathrm{c}}^0)/4, \end{aligned} \tag{4.34}$$

where the up-wind, σ_{u}^0, down-wind, σ_{d}^0, and cross-wind, σ_{c}^0, mean backscatter values are in turn determined by the following relationship, which captures the influence of both the wind speed and grazing angle:

$$\sigma_{\mathrm{dB}}^0 = d_{\mathrm{IRSG}} + e_{\mathrm{IRSG}}\phi + f_{\mathrm{IRSG}} \log_{10} U. \tag{4.35}$$

The coefficients d_{IRSG}, e_{IRSG} and f_{IRSG} are given in Table 4.5 for both the co- and cross-polarised returns.

Figures 4.1–4.3 show the model validation results* for the IRSG model compared to data points from the Ingara X-band data set [12]. This is contrasted to model outputs

Table 4.5 IRSG mean backscatter model coefficients [21]

Pol.	Wind direction	d_{IRSG}	e_{IRSG}	f_{IRSG}
HH	Up-wind	–60.03	23.39	22.65
	Down-wind	–67.80	28.58	23.92
	Cross-wind	–67.09	23.12	24.71
HV	Up-wind	–66.65	9.72	25.46
	Down-wind	–68.74	14.17	24.47
	Cross-wind	–73.09	14.85	26.66
VV	Up-wind	–50.18	12.41	25.15
	Down-wind	–50.16	12.30	23.92
	Cross-wind	–52.60	12.30	22.09

*In the initial publication, the results were scaled by a constant.

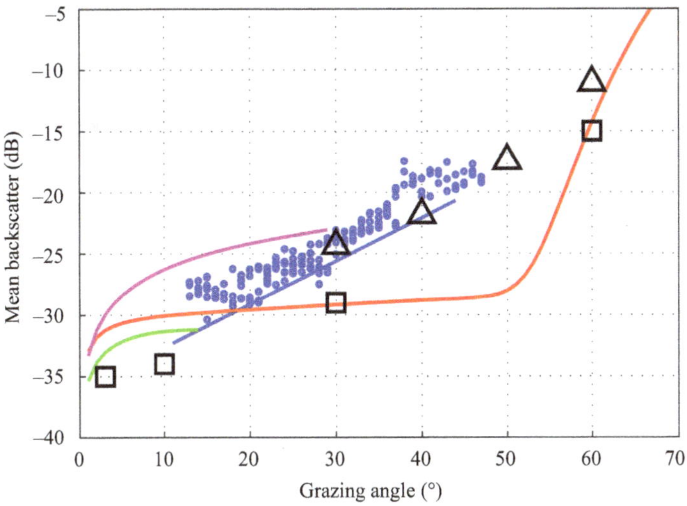

Figure 4.1 *Mean backscatter variation with grazing angle. Data is from the horizontal polarisation, 9.4 m/s wind speed and up-wind. Model/data represented include the following: (—) IRSG-lin, (—) hybrid, (—) TSC, (—) GIT, △ Masuko, □ Nathanson, (•) Ingara data. Adapted from [12].*

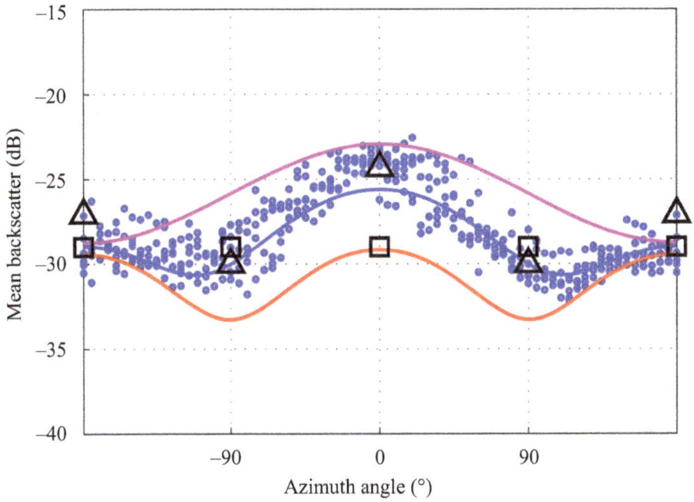

Figure 4.2 *Mean backscatter variation with azimuth angle. Data is from the horizontal polarisation, 30° grazing angle and wind speed 9.4 m/s. Model/data represented include the following: (—) IRSG-lin, (—) hybrid, (—) TSC, △ Masuko, □ Nathanson, (•) Ingara data. Adapted from [12].*

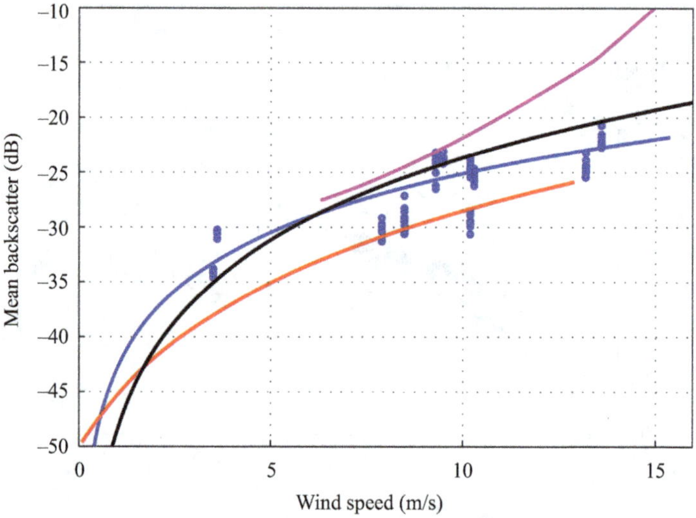

Figure 4.3 Mean backscatter variation with wind speed. Data is from the horizontal polarisation, 30° grazing angle and up-wind. Model/data represented include the following: (—) IRSG-lin, (—) hybrid, (—) TSC, (—) Masuko, (•) Ingara data. Adapted from [12]. © Commonwealth of Australia 2008

from the GIT, hybrid and TSC models with data points shown from Masuko's models and Nathanson's tables. As expected, the IRSG model shows a good fit to the data, while many of the other models show significant mismatch in azimuth, grazing or wind speed.

4.2.8 Naval Research Laboratory model

The NRL model [9] is an empirical sea clutter model based on fits to Nathanson's tables [6]. As a result, the model covers both horizontal and vertical polarisations, frequencies from 0.5 to 35 GHz, sea states 2–6 and grazing angles from 0.1° to 60°. The model can be summarised by the following relationship with the model parameters given in Table 4.6,

$$\sigma^0_{\text{dB}} = a_{\text{NRL}} + b_{\text{NRL}} \log_{10}(\sin\phi) + \frac{(27.5 + c_{\text{NRL}}\phi_{\text{deg}})\log_{10} f_{\text{RF,GHz}}}{1 + 0.95\phi_{\text{deg}}}$$
$$+ d_{\text{NRL}}(1 + S_{\text{SS}})^{1/(2+0.085\phi_{\text{deg}}+0.033S_{\text{SS}})} + e_{\text{NRL}}\phi^2_{\text{deg}}. \tag{4.36}$$

where a_{NRL}, b_{NRL}, c_{NRL}, d_{NRL} and e_{NRL} are the model parameters. The first term is a constant, while the second term represents a logarithmic dependence on the grazing angle. The third term accounts for the effect of radar frequency with an additional empirical correction for the grazing angle, while the fourth term adds the effect of sea state, again with an additional empirical correction for the grazing angle and the fifth

Table 4.6 NRL mean backscatter model coefficients [9]

Pol.	a_{NRL}	b_{NRL}	c_{NRL}	d_{NRL}	e_{NRL}
HH	–73.0	20.78	7.351	25.65	0.0054
VV	–50.79	25.93	0.7093	21.58	0.0021

term adds the rapid increase in reflectivity as the grazing angle approaches vertical incidence. To include a variation with the wind direction, Whitrow [22] has proposed the addition of the d_{hyb} term from the hybrid model.

4.2.9 DSTG continuous mean backscatter model

The DSTG continuous X-band mean backscatter model [13] utilises the GIT model, $\sigma^0_{\text{GIT}}(\cdot)$ in the low grazing angle region, 0.1°–3° and a medium grazing angle model, $\sigma^0_{\text{med}}(\cdot)$ over 20°–45°. To ensure continuity, a linear function, $\sigma^0_{\text{lin}}(\cdot)$ is used to bridge between the two model components and to make sure that the mean backscatter is always increasing in grazing angle, the maximum of the linear and medium grazing angle models is used in the grazing region $20° \le \phi_{\text{deg}} \le 45°$. The complete model is given by

$$\sigma^0(\theta,\phi) = \begin{cases} \sigma^0_{\text{GIT}}(\theta,\phi), & 0.1° \le \phi_{\text{deg}} < 3° \\ \sigma^0_{\text{lin}}(\theta,\phi), & 3° \le \phi_{\text{deg}} < 20° \\ \text{Max}\left[\sigma^0_{\text{med}}(\theta,\phi), \sigma^0_{\text{lin}}(\theta,20)\right], & 20° \le \phi_{\text{deg}} \le 45° \end{cases} \tag{4.37}$$

where Max[·] is the maximum value,

$$\sigma^0_{\text{lin}}(\theta,\phi) = \sigma^0_{\text{GIT}}(\theta,3) + K_{\text{cont}}(\theta)\left(\phi_{\text{deg}} - 3\right) \tag{4.38}$$

and the slope of the line $K_{\text{cont}}(\theta)$ is always positive and spans from 3° to 20°:

$$K_{\text{cont}}(\theta) = \left|\sigma^0_{\text{med}}(\theta,20) - \sigma^0_{\text{GIT}}(\theta,3)\right|/17. \tag{4.39}$$

For the medium grazing angle component, $\sigma^0_{\text{med}}(\cdot)$, there are two relationships which are important. The first captures the azimuth and grazing variations:

$$\sigma^0(\theta,\phi) = \tilde{a}_{\text{cont}}\phi^{\gamma_{\text{cont}}}\left(1 + \tilde{b}_{\text{cont}}\cos\theta + \tilde{c}_{\text{cont}}\cos(2\theta)\right) \tag{4.40}$$

where $\tilde{a}_{\text{cont}}$, $\tilde{b}_{\text{cont}}$, $\tilde{c}_{\text{cont}}$ and γ_{cont} are the geometric model coefficients. Representing the ocean conditions is difficult, so the model uses only the wind speed, U and significant wave height $H_{1/3}$,

$$Y_{\text{cont}} = d_{\text{cont}} + e_{\text{cont}}\log_{10}U + f_{\text{cont}}H_{1/3} \tag{4.41}$$

where d_{cont}, e_{cont} and f_{cont} are the environmental model coefficients. To relate these two models, the coefficients in (4.40) must be altered so they are independent of grazing angle. This is achieved by introducing a normalisation factor, ϕ_0, and then redefining (4.40) as

$$\sigma^0_{\text{med}}(\theta, \phi) = \left(\frac{\phi}{\phi_0}\right)^{\gamma_{\text{cont}}} (a_{\text{cont}} + b_{\text{cont}} \cos\theta + c_{\text{cont}} \cos(2\theta)) \tag{4.42}$$

where the new coefficients are given by

$$a_{\text{cont}} = \tilde{a}_{\text{cont}} \phi_{0,\text{deg}}^{\gamma_{\text{cont}}}, \quad b_{\text{cont}} = \tilde{a}_{\text{cont}} \tilde{b}_{\text{cont}} \phi_{0,\text{deg}}^{\gamma_{\text{cont}}}, \quad c_{\text{cont}} = \tilde{a}_{\text{cont}} \tilde{c}_{\text{cont}} \phi_{0,\text{deg}}^{\gamma_{\text{cont}}}. \tag{4.43}$$

The model is then implemented by equating each coefficient $a_{\text{cont}}, b_{\text{cont}}, c_{\text{cont}}, \gamma_{\text{cont}}$ to the model Y_{cont} in (4.41). For the model published in [13], the normalisation parameter was fixed at $\phi_{0,\text{deg}} \equiv 30°$ and the coefficients are given in Table 4.7.

Figure 4.4 shows the model data and fits for the mean backscatter horizontal and vertical polarisations where the wind speed and wave height are 10.2 m/s and 1.21 m, respectively. For the low grazing angle region, 0.1°–10°, the GIT model has been used as a data substitute. The model fit shows a continuous variation across both azimuth and grazing angles, with the transition from a single to a double azimuth peak at a grazing angle of approximately 20° for the horizontal polarisation and 10° for vertical.

To better visualise the fit results, azimuth slices of the continuous model are shown in Figure 4.5 along the main wind directions. Also shown is the GIT model in the low grazing angle region, the Ingara data and the IRSG model in the medium grazing angle model region. The continuous model covers the entire region and provides an extremely good fit for each wind direction. The continuous mean backscatter model was also compared to the GO-SSA extended model [13] along the up-wind direction. Figure 4.6 shows the results for sea states 2, 4 and 6 and both HH and VV polarisations. The two models show a reasonable match over the range of grazing angles with only a few dB mismatch.

Table 4.7 DSTG continuous mean backscatter coefficients for the medium grazing angle model, σ^0_{med} [13]

Pol.		d_{cont}	e_{cont}	f_{cont}
	a_{cont}	−78.74	49.42	−0.20
HH	b_{cont}	11.72	−9.14	−0.012
	c_{cont}	−0.035	2.21	−0.15
	γ_{cont}	−1.21	0.64	0.0082
	a_{cont}	−47.78	27.15	−0.56
VV	b_{cont}	−0.032	0.69	0.023
	c_{cont}	0.51	2.31	−0.21
	γ_{cont}	0.16	−0.57	0.050

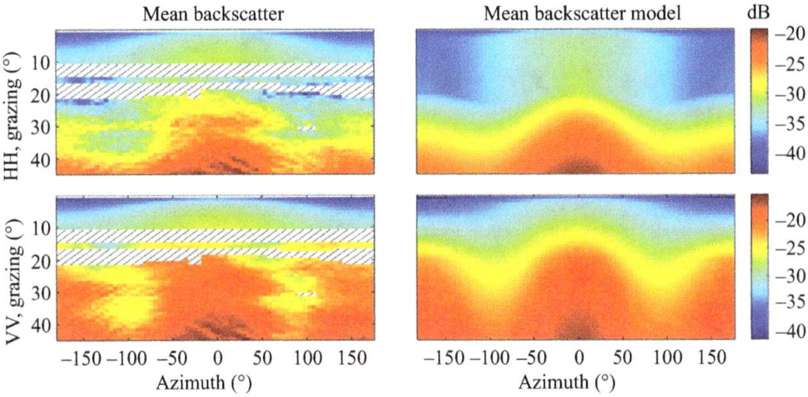

Figure 4.4 Mean backscatter – data on left and model fit on right. Data comprises the GIT model for grazing angles 0.1°–10° *and the Ingara data from* 15° *to* 45°*. The wind speed and wave height are 10.2 m/s and 1.21 m, respectively [13].* © *Commonwealth of Australia 2017*

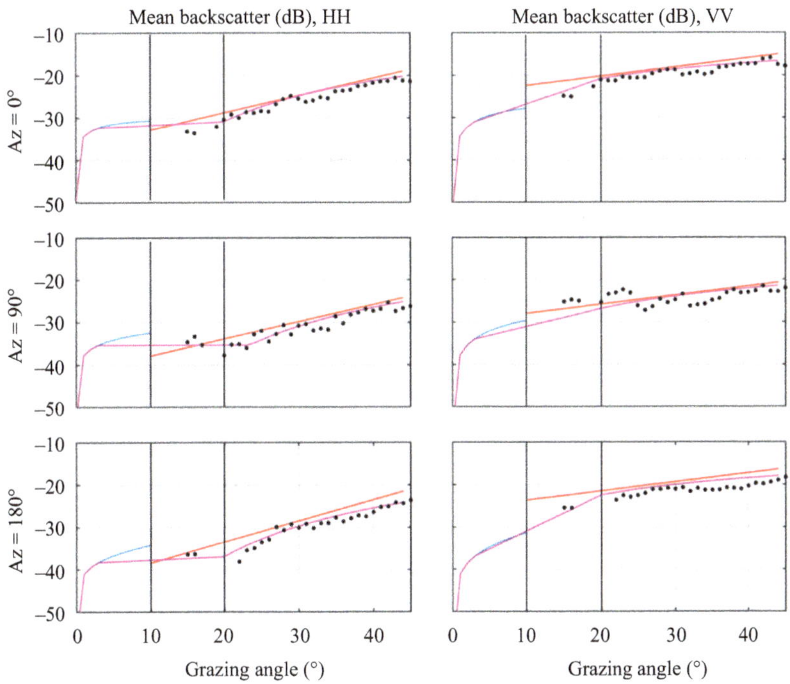

Figure 4.5 Mean backscatter: (—) GIT model, (—) IRSG model, (—) DSTG continuous model, (·) Ingara data, vertical lines indicate model limits. The wind speed and wave height are 10.2 m/s and 1.21 m, respectively [13]. © *Commonwealth of Australia 2017*

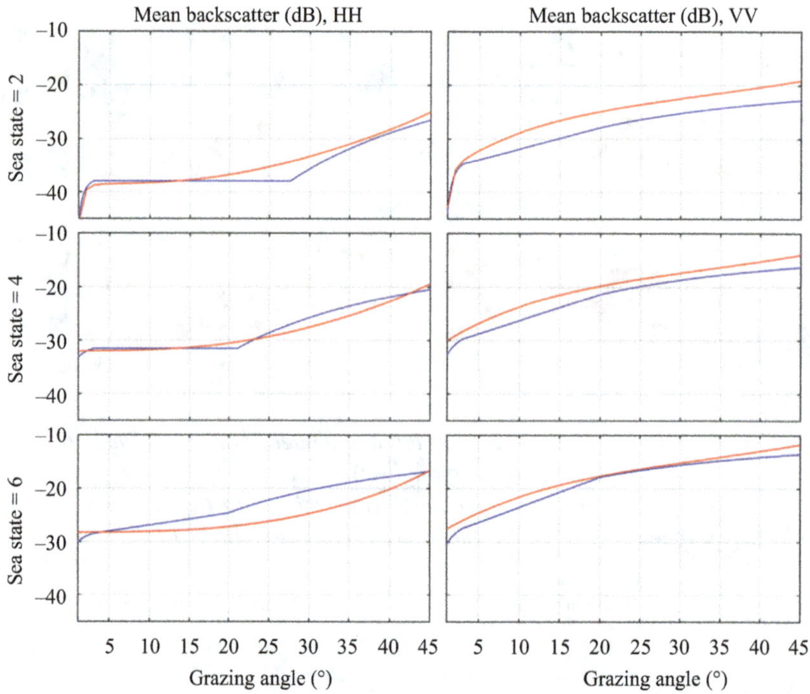

Figure 4.6 Mean backscatter with different sea states: (—) DSTG continuous model, (—) GO-SSA extended model [13]. © Commonwealth of Australia 2017

4.3 Bistatic mean reflectivity

As discussed in Chapter 3, very few measurements of bistatic sea clutter have been made and consequently few empirical models have been developed. However, some limited progress has been made for modelling bistatic mean reflectivity. These models are discussed next.

4.3.1 In-plane scattering

Domville [23,24] proposed a reflectivity model for X-band in-plane scattering based on his measurements. The empirical model covers both backscatter and forward scatter collected with wind speeds in the range 4–10 m/s and using the model with values outside of this region may not give reliable results. Also, for horizontally polarised scattering, the modelled value of the normalised bistatic mean reflectivity, σ_B^0, shows very little dependence on wind speed. The original model was presented as a graphical flow diagram, which was transcribed into algorithmic form in [1] and is reproduced here. First, define the following angles which relate to the incident and reflected angles, ϕ_1 and ϕ_2:

$$\begin{aligned} \phi_{11} &= |\phi_1 - \pi/2|, \\ \phi_{21} &= |\phi_2 - \pi/2|, \\ \phi_n &= \phi_{11} + \phi_{21}. \end{aligned} \tag{4.44}$$

The values of σ_B^0 for in-plane backscatter, $\psi_B = \pi$, and three different grazing angle regimes are then given as

1. Low grazing angles, $\phi_{11,\text{deg}} > 88.2°$ or $\phi_{21,\text{deg}} > 88.2°$:

$$\sigma^0_{B,dB} = \begin{cases} -50, & U \leq 5 \\ -50 + 114\phi(3.8U - 23), & U > 5, \phi < 0.0087 \\ -67 + 3.4U + (0.4U - 6)(1.33 - 38\phi), & U > 5, \phi \geq 0.0087 \end{cases} \tag{4.45}$$

where

$$\phi = \pi/2 - \text{Max}[\phi_{11}, \phi_{21}], \quad \phi_{11,\text{deg}} > 84.8° \text{ or } \phi_{21,\text{deg}} > 84.8° \tag{4.46}$$

2. Medium grazing angles, $84.8° < \phi_{11,\text{deg}} \leq 88.2°$ or $84.8° < \phi_{21,\text{deg}} \leq 88.2°$:

$$\sigma^0_{B,dB} = \begin{cases} -67 + 3.4U + (\sigma_2 + 67 - 3.4U)(19\phi - 0.67), & U > 5 \\ -50 + (\sigma_2 + 50)(19\phi - 0.67), & U \leq 5 \end{cases} \tag{4.47}$$

where

$$\sigma_2 = m_B\phi_n + c_B, \quad \phi_{11,\text{deg}} \leq 84.8°, \ \phi_{21,\text{deg}} \leq 84.8° \tag{4.48}$$

and the values of m_B and c_B are given in Table 4.8 for different values of ϕ_n and wind velocity, U.

Table 4.8 Domville bistatic mean reflectivity model: values of m_B and c_B for in-plane backscatter ($\psi_B = \pi$)

Pol.	ϕ_n	m_B	c_B
VV	$\phi_n \geq 2 - 0.1U$	−10.18	$-11 + 0.8U$
VV	$\phi_n < 2 - 0.1U$	−17.63	4
HH	$\phi_n \geq 1.6 - 0.008U$	−6.69	$-13.8 + 0.08U$
HH	$\phi_n < 1.6 - 0.008U$	−17.63	4

3. High grazing angles, $\phi_{11,\text{deg}} \leq 84.8°$ and $\phi_{21,\text{deg}} \leq 84.8°$:

$$\sigma^0_{\text{B,dB}} = \sigma_2. \tag{4.49}$$

For the region of strong specular reflection (forward scatter), $\psi_{\text{B}} = 0$, and the values for σ^0_{B} are given in Table 4.9, where the following angles are defined as

$$\begin{aligned} \phi_s &= \sqrt{\phi_{11}^2 + \phi_{21}^2}, \\ \phi_b &= |\phi_{11} - \phi_{21}|, \\ \phi_d &= \text{Max}[\phi_{11}, \phi_{21}]. \end{aligned} \tag{4.50}$$

For other regions of forward scatter, the following values are used:

$$\sigma^0_{\text{B,dB}} = \sigma_1 - \frac{\phi_c(\sigma_1 - \sigma_2)}{\phi_a} \tag{4.51}$$

where the variables $\sigma_1, \sigma_2, \phi_a$ and ϕ_c are given in Tables 4.9–4.11 for different values of ϕ_s, ϕ_b, ϕ_d and U.

A typical plot from this model for the horizontal polarisation and a wind speed of 10 m/s is shown in Figure 4.7. The angles ϕ_1 and ϕ_2 represent the two grazing angles, with $\phi_{2,\text{deg}} > 90°$ representing the forward scatter region. The bistatic angle is given by $\beta = |\phi_1 - \phi_2|$.

Domville [23,24] produced an empirical fit to Pidgeon's C-band data (see Chapter 3). This data was collected for sea states 1–3 and transmitter grazing

Table 4.9 Domville bistatic mean reflectivity model: values of σ^0_B for in-plane forward scatter $\psi_B = 0$

Pol.	ϕ_s	ϕ_b	σ_1	$\sigma^0_{\text{B,dB}}$
HH	$\phi_s > 1.40$	$\phi_b \leq 0.345$		14
VV	$\phi_s > 1.22$	$\phi_b < 0.087$	$18.8 - 7.16\phi_s$	$\sigma_1 - \phi_b(\sigma_1 - 10)/0.0873$
VV	$\phi_s > 1.22$	$0.087 \leq \phi_b \leq 0.345$		10

Table 4.10 Domville bistatic mean reflectivity model: values of σ_1, ϕ_a, ϕ_c for forward scatter, as a function of ϕ_s and ϕ_b

Pol.	ϕ_s, ϕ_b	σ_1	ϕ_a	ϕ_c
HH	$\phi_s \leq 1.40$	$4 + 7.16\phi_s$	ϕ_s	ϕ_b
HH	$\phi_s > 1.42$ and $\phi_b > 0.345$	14	$\phi_s - 0.349$	$\phi_b - 0.349$
VV	$\phi_s \leq 1.22$	$4 + 4.91\phi_s$	ϕ_s	ϕ_b
VV	$\phi_s > 1.22$ and $\phi_b > 0.345$	10	$\phi_s - 0.349$	$\phi_b - 0.349$

Table 4.11 Domville bistatic mean reflectivity model: values of σ_2 for forward scatter as a function of ϕ_d and U

Pol.	ϕ_d	U	σ_2
HH	$\phi_d < 1.48$	All U	$-17.5\phi_d + 4$
HH	$\phi_d \geq 1.48$	$U > 5$	$\text{Max}\,[-50, -70 + 3.34U + 11.4(47 - 3.34U)(\pi/2 - \phi_d)]$
HH	$\phi_d \geq 1.48$	$U \leq 5$	$\text{Max}[-50, -70 + 535.8(\pi/2 - \phi_d)]$
VV	$\phi_d < 1.48$	All U	$-17.5\phi_d + 4 + 0.26U\phi_d$
VV	$\phi_d \geq 1.48$	$U > 5$	$\text{Max}[-50, -70 + 3.34U + 11.4(47 - 2.94U)(\pi/2 - \phi_d)]$
VV	$\phi_d \geq 1.48$	$U \leq 5$	$\text{Max}[-50, -70 + 11.4(47 - 0.4U)(\pi/2 - \phi_d)]$

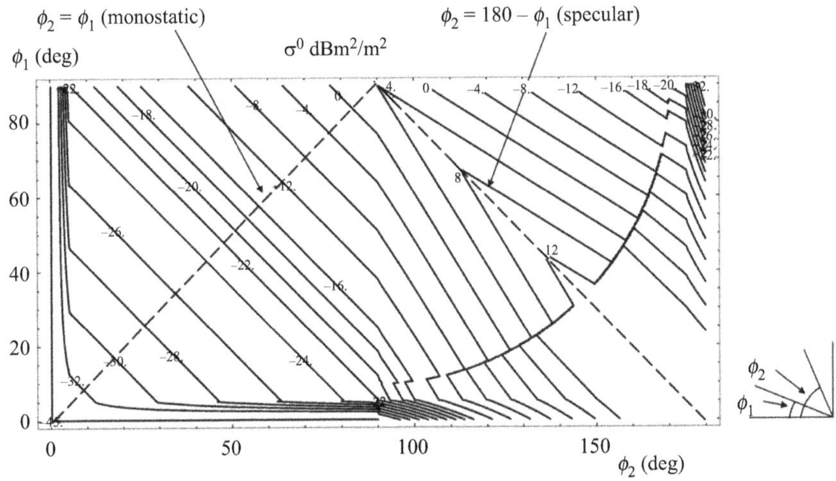

Figure 4.7 In-plane bistatic reflectivity from Domville's model with 10 m/s wind speed and HH polarisation

angles, $0.1° \leq \phi_{1,\text{deg}} \leq 3°$. The receiver grazing angle values were in the region $10° \leq \phi_{2,\text{deg}} \leq 90°$, and it was found that the reflectivity was determined only by the transmitter grazing angles. The fit to the vertical reflectivity was given as

$$\sigma_B^0 = \frac{2.4 \times 10^9}{0.3a_D} U_{kn}^{1.7} 10^{0.16} H_{in}^{-\left(1.2\sin\left((\phi_{1,\text{deg}}+\phi_{2,\text{deg}})/2+0.25\right)(5/U_{kn}+1.1)\right)} \tag{4.52}$$

where $a_D = \left(1/\phi_{1,\text{deg}} + 1/\phi_{2,\text{deg}}\right)$, U_{kn} is the wind speed in knots and H_{in} is the wave height in inches.

In [25], building on the work by Willis [26], it was proposed that for in-plane scattering, σ_B^0 could be approximated by

$$\sigma_B^0 = \sqrt{\sigma^0(\phi_1)\sigma^0(\phi_2)}, \quad \text{m}^2/\text{m}^2 \tag{4.53}$$

where $\sigma^0(\phi_1)$ and $\sigma^0(\phi_2)$ are the monostatic values of reflectivity at grazing angles ϕ_1 and ϕ_2, respectively. Using values of $\sigma^0(\phi_1)$ and $\sigma^0(\phi_2)$ from the GIT model, this approach was found to give satisfactory agreement with the results of both Domville's and Willis's models. It should be noted that the GIT model is only valid for $\phi_{\text{deg}} < 10°$ and for larger values of grazing angle, the models discussed in Section 4.2 could be used. Also, the modelling approach proposed in (4.53) does not accurately predict the very low values of σ_{B}^0 reported by Domville when one of the grazing angles is extremely small and the other is much larger.

4.3.2 Out-of-plane scattering

The bistatic geometry was described in Chapter 3 with out-of-plane scattering when the azimuth angle, $\psi_{\text{B}} \neq 0$ or π radians. As described in [25], there are two effects of out-of-plane scattering that must be modelled. The first is the reduction of the co-polarised scattering component with increasing azimuth angle. The second is skew polarisation which is the projection of the scattered electromagnetic fields from the frame of the transmitting antenna to the frame of the receiving antenna. As the azimuth angle increases, this effect will reduce the co-polar component and increase the cross-polar component seen in the receiving antenna. In order to capture these effects, the following model has been proposed [25]:

$$\sigma_{\text{B}}^0 = \sqrt{\sigma^0(\phi_1, 0)\sigma^0(\phi_2, \psi_{\text{B}})}\,|\cos\psi_{\text{B}}|^{m_c} + k_c\sqrt{\sigma_{\text{HV}}^0(\phi_1, 0)\sigma_{\text{HV}}^0(\phi_2, \psi_{\text{B}})}\,|\sin\psi_{\text{B}}|^{n_c} \tag{4.54}$$

where $\sigma^0(\phi, \psi_{\text{B}})$ is the monostatic co-polar mean backscatter (HH or VV) with grazing angle ϕ and azimuth direction ψ_{B}, $\sigma_{\text{HV}}^0(\phi, \psi_{\text{B}})$ is the equivalent cross-polar reflectivity and $m_c, n_c, k_c > 0$ are scaling factors adjusted to fit the data. A suitable model for the cross-polar reflectivity is given by Long [27] who modelled a dependence on wind speed and direction using a C-band data set. The grazing angles were in the range 1.5°–4.0° and the wave heights (mean surface to wave crest height) varied from 0.18 to 0.6 m, which is equivalent to sea states 2–4. The model averaged over all grazing the angles is given by

$$\sigma_{\text{HV, dB}}^0 = 29.8\log_{10}(0.5144U) + 6\cos\theta - 84.7. \tag{4.55}$$

The model in (4.55) was also verified by Griffiths *et al.* [25] who investigated the fit of (4.54) for the cross-polar monostatic reflectivity and by selecting appropriate values of m_c, n_c and k_c.

Other measurements of cross-polar σ_{HV}^0 are given in [28,29]. A common observation is that the cross-polar reflectivity is much less than the equivalent co-polar result. The results in [28] suggest that the reduction is about 5 dB when compared to the horizontal polarisation at 15° grazing angle (X-band, 7.7 m/s wind speed), and up to about 18 dB when approaching vertical incidence. According to [28], this implies that the out-of-plane σ_{B}^0 will be lower than the monostatic σ^0 for the same grazing angles.

This conclusion is supported by results in [23,24] and by Ewell and Zehner [30] who published values for the ratio of σ_B^0/σ^0.

4.4 K-distribution shape

The K-distribution is the most common amplitude distribution model for representing the sea clutter amplitude fluctuation. At low grazing angles ($<10°$), there are a few models for the K-distribution shape, ν, based on a data set summarised in [1]. This includes the original shape model presented by Ward [31] and the updated fits by Watts and Wicks [32]. At medium grazing angles, there are a number of models based on the Ingara X-band data set. The first was designed to model only the geometry for an individual day [33]. This was then extended by Rosenberg *et al.* [34] to include extra components for the environmental conditions. Finally, a continuous model over the grazing angle range 0.1°–45° was presented in [13] using a modified version of the Watts and Wicks model to cover all swell directions combined with the medium grazing angle model from [34].

4.4.1 Low grazing angle models

The most popular low grazing angle shape model was proposed by Ward [31] and spans grazing angles from 0.1° to 10°. While there were over 300 data sets used in the development of the model, the data was insufficient to get reliable trends for many parametric variations such as the wind speed or sea state. Also, although the data was collected at X-band, it is believed to work reasonably well from 5 to 35 GHz [1]. The logarithm of the shape is given by

$$\log_{10}\nu = \frac{2}{3}\log_{10}\phi_{\mathrm{deg}} + \frac{5}{8}\log_{10}A_{\mathrm{cl}} - k_{\mathrm{WW}} - \frac{\cos(2\theta_{\mathrm{sw}})}{3} \tag{4.56}$$

where θ_{sw} is the swell direction, A_{cl} is the radar resolved area, k_{WW} is a polarisation-dependent parameter (1.39 for VV and 2.09 for HH). This model was developed using data spanning radar resolved areas, $630 \leq A_{\mathrm{cl}} \leq 1\,260$ m^2, with a range resolution of 4.2 m.

Using additional data, Watts and Wicks [32] subsequently improved the fits for the up-swell (USW) and cross-swell (CSW) directions, and for horizontally and vertically polarised data. Their model is given as

$$\log_{10}\nu = a_{\mathrm{WW}}\log_{10}\phi_{\mathrm{deg}} + b_{\mathrm{WW}}\log_{10}A_{\mathrm{cl}} + c_{\mathrm{WW}} \tag{4.57}$$

with the coefficients a_{WW}, b_{WW} and c_{WW} given in Table 4.12. This model can easily be extended to include a sinusoidal variation over the swell direction with maximum values in the up and down swell directions [13]. If the sum and difference shape values are defined as

$$\begin{aligned}\nu_{\text{sum}}(\phi) &= (\nu_{\text{USW}}(\phi) + \nu_{\text{CSW}}(\phi))/2,\\ \nu_{\text{diff}}(\phi) &= (\nu_{\text{USW}}(\phi) - \nu_{\text{CSW}}(\phi))/2\end{aligned} \tag{4.58}$$

then the modified model can be written as

$$\nu = \nu_{\text{sum}}(\phi) + \nu_{\text{diff}}(\phi)\cos(2\theta_{\text{sw}}). \tag{4.59}$$

Figure 4.8 shows the model fit at grazing angles 1°, 5° and 10° with crosses showing the original data points. The main trends observed with this model are peaks in the cross swell directions and increasing shape as the grazing increases. The vertically

Table 4.12 Watts and Wicks shape parameter coefficients

Pol.	**Swell**	a_{WW}	b_{WW}	c_{WW}
HH	Up/down	0.40	0.46	−1.52
HH	Cross	0.25	0.79	−2.22
VV	Up/down	0.83	0.71	−1.69
VV	Cross	1.04	1.23	−2.81

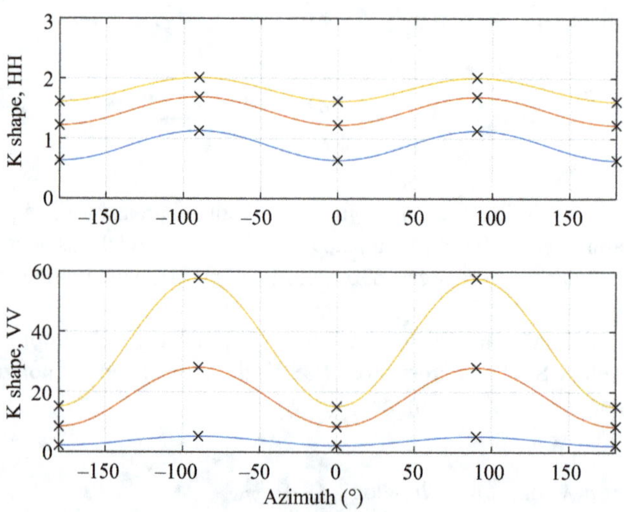

Figure 4.8 Modified Watts and Wicks shape model with area $A_{\text{cl}} = 756\ m^2$ at grazing angles: (—) 1°, (—) 5°, (—) 10°, (×) original data points [13].

polarised shape is also much higher than horizontal with Gaussian-like statistics when $\nu \gtrsim 30$.

4.4.2 DSTG continuous shape model

The DSTG continuous shape model does not directly use the modified Watts and Wicks model in the low grazing angle region, 0.1°–10°. Rather, a power law fit is introduced to match the trend at low grazing regions. To ensure continuity, a transition point has been chosen at $\phi_{\text{deg}} = 12°$ with a medium grazing angle model, ν_{med} used from 12° to 45° grazing. This model has the same form as σ^0_{med} in the DSTG continuous mean backscatter model in Section 4.2.9, except that the variation in azimuth is related to the swell direction rather the wind direction. It is given by

$$\nu_{\text{med}}(\theta_{\text{sw}}, \phi) = \left(\frac{\phi}{\phi_0}\right)^{\gamma_{\text{cont}}} (a_{\text{cont}} + b_{\text{cont}} \cos\theta_{\text{sw}} + c_{\text{cont}} \cos(2\theta_{\text{sw}})) \tag{4.60}$$

where θ_{sw} is the azimuth swell direction relative to up-swell, the coefficients a_{cont}, b_{cont}, c_{cont} and γ_{cont} are the geometric model coefficients and ϕ_0 is the reference grazing angle. Representing the ocean conditions is achieved using the wind speed, U and the significant wave height $H_{1/3}$,

$$Y_{\text{cont}} = d_{\text{cont}} + e_{\text{cont}} \log_{10} U + f_{\text{cont}} H_{1/3} \tag{4.61}$$

where d_{cont}, e_{cont} and f_{cont} are the environmental model coefficients. The model is then implemented by equating each coefficient a_{cont}, b_{cont}, c_{cont} and γ_{cont} to the model Y_{cont} in (4.61).

To determine the variation of the K-distribution shape with radar-resolved area, A_{cl}, the Ingara data range resolution was coarsened to achieve radar-resolved areas of 50–1,200 m^2. The estimated shape revealed no strong trends along grazing, azimuth or sea state. Consequently, each of the shape values was averaged and a power law was fitted to the result. The data and model fit are shown in Figure 4.9 for both horizontal and vertical polarisations where the area is normalised by the smallest resolved area of the Ingara data, $A_0 = 47.3$ m^2. The model can then be described as

$$\nu(\theta_{\text{sw}}, \phi) = \begin{cases} \nu_{\text{med}}(\theta_{\text{sw}}, 12)\left(\dfrac{A_{\text{cl}}}{A_0}\right)^{\delta_{\text{cont}}}\left(\dfrac{\phi_{\text{deg}}}{12}\right)^{\eta_{\text{cont}}}, & 0.1° \le \phi_{\text{deg}} < 12°, \\ \nu_{\text{med}}(\theta_{\text{sw}}, \phi)\left(\dfrac{A_{\text{cl}}}{A_0}\right)^{\delta_{\text{cont}}}, & 12° \le \phi_{\text{deg}} < 45° \end{cases} \tag{4.62}$$

with the parameter values given in Table 4.13. As an example, Figure 4.10 has a radar resolved areas of 756 m^2 and the modified Ward and Wicks model has been used as a data substitute at low grazing angles. Both HH and VV polarisations are shown and the data is from 30° grazing, with a wind speed of 10.2 m/s and wave height of 1.2 m. The model clearly shows peaks in the cross swell direction and shape values

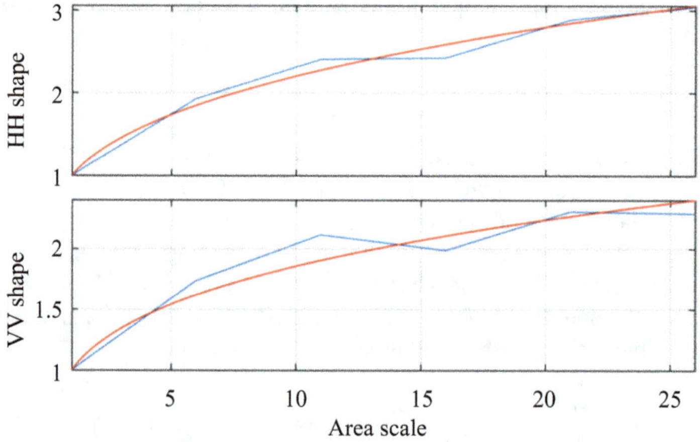

Figure 4.9 *DSTG continuous model showing K-distribution shape variation with area scale (A_{cl}/A_0): (—) averaged Ingara data, (—) model fit. The wind speed and wave height are 10.2 m/s and 1.21 m, respectively, [13].*

Table 4.13 *DSTG continuous shape model coefficients*

Pol.		d_{cont}	e_{cont}	f_{cont}
HH	γ_{cont}	1.39	–0.78	0.16
	a_{cont}	3.11	–1.85	0.35
	b_{cont}	–0.14	0.19	–0.030
	c_{cont}	–0.51	0.23	–0.049
	η_{cont}	0.30		
	δ_{cont}	0.34		
VV	γ_{cont}	0.61	–0.35	–0.035
	a_{cont}	25.19	–6.96	–0.53
	b_{cont}	–0.014	0.012	0
	c_{cont}	–7.75	2.18	0.16
	η_{cont}	0.55		
	δ_{cont}	0.27		

that increase with grazing angle. Figures 4.11 and 4.12 then show slices through the model for radar-resolved areas of 47.3 and 756 m^2, respectively. These correspond to the smallest radar-resolved area for the Ingara data and a value where the modified Ward and Wicks model is valid. The transition between the two regions looks very smooth and the model matches the trends in the data quite well. Also shown in these figures are model fits for the generic parameter model described in Section 4.5.3 and the modified Watts and Wicks model.

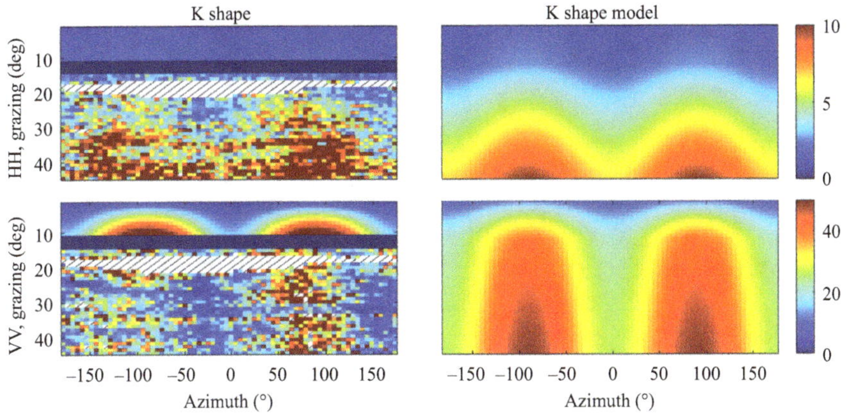

Figure 4.10 *DSTG continuous model for the K-distribution shape showing a real data comparison with area, $A_{\mathrm{cl}} = 756\ m^2$ and HH and VV polarisations. The wind speed and wave height are 10.2 m/s and 1.21 m, respectively [13]. © Commonwealth of Australia 2017*

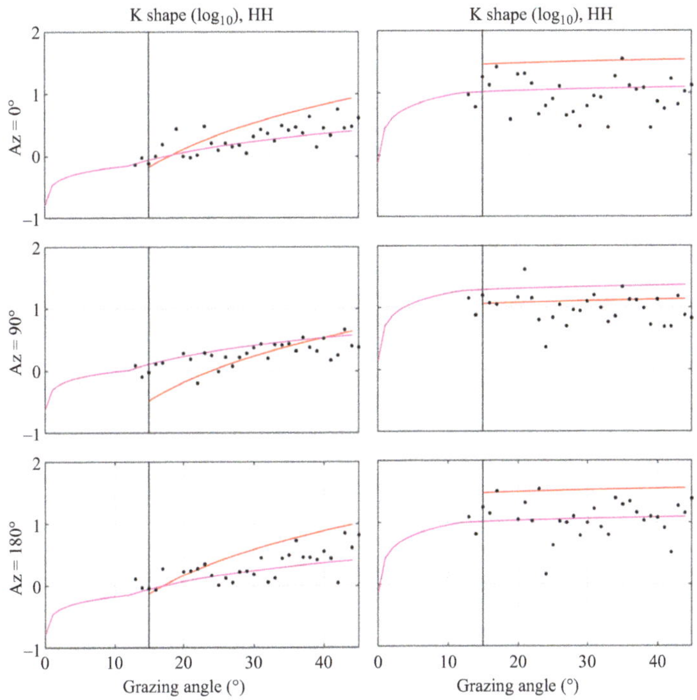

Figure 4.11 *DSTG continuous model showing K-distribution shape with area, $A_{\mathrm{cl}} = 47.5\ m^2$, azimuth look directions of 0°, 90° and 180° and HH and VV polarisations. The wind speed and wave height are 10.2 m/s and 1.21 m, respectively. (—) Medium grazing angle shape model, (—) continuous model, (·) Ingara data; vertical lines indicate model limits [13]. © Commonwealth of Australia 2017*

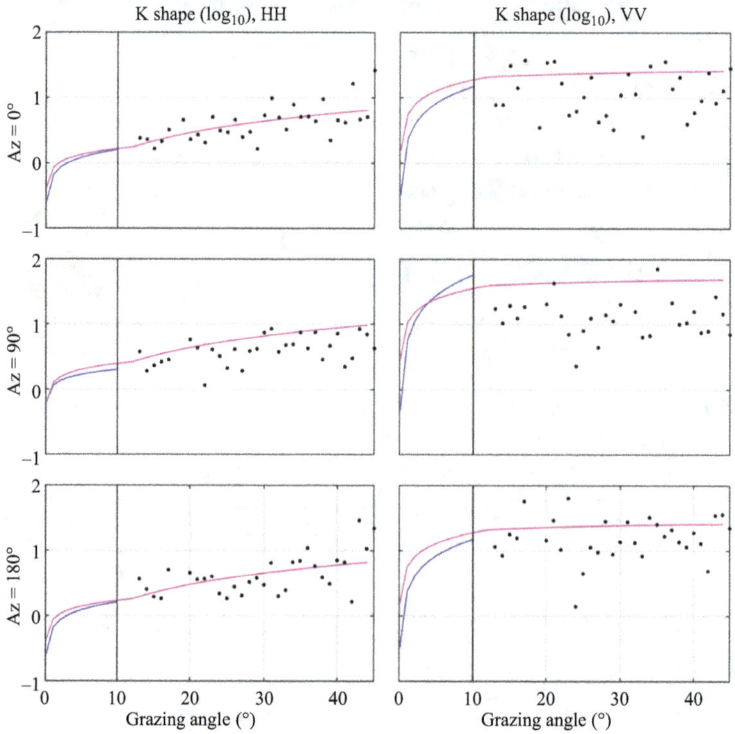

Figure 4.12 *DSTG continuous model showing K-distribution shape with area $A_{cl} = 756\ m^2$ for azimuth look directions of 0°, 90° and 180° and for HH and VV polarisation. The wind speed and wave height are 10.2 m/s and 1.21 m, respectively. (—) Modified Watts and Wicks model, (—) continuous model, (·) Ingara data; vertical lines indicate model limits [13]. © Commonwealth of Australia 2017*

4.5 Doppler spectrum

While there has been a lot of study into the Doppler characteristics of sea clutter, there are very few parametric models for the Doppler spectrum. As discussed in Chapter 2, this is due to the difficulty in modelling the time- and range-varying characteristics.

4.5.1 Mean Doppler spectrum models

To model the mean Doppler spectrum, a single Gaussian model is often used [29]. This model is based on low grazing angle data in the up-wind direction and relates the mean velocity of the Doppler spectrum to the wind speed:

$$\begin{aligned} \bar{v}_{HH} &= 0.25 + 0.25U, \\ \bar{v}_{VV} &= 0.25 + 0.18U \end{aligned} \tag{4.63}$$

where $\bar{v}_{\mathrm{VV}}$ and $\bar{v}_{\mathrm{HH}}$ denote the mean velocities of the vertical and horizontal polarisations, respectively. The half-power (HP) width is quite variable, with an approximate relationship,

$$\bar{w}_{\mathrm{HP}} = 0.24U. \tag{4.64}$$

In the up-wind direction as the grazing angle increases, the mean Doppler velocity reduces approximately in proportion to $\cos\phi$ and returns from the horizontal and vertical polarisations become similar. These two additions are captured in the model proposed by Stove and published in [1]:

$$\bar{v}_{\mathrm{HH}} = \begin{cases} (0.25 + 0.18U + 0.07U\cos(2\phi))\cos\phi, & \phi_{\mathrm{deg}} \leq 45^\circ \\ (0.25 + 0.18U + 0.07U)\cos\phi & \phi_{\mathrm{deg}} > 45^\circ \end{cases} \tag{4.65}$$

$$\bar{v}_{\mathrm{VV}} = (0.25 + 0.18U + 0.07U)\cos\phi. \tag{4.66}$$

For both of these mean Doppler spectra models, it can be assumed that the mean Doppler shift of the clutter will vary with the azimuth look direction.

A model for the RMS clutter widths, $\bar{w}_{\mathrm{RMS}}$ in X- and Ku-bands, is given by Sittrop [4] for VV and HH polarisations, wind speeds in the range 2–20 m/s and up- and cross-wind directions:

$$\log_{10}(\bar{w}_{\mathrm{RMS}}) = \log_{10}(a_{\mathrm{Sitv}}) + \frac{b_{\mathrm{Sitv}}\log_{10}\left(\frac{\phi}{\phi_0}\right) + \left(c_{\mathrm{Sitv}}\log_{10}\left(\frac{\phi}{\phi_0}\right) + d_{\mathrm{Sitv}}\right)\log_{10}\left(\frac{U}{U_0}\right)}{10} \tag{4.67}$$

where the reference grazing angle and wind velocity are given by $\phi_{0,\mathrm{deg}} = 0.5^\circ$ and $U_0 = 5$ m/s, respectively. The X- and Ku-band model parameters, a_{Sitv}, b_{Sitv}, c_{Sitv} and d_{Sitv}, are given in Table 4.14.

Table 4.14 Sittrop's RMS spectrum width X- and Ku-band model coefficients

Pol.	Freq. band	Wind direction	a_{Sitv}	b_{Sitv}	c_{Sitv}	d_{Sitv}
HH	X	Up-wind	60	0	−1.7	7
HH	X	Cross-wind	32	2.6	−9.6	15
VV	X	Up-wind	63	0	−7.4	8
VV	X	Cross-wind	50	−2.2	−1.6	9.2
HH	Ku	Up-wind	130	−3	3.6	3.4
HH	Ku	Cross-wind	160	−3	3.3	2
VV	Ku	Up-wind	125	−4.1	6	4
VV	Ku	Cross-wind	150	−5.4	7	3.4

A parametric model for the mean Doppler spectrum at low grazing angles has been presented by Whitrow [22] and is based on Walker's mean Doppler model [35]. The model uses data collected at Ku-band (14 GHz) with wind speeds from 2 to 12 m/s implying sea states from 2 to 6 and grazing angles from 6° to 12°. Interestingly, the model results were validated using X-band data with a 2° grazing angle, implying that the model is valid beyond the experimental conditions.

Walker's model comprises two Gaussian components for the vertical polarisation and three for the horizontal. Common to both are the Bragg and persistent whitecap components, while only the horizontal polarisation has a discrete sea-spike component. The mean Doppler spectra for vertical and horizontal polarisations are given by $G_{VV}(f)$ and $G_{HH}(f)$, respectively,

$$G_{VV}(f) = A_{Br,VV} \exp\left(-\frac{(f-\bar{f}_{Br})^2}{\bar{w}_{Br}}\right) + A_W\left(-\frac{(f-\bar{f}_W)^2}{\bar{w}_W}\right)$$

$$G_{HH}(f) = A_{Br,HH} \exp\left(-\frac{(f-\bar{f}_{Br})^2}{\bar{w}_{Br}}\right) + A_W\left(-\frac{(f-\bar{f}_W)^2}{\bar{w}_W}\right) + A_S\left(-\frac{(f-\bar{f}_W)^2}{\bar{w}_S}\right) \tag{4.68}$$

where $A_{Br,HH}$ and $A_{Br,VV}$ are the Bragg magnitudes for the horizontal and vertical polarisations, respectively, and A_W and A_S are the whitecap and sea-spike magnitudes. The Bragg and whitecap centre points, $\bar{f}_{Br}$ and $\bar{f}_W$, are common to both polarisations, and the whitecap and sea-spike components have the same centre frequency. The widths for each component are $\bar{w}_{Br}$, $\bar{w}_W$ and $\bar{w}_S$, respectively.

To determine models for each parameter, Whitrow deduced relationships based on wave tank measurements by Rozenberg [36,37] with the centre point and widths scaled to match the measured results by Walker [35]. The parametric model requires the radar wavelength, λ, azimuth wind direction, θ, wind speed, U, and a mean backscatter model, σ^0. For the Bragg centre frequency,

$$\bar{f}_{Br} = \begin{cases} \dfrac{0.0273\cos\theta}{\lambda}\left(17.36 + 10.59U^{0.29} + 0.0153U^{3.05}\right), & 0 \le \theta \le \pi/2 \\ \dfrac{0.0273\cos\theta}{\lambda}(22.83 + 2.84U), & \pi/2 < \theta \le \pi \end{cases} \tag{4.69}$$

while for the whitecap centre frequency,

$$\bar{f}_W = \begin{cases} \dfrac{0.0567\cos\theta}{\lambda}\left(-39.43 + 57.48\sqrt{U} - 5.69U\right), & 0 \le \theta \le \pi/2 \\ \dfrac{0.0483\cos\theta}{\lambda}(22.83 + 2.84U), & \pi/2 < \theta \le \pi. \end{cases} \tag{4.70}$$

For the Bragg width,

$$\bar{w}_{\mathrm{Br}} = \frac{0.0273}{\lambda}\left(4.6 + 0.68U\cos\theta\right), \tag{4.71}$$

and for the whitecap and sea spike widths,

$$\bar{w}_{\mathrm{W}} = \begin{cases} \dfrac{0.0567U}{\lambda}\left(5.035 + 1.115\cos\theta\right), & U < 5.97 \\ \dfrac{0.0567}{\lambda}\left(18.35 + 1.96U + (18.35 - 1.96U)\cos\theta\right), & U > 5.97 \end{cases} \tag{4.72}$$

$$\bar{w}_{\mathrm{S}} = 1.13\bar{w}_{\mathrm{W}}. \tag{4.73}$$

For the magnitudes, there are two different relationships that produce similar results. In the first approach, ratios of the sea-spike and whitecap magnitudes to the horizontal Bragg component are given by

$$\frac{A_{\mathrm{W}}}{A_{\mathrm{Br,HH}}} = 1.41U\left(\frac{1+\cos\theta}{2}\right) + 0.15U\left(\frac{1-\cos\theta}{2}\right), \tag{4.74}$$

$$\frac{A_{\mathrm{S}}}{A_{\mathrm{Br,HH}}} = \mathrm{Max}\left[0, 19.1\left(\frac{U-4}{6}\right)\right]\left(\frac{1+\cos\theta}{2}\right). \tag{4.75}$$

An appropriate mean backscatter model for the horizontal polarisation, σ^0_{HH}, can then be used to determine the Bragg magnitude,

$$A_{\mathrm{Br,HH}} = \frac{\sigma^0_{\mathrm{HH}}}{\sqrt{\pi}\left(\bar{w}_{\mathrm{Br}} + \bar{w}_{\mathrm{W}}\dfrac{A_{\mathrm{W}}}{A_{\mathrm{Br,HH}}} + \bar{w}_{\mathrm{S}}\dfrac{A_{\mathrm{S}}}{A_{\mathrm{Br,HH}}}\right)}. \tag{4.76}$$

with the magnitude values for A_{W} and A_{S} then determined from (4.74) and (4.75). Similarly, the vertically polarised Bragg magnitude can be determined using the corresponding mean backscatter, σ^0_{VV},

$$A_{\mathrm{Br,VV}} = \frac{\sigma^0_{\mathrm{VV}} - \sqrt{\pi}\,\bar{w}_{\mathrm{W}}A_{\mathrm{W}}}{\sqrt{\pi}\,\bar{w}_{\mathrm{Br}}}. \tag{4.77}$$

The second method uses the magnitude ratio of the whitecap and vertical Bragg components,

$$\frac{A_{\mathrm{W}}}{A_{\mathrm{Br,VV}}} = 0.0437U\left(\frac{1+\cos\theta}{2}\right) + 0.00595U\left(\frac{1-\cos\theta}{2}\right) \tag{4.78}$$

with

$$A_{\mathrm{Br,VV}} = \frac{\sigma^0_{\mathrm{VV}}}{\sqrt{\pi}\left(\bar{w}_{\mathrm{Br}} + \bar{w}_{\mathrm{W}}\dfrac{A_{\mathrm{W}}}{A_{\mathrm{Br,VV}}}\right)}. \tag{4.79}$$

The whitecap and horizontal Bragg magnitudes, A_{W} and $A_{\mathrm{Br,HH}}$, can then be determined from (4.78) and (4.74), respectively. The sea-spike magnitude follows from

$$A_{\mathrm{S}} = \frac{\sigma^0_{\mathrm{HH}} - \sqrt{\pi}\bar{w}_{\mathrm{Br}}A_{\mathrm{Br,HH}} - \sqrt{\pi}\bar{w}_{\mathrm{W}}A_{\mathrm{W}}}{\sqrt{\pi}\bar{w}_{\mathrm{S}}}. \tag{4.80}$$

4.5.2 Temporal decorrelation model

As the sea state becomes greater, the wind speed, wave height, wave length and the Doppler bandwidth all increase. Regarding the latter, Hicks *et al.* [38] have shown experimentally that the Doppler spread at high grazing angles is proportional to the wave height divided by the wave period. This relationship has been verified with the Ingara X-band data set [39]. Further analysis of the data has allowed a model to be constructed for the mean temporal correlation, T_{temp}, as measured at the $1/e$ point. The analysis found only minor trends in azimuth and grazing angle and hence the model is only a function of wind speed, U, and significant wave height, $H_{1/3}$. It is described by three sets of coefficients, one for each polarisation,

$$T_{\mathrm{temp}} = (a_{\mathrm{temp}} + b_{\mathrm{temp}}U + c_{\mathrm{temp}}H_{1/3}) \times 10^{-3} \tag{4.81}$$

which can be related to an autocorrelation model represented as either a Gaussian or negative exponential, with the latter is given by

$$\rho_n = \exp\left(-\frac{|n|}{T_{\mathrm{temp}}f_{\mathrm{r}}}\right) \tag{4.82}$$

where $n > 0$ is the lag. The model coefficients for the three polarisations are described in Table 4.15 with validity over grazing angles 15°–45°. Note that this model only considers the real part of the correlation and hence assumes that the Doppler spectrum is centred at 0 Hz. An example of the model fit for the horizontal polarisation is shown in Figure 4.13 with the measured mean correlation values for each day marked with red lines below the model and blue for above.

4.5.3 Evolving Doppler spectrum model

The model for evolving Doppler spectra was presented in Chapter 2 and takes the form

$$G(f) = \frac{\tau}{\sqrt{2\pi}w}\exp\left(-\frac{(f - m_{\mathrm{f}}(\tau))^2}{2w^2}\right) \tag{4.83}$$

Table 4.15 Temporal correlation model coefficients

Pol.	a_{temp}	b_{temp}	c_{temp}
HH	6.34	–0.20	–0.23
HV	5.58	–0.10	–0.21
VV	5.83	–0.10	–0.21

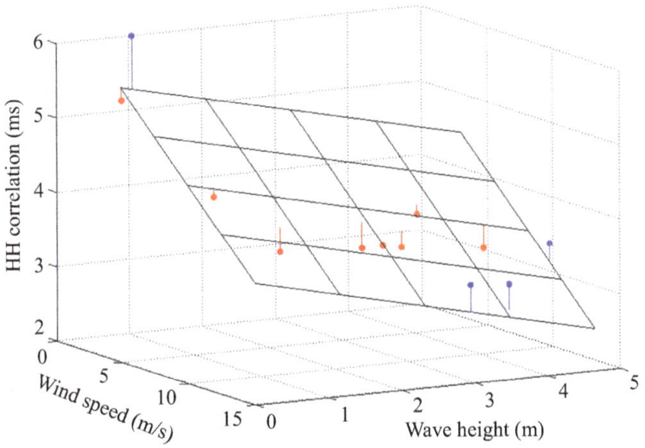

Figure 4.13 Horizontal mean temporal correlation (ms) as a function of wind speed and wave height [40]. © Commonwealth of Australia 2012

where $G(f)$ is the mean power spectral density for Doppler frequency, f, and τ is the local clutter intensity or texture, having a gamma-distributed PDF if the clutter has K-distribution. The mean Doppler frequency is related to the normalised mean speckle power, $\bar{\tau} = \tau / \langle \tau \rangle$, with

$$m_f(\tau) = A_c + B_c\bar{\tau} + f_{lin} \tag{4.84}$$

where the term f_{lin} was added to the model in [34] and captures the fluctuation around the model fit. It is described by a Gaussian random variable with zero mean and standard deviation, σ_{lin}. The final term in (4.83) is the spectrum width, w, which is a random variable with mean, m_w, and standard deviation, σ_w.

A parametric model for the evolving Doppler spectrum was presented in [41] and later extended for the bimodal case in [34]. This model is for medium grazing angles and is based on relationships in [10] using the Ingara X-band data set with an integration time of 0.22 s ($N = 64$ pulses) and the finest radar-resolved area,

$A_{\mathrm{cl}} = 47.25\ \mathrm{m}^2$. It is a generic model that includes both the wind and swell directions and is an extension of the continuous mean backscatter and K-distribution shape parameter models presented earlier.

If X is a generic parameter to be modelled, then the geometric relationship can be written as the weighted sum of two components which represent the wind and swell directions:

$$X(\theta, \phi | U, H_{1/3}) = \left(\frac{\phi}{\phi_0}\right)^{\gamma_{\mathrm{gen}}} \big(a_{\mathrm{gen}} + b_{\mathrm{gen}} \cos\theta + c_{\mathrm{gen}} \cos(2\theta) + d_{\mathrm{gen}} \cos(\theta - \theta_{\mathrm{sw}}) + e_{\mathrm{gen}} \cos(2(\theta - \theta_{\mathrm{sw}}))\big) \quad (4.85)$$

where θ and θ_{sw} are the azimuth and swell angles, both defined relative to up-wind, ϕ_0 is the reference grazing angle fixed at 10°, and the model coefficients are $a_{\mathrm{gen}}, b_{\mathrm{gen}}, c_{\mathrm{gen}}, d_{\mathrm{gen}}, e_{\mathrm{gen}}$ and γ_{gen}. The ocean conditions are related by the wind speed, U, and the significant wave height $H_{1/3}$,

$$Y = f_{\mathrm{gen}} + g_{\mathrm{gen}} \log_{10}(U) + h_{\mathrm{gen}} H_{1/3} \quad (4.86)$$

where $f_{\mathrm{gen}}, g_{\mathrm{gen}}$ and h_{gen} are the model coefficients. Note that each coefficient in (4.85) depends on the ocean conditions and is modelled as (4.86) with individual parameters, $f_{\mathrm{gen}}, g_{\mathrm{gen}}$ and h_{gen}.

Consider an example day from the Ingara data set with 30° grazing, up-wind direction, relative swell angle of 135°, wind speed of 10.2 m/s and a wave height of 1.2 m. Figures 4.14 and 4.15 show the estimated and modelled results for A_{c} and B_{c} with the model coefficients given in Table 4.16. For the horizontal and cross polarisations, all 12 flights result in a peak at up-wind with small variation across grazing angle, confirming that the scattering from these channels is primarily wind driven. However, for the vertical channel of A_{c}, there is a peak smeared around the up-swell direction. This complex interaction has been captured by the model in (4.85) which was run independently for each polarisation. The modelled results use the same input sea state conditions and clearly capture the significant trends, although some of the peaks in the vertical polarisation are not quite as strong as the data. This is due to the low wave height used for the modelling which has been balanced over the 12 flights. Results for the parameter σ_{lin} are shown in Figure 4.16 with the model coefficients given in Table 4.17. These show similar trends to the A_{c} vertical polarisation with peaks located in the up-swell location.

Figures 4.17 and 4.18 then show the mean, m_{w}, and standard deviation, σ_{w}, of the width, w, with the coefficients given in Table 4.18. The first observation is that the trends match for both parameters and for all three polarisations. Peaks are present in the vicinity of the up-wind and down-swell directions with strengths related to the wind speed and swell strength (wave height). The modelled results reveal a good match

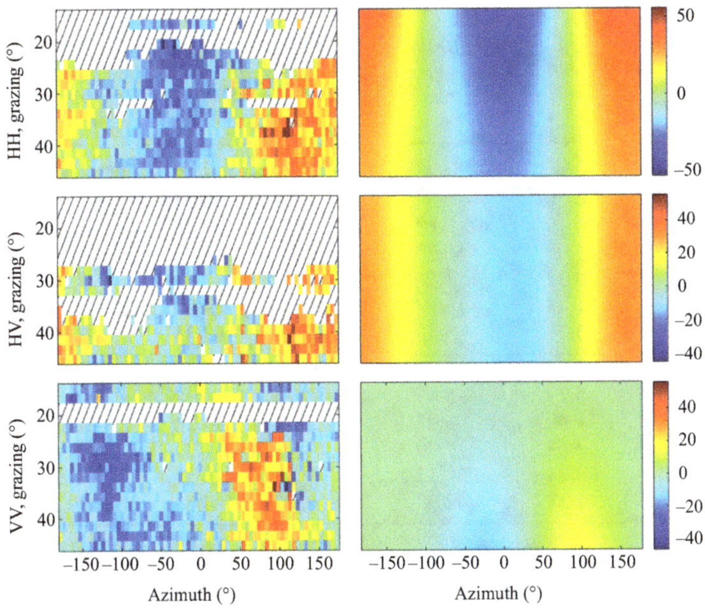

Figure 4.14 *Variation of linear parameter, A_c, with azimuth and grazing angle. Data on left, model on right. Top plot: HH, middle plot: HV, bottom plot: VV. The wind speed and wave height are 10.2 m/s and 1.21 m, respectively. Adapted from [42]. © IEEE 2014*

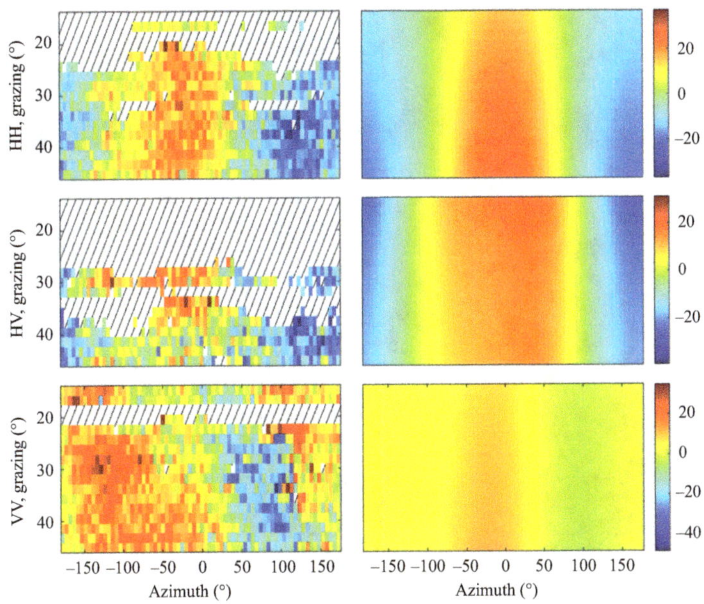

Figure 4.15 *Variation of linear parameter, B_c, with azimuth and grazing angle. Data on left, model on right. Top plot: HH, middle plot: HV, bottom plot: VV. The wind speed and wave height are 10.2 m/s and 1.21 m, respectively. Adapted from [42]. © IEEE 2014*

Table 4.16 Evolving Doppler model linear coefficients for A_c and B_c

Pol.		A_c			B_c		
		f_{gen}	g_{gen}	h_{gen}	f_{gen}	g_{gen}	h_{gen}
HH	γ_{gen}	6.60	−6.87	−0.012	−0.055	0.26	0.084
	a_{gen}	−0.19	0.64	−0.054	12.10	−10.81	0.17
	b_{gen}	15.98	−58.21	5.05	−2.83	30.86	−3.84
	c_{gen}	−0.32	1.22	−0.62	3.99	−4.30	0.47
	d_{gen}	1.43	−16.24	6.01	−19.96	28.86	−3.60
	e_{gen}	5.77	−5.17	0.71	7.31	−6.90	−0.44
HV	γ_{gen}	−1.10	0.96	0.062	1.41	−1.74	0.035
	a_{gen}	6.29	−5.79	2.52	0.73	−0.38	−0.054
	b_{gen}	−51.49	5.97	8.75	30.24	4.04	−6.33
	c_{gen}	2.64	−7.79	2.75	−44.32	49.16	−3.79
	d_{gen}	−60.24	50.13	2.93	6.37	−1.32	−1.76
	e_{gen}	41.78	−42.92	1.96	−4.63	6.24	−1.52
VV	γ_{gen}	−3.05	6.59	−0.69	−1.88	3.06	−0.36
	a_{gen}	0.14	−0.044	0.016	7.58	−6.77	0.73
	b_{gen}	1.42	−21.46	8.14	−2.15	20.41	−7.15
	c_{gen}	−6.27	6.70	−1.23	1.88	−1.34	0.86
	d_{gen}	2.48	−4.63	3.25	−2.88	5.70	−2.92
	e_{gen}	−2.57	4.34	−1.024	0.69	−2.41	0.83

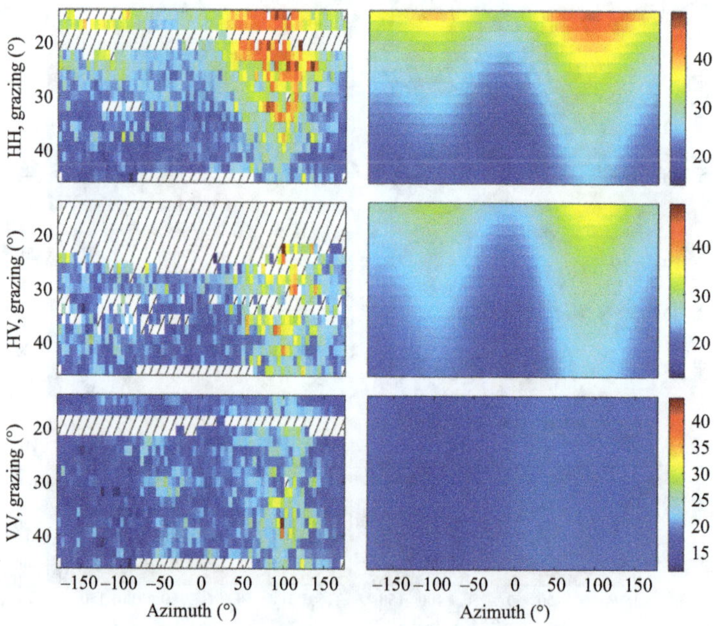

Figure 4.16 Variation of linear parameter, σ_{lin}, with azimuth and grazing angle. Data on left, model on right. Top plot: HH, middle plot: HV, bottom plot: VV. The wind speed and wave height are 10.2 m/s and 1.21 m, respectively.

Table 4.17 Evolving Doppler model linear coefficients for σ_{lin}

Pol.		σ_{lin}		
		f_{gen}	g_{gen}	h_{gen}
HH	γ_{gen}	−1.07	0.27	0.12
	a_{gen}	38.31	−16.22	1.42
	b_{gen}	−10.37	9.67	0.63
	c_{gen}	9.76	−12.71	0.27
	d_{gen}	−12.38	13.83	0.92
	e_{gen}	5.48	−4.54	−0.33
HV	γ_{gen}	−0.0016	−0.39	0.055
	a_{gen}	25.65	−5.35	1.31
	b_{gen}	4.83	−5.94	0.60
	c_{gen}	0.72	−3.57	0.24
	d_{gen}	−6.00	4.90	1.46
	e_{gen}	15.33	−14.05	−0.32
VV	γ_{gen}	−0.22	0.23	0.0043
	a_{gen}	10.14	1.28	2.50
	b_{gen}	1.56	−1.34	0.58
	c_{gen}	−0.94	1.63	0.22
	d_{gen}	−1.77	2.98	0.51
	e_{gen}	1.79	−3.73	0.76

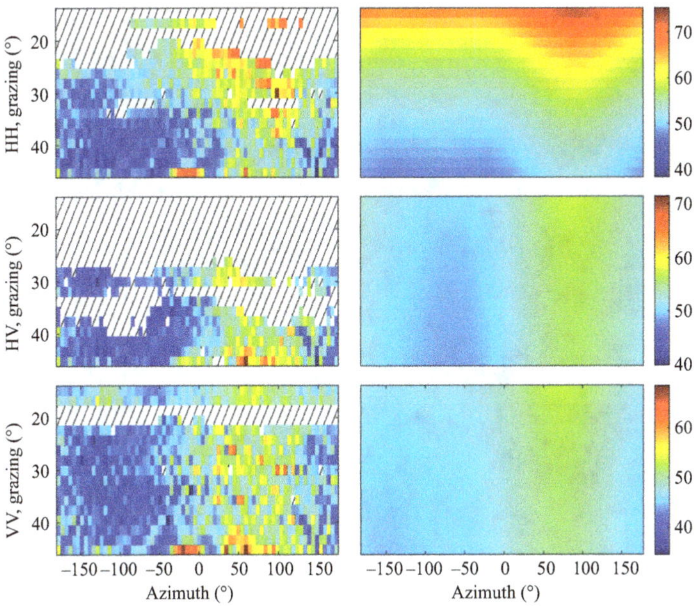

Figure 4.17 Variation of linear parameter, m_w, with azimuth and grazing angle. Data on left, model on right. Top plot: HH, middle plot: HV, bottom plot: VV. The wind speed and wave height are 10.2 m/s and 1.21 m, respectively. Adapted from [42]. © IEEE 2014

Table 4.18 Evolving Doppler model coefficients for m_w and σ_w

Pol.		m_w			σ_w		
		f_{gen}	g_{gen}	h_{gen}	f_{gen}	g_{gen}	h_{gen}
HH	γ_{gen}	0.49	−0.79	0.013	0.95	−1.27	0.056
	a_{gen}	22.06	31.27	0.59	14.24	−2.23	0.84
	b_{gen}	11.57	−12.50	1.30	−13.31	13.43	0.60
	c_{gen}	−5.29	3.80	0.20	4.71	−4.60	0.026
	d_{gen}	3.74	−3.73	1.56	−2.21	3.24	0.54
	e_{gen}	−2.96	2.51	0.23	0.53	−1.15	0.20
HV	γ_{gen}	1.14	−1.15	0.0030	2.28	−2.50	0.085
	a_{gen}	37.44	13.76	0.28	4.61	4.91	0.88
	b_{gen}	11.89	−11.65	1.01	7.95	−8.87	0.54
	c_{gen}	−5.30	5.17	−0.17	−2.23	2.25	0.022
	d_{gen}	−1.35	2.65	1.47	−4.33	4.12	0.65
	e_{gen}	4.12	−4.73	0.21	3.78	−4.34	0.26
VV	γ_{gen}	−0.20	0.16	0.0022	0.018	0.10	0.0016
	a_{gen}	46.05	−0.16	0.68	4.91	3.52	1.28
	b_{gen}	−4.19	6.26	0.61	0.39	0.071	0.35
	c_{gen}	−1.24	0.62	−0.081	−1.03	1.92	0.13
	d_{gen}	−4.04	6.70	0.59	−1.36	2.33	0.31
	e_{gen}	0.32	−0.26	−0.14	0.21	−1.094	0.34

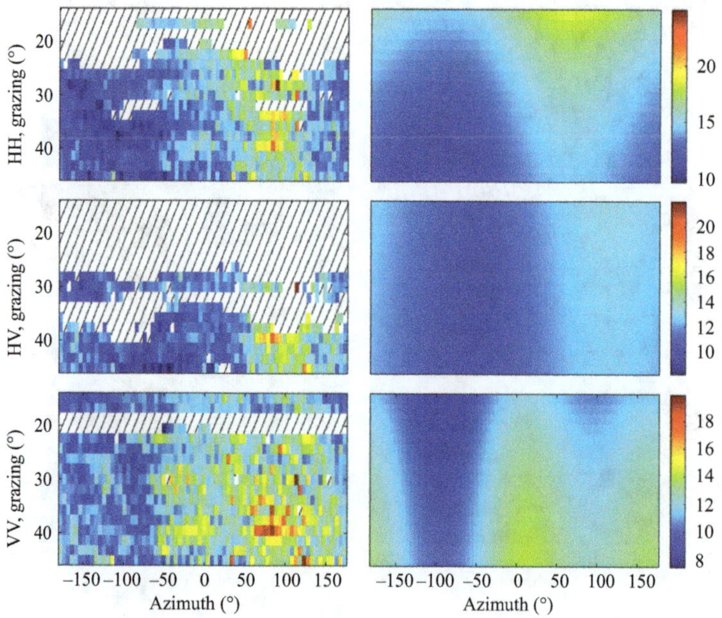

Figure 4.18 Variation of linear parameter, σ_w, with azimuth and grazing angle. Data on left, model on right. Top plot: HH, middle plot: HV, bottom plot: VV. The wind speed and wave height are 10.2 m/s and 1.21 m, respectively. Adapted from [42]. © IEEE 2014

with the data. The strong values which are in the horizontal low grazing region are present in a number of other data sets and are a real characteristic of the data. Further results for the bimodal evolving Doppler spectrum model are given in Appendix B.

A follow-on paper [43] proposed an 'extended' Doppler parametric model that accounts for changes in the integration time, $T_{\rm I}$, and radar-resolved area, $A_{\rm cl}$. It was carefully designed to work in conjunction with the original model in (4.85) as an additive term. The model was characterised using the Ingara X-band data set, with the range of integration times covering 0.053–0.43 s and the radar-resolved area covering 47.25–236.25 m^2. However, as demonstrated in [43], the extended model can produce realistic values that go beyond the range of the Ingara data, making it more useful in a broader range of scenarios.

The model is applied independently to each polarisation and is a two-dimensional power law model with independent terms for the integration time and radar-resolved area,

$$W_{\rm ext}(A_{\rm cl}, T_{\rm I}) = a_{\rm ext} + b_{\rm ext} T_{\rm I}^{c_{\rm ext}} + d_{\rm ext} A_{\rm cl}^{e_{\rm ext}} \tag{4.87}$$

with model coefficients $a_{\rm ext}, b_{\rm ext}, c_{\rm ext}, d_{\rm ext}$ and $e_{\rm ext}$. The extended Doppler model is then

$$X_{\rm ext}(\theta, \phi, A_{\rm cl}, T_{\rm I} | U, H_{1/3}) = X(\theta, \phi | U, H_{1/3}) + W_{\rm ext}(A_{\rm cl}, T_{\rm I}). \tag{4.88}$$

Figure 4.19 shows a sub-set of the model fits with the red and blue stems showing the averaged results for each value of integration time and radar-resolved area that are below and above the model fit, respectively. There are clear trends observed for the majority of parameters and polarisations, but some are clearly stronger than others, particularly in the HV polarisation. Using a variation metric to determine the significance of the model fit, the final model values have zeros for the appropriate coefficient if the variation is not significant. Also, as the original model was determined with an integration time of 0.22 s and a radar-resolved area of 47.25 m^2, the model coefficients (i.e. $a_{\rm ext}$) have been adjusted so that $W_{\rm ext}(\cdot)$ has a value of 0 given this input. The linear model values are given in Table 4.19, with the bimodal coefficients given in Appendix B.

4.6 Spatial correlation

Spatial correlation is a key component in representing the sea clutter texture. However, it is difficult to model due to the non-stationarity of sea clutter. An example highlighting the variability from the Ingara X-band data set is shown in Figure 4.20 where the correlation is averaged over 2–2.2 s and covers a 2° grazing span or approximately 200 m in range. The result show an initial decay and a number of modes present due to the interaction of the waves. Typically, the correlation extent or decorrelation length, $R_{\rm L}$, is used to characterise the initial decay and is measured at the point where the normalised correlation reaches $1/e$. Using this value, either a negative exponential or

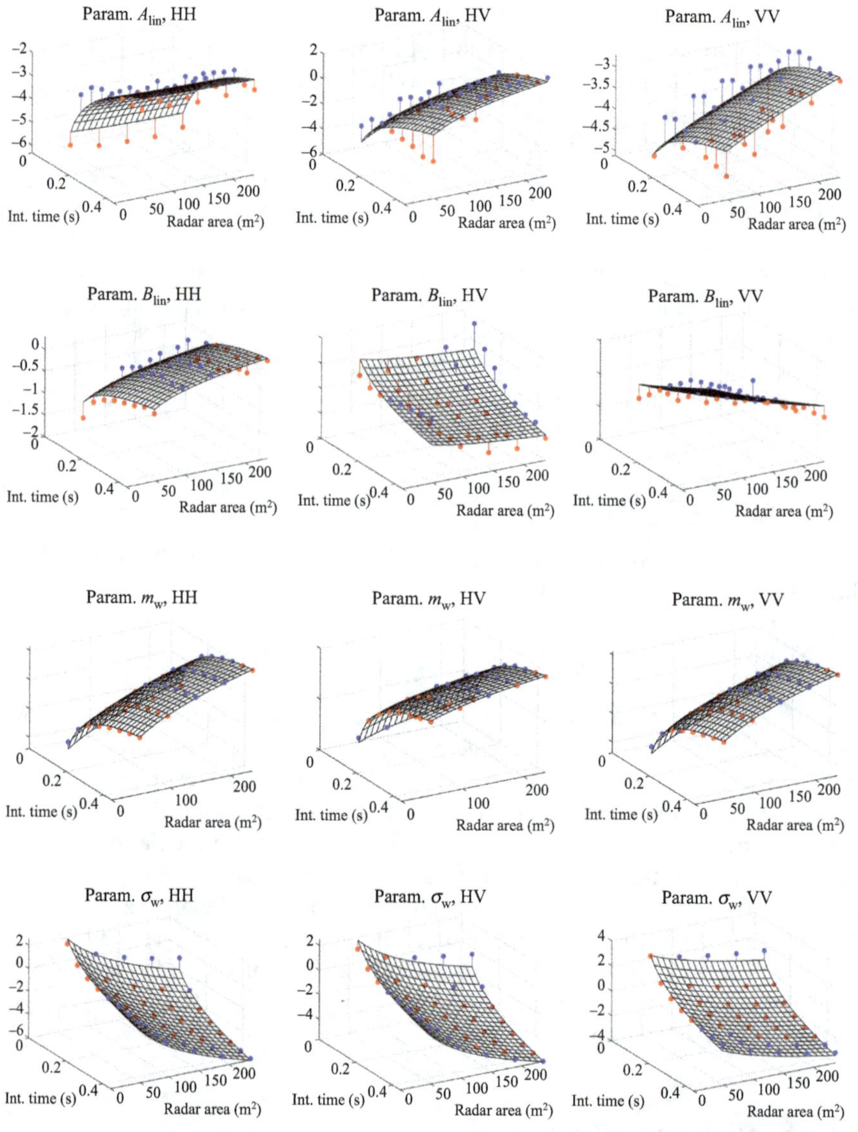

Figure 4.19 Extended Doppler model fits: variation with integration time and radar area. The red and blue points indicate the data points which are below and above the model fit [43]. © Commonwealth of Australia 2008

Gaussian function can be used to model the correlation. An example of the latter is given by

$$\rho_{\text{spat}}(\tau) = \exp\left(-\frac{\tau^2}{R_L^2}\right). \tag{4.89}$$

Table 4.19 Extended Doppler model linear coefficients

Pol.		A_{lin}	B_{lin}	σ_{lin}	m_{w}	σ_{w}
HH	a_{ext}	0.4024	0	–20.94	–0.81	–50.98
	b_{ext}	–0.086	0	1.33	–1.51	1.33
	c_{ext}	–1.32	0	–0.57	–0.48	–0.54
	d_{ext}	0	0	44.89	0.99	61.22
	e_{ext}	0	0	–0.24	0.37	–0.06
HV	a_{ext}	8.24	–21.85	–9.30	0.027	–34.22
	b_{ext}	–2.85	18.11	0.51	–3.04	3.19
	c_{ext}	–0.35	–0.076	–0.80	–0.44	–0.34
	d_{ext}	–100	100	28.14	0	41.52
	e_{ext}	–0.88	–1.08	–0.34	0	–0.094
VV	a_{ext}	7.74	–1.74	–37.08	–0.81	–34.81
	b_{ext}	0	–26.62	1.57	–1.51	1.46
	c_{ext}	–5.81	–0.02	–0.59	–0.48	–0.54
	d_{ext}	–11.54	32.19	39.61	0.99	37.12
	e_{ext}	–0.10	–0.026	–0.045	0.36	–0.042

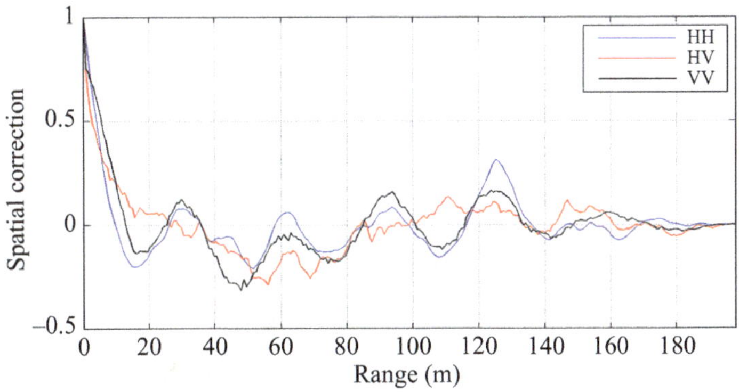

Figure 4.20 Spatial correlation example. Blue: HH, red: HV, black: VV [42]. © IEEE 2014

At low grazing angles, a suitable model for the spatial decorrelation length is presented in [44],

$$R_{\text{L}} = \frac{\pi U^2}{2g}\sqrt{3\cos^2\theta + 1} \tag{4.90}$$

where g is the gravitational acceleration.

For medium grazing angles, the generic model presented in Section 4.5.3 has been applied to the Ingara X-band data set [34] with model coefficients given in Table 4.20. The result in Figure 4.21 shows an example of the spatial decorrelation. The main feature is the higher decorrelation length in the cross-polarised up-wind

Table 4.20 Spatial correlation model coefficients for R_L

Pol.		R_L		
		f_{gen}	g_{gen}	h_{gen}
HH	γ_{gen}	1.24	−1.32	0.11
	a_{gen}	−1.14	3.62	0.47
	b_{gen}	0.28	−0.098	0.093
	c_{gen}	−0.67	0.89	0.20
	d_{gen}	0.33	−0.50	−0.054
	e_{gen}	0.30	−0.88	0.26
HV	γ_{gen}	−1.65	1.66	−0.11
	a_{gen}	−3.89	6.59	0.70
	b_{gen}	2.99	−3.61	0.62
	c_{gen}	−3.58	3.67	0.12
	d_{gen}	−3.10	2.58	0.082
	e_{gen}	−7.44	8.23	−0.45
VV	γ_{gen}	0.45	−0.57	−0.12
	a_{gen}	2.52	0.66	1.76
	b_{gen}	1.55	−1.88	−0.34
	c_{gen}	1.07	−1.92	0.011
	d_{gen}	−0.30	−0.32	0.0043
	e_{gen}	−0.70	1.78	−0.86

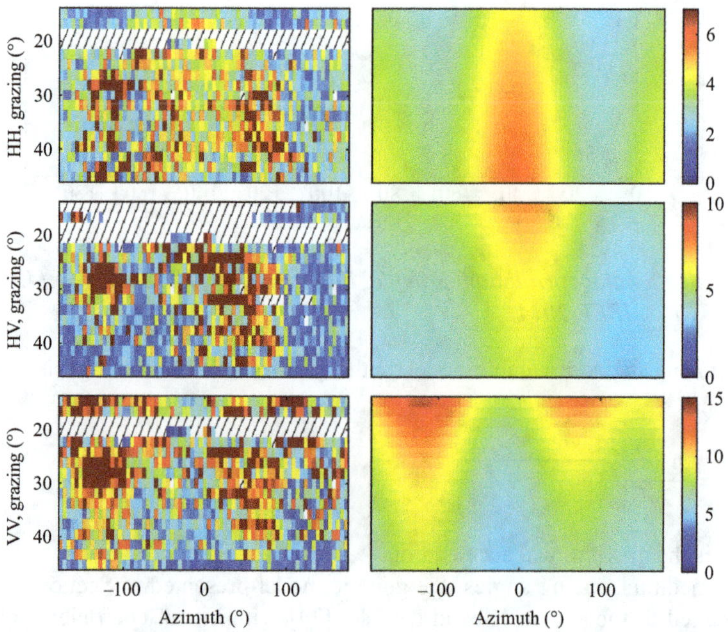

Figure 4.21 Spatial decorrelation lengths (m), left: data, right: model. Top plot: HH, middle plot: HV, bottom plot: VV [42]. © IEEE 2014

direction and around the up- and down-swell angles for the vertical polarisation. Other data sets show similar results for vertical polarisation but also reveal peaks in the up-wind direction for the horizontal and cross-polarised channels. The model result has captured the main trends in each of the polarisations quite successfully.

References

[1] K. D. Ward, R. J. A. Tough, and S. Watts, *Sea clutter: Scattering, the K-distribution and radar performance*, 2nd ed. The Institute of Engineering Technology, London, 2013.

[2] S. Angelliaume, V. Fabbro, G. Soriano, and C. A. Guerin, "The GO-SSA extended model for all-incidence sea clutter modelling," *IEEE Geoscience and Remote Sensing Symposium*, 2014, pp. 5017–5020.

[3] Z. Guerraou, S. Angelliaume, L. Rosenberg, and C. Guérin, "Investigation of azimuthal variations from X-band medium-grazing-angle sea clutter," *IEEE Transactions on Geoscience and Remote Sensing*, vol. 54, no. 10, pp. 6110–6118, 2016.

[4] H. Sittrop, "On the sea-clutter dependency on windspeed," *IEE International Radar Conference*, 1977, pp. 110–114.

[5] M. M. Horst, F. B. Dyer, and M. T. Tuley, "Radar sea clutter model," *IEEE International Conference of Antennas and Propagation*, no. 169, pp. 5–10, 1978.

[6] F. E. Nathanson, J. P. Reilly, and M. N. Cohen, *Radar design principles*, 2nd ed. McGraw-Hill, New York, 1991.

[7] I. Antipov, "Simulation of sea clutter returns," Defence Science Technology Organisation, Technical Report DSTO-TR-0679, Jun. 1998.

[8] Technology Service Corporation, "Backscatter from sea," *Radar Workstation*, vol. 2, pp. 177–186, 1990.

[9] V. Gregers-Hansen and R. Mittal, "An improved empirical model for radar sea clutter reflectivity," *IEEE Transactions on Aerospace and Electronic Systems*, vol. 48, no. 4, pp. 3512–3524, 2012.

[10] F. T. Ulaby, R. K. Moore, and A. K. Fung, *Microwave remote sensing: Active and passive, Volume II: Radar remote sensing and surface scattering and emission theory*. Addison-Wesley, Boston, MA, 1982.

[11] H. Masuko, K. Okamoto, M. Shimada, and S. Niwa, "Measurement of microwave backscattering signatures of the ocean surface using X band and Ka band airborne scatterometers," *Geophysical Research*, vol. 91, no. C11, pp. 13 065–13 083, 1986.

[12] D. J. Crisp, R. Kyprianou, L. Rosenberg, and N. J. Stacy, "Modelling X-band sea clutter at moderate grazing angles," *International Radar Conference*, 2008, pp. 596–601.

[13] L. Rosenberg and S. Watts, "Continuous sea clutter models for the mean backscatter and K-distribution shape," *International Radar Conference*, 2017, pp. 1–6.

[14] J. P. Reilly and G. D. Dockery, "Influence of evaporation ducts on radar sea return," *IEE Proceedings*, vol. 137F, no. 2, pp. 80–88, 1990.
[15] J. P. Reilly, "Clutter models for shipboard radar applications, 0.5 to 70 GHz, multi-sensor propagation data and clutter modelling," NATO AAW System Program Office, Technical Report F2A-88-0-307R2, NAAW-88-062R2, 1998.
[16] F. E. Nathanson, *Radar design principles*, 1st edn. McGraw-Hill, New York, 1969.
[17] J. C. Daley, J. T. Ransome, J. A. Burkett, and J. R. Duncan, "Sea clutter measurements on four frequencies," *Naval Research Laboratory Report*, vol. 6806, 1968.
[18] S. P. Zehner, "GIT model errata," 1988, unpublished.
[19] P. Beckmann, *The scattering of electromagnetic waves from rough surfaces*. Pergamon Press Ltd., London, 1963.
[20] B. Spaulding, D. Horton, and H. Pham, "Wind aspect factor in sea clutter modeling," *International Radar Conference*, 2005, pp. 89–92.
[21] L. Rosenberg and S. Watts, "High grazing angle sea-clutter literature review," Defence Science Technology Organisation, General Document DSTO-GD-0736, 2013.
[22] J. L. Whitrow, "A model for low grazing angle sea clutter for coherent radar performance analysis," Defence Science and Technology Organisation, Tech. Rep. DSTO-TR-2864, 2013.
[23] A. R. Domville, "The bistatic reflection from land and sea of X-band radio waves, part I," GEC (Electronics) Ltd, Stanmore, England, Tech. Rep., Jul. 1967, memorandum SLM1802.
[24] A. R. Domville, "The bistatic reflection from land and sea of X-band radio waves, part II," GEC (Electronics) Ltd, Stanmore, England, Tech. Rep., Jul. 1968, memorandum SLM2116.
[25] H. D. Griffiths, W. A. Al-Ashwal, K. D. Ward, R. J. A. Tough, C. J. Baker, and K. Woodbridge, "Measurement and modelling of bistatic radar sea clutter," *IET Radar, Sonar and Navigation*, vol. 4, no. 2, pp. 280–292, 2010.
[26] N. J. Willis, *Bistatic radar*. SciTech Publishing Inc., London, 1995.
[27] M. W. Long, "On the polarization and the wavelength dependence of sea echo," *IEEE Transactions on Antennas and Propagation*, vol. AP-13, no. 5, pp. 749–754, 1965.
[28] J. C. Wiltse, S. P. Schlesinger, and C. M. Johnson, "Back-scattering characteristics of the sea in the region from 10 to 50 KMC," *Proceedings of the IRE*, vol. 45, no. 2, pp. 220–228, 1957.
[29] L. B. Wetzel, *Radar handbook*, 3rd ed. McGraw Hill, New York, 2008, ch. Sea Clutter.
[30] G. W. Ewell and S. P. Zehner, "Bistatic sea clutter return near grazing incidence," *IEE Conference*, no. 216, pp. 188–192, 1982.
[31] K. D. Ward, "A radar sea clutter model and its application to performance assessment," *International Radar Conference*, 1982, pp. 203–207.
[32] S. Watts and D. C. Wicks, "Empirical models for detection prediction in K-distribution sea clutter," *IEEE International Radar Conference*, 1990, pp. 189–194.

[33] L. Rosenberg, D. J. Crisp, and N. J. Stacy, "Statistical models for medium grazing angle X-band sea-clutter," *Defence Applications of Signal Processing*, 2009, pp. 1–6.

[34] S. Watts, L. Rosenberg, S. Bocquet, and M. Ritchie, "The Doppler spectra of medium grazing angle sea clutter; part 2: Exploiting the models," *IET Radar Sonar and Navigation*, vol. 10, no. 1, pp. 32–42, 2016.

[35] D. Walker, "Doppler modelling of radar sea clutter," *IEE Proceedings of Radar, Sonar and Navigation*, vol. 148, no. 2, pp. 73–80, 2001.

[36] A. Rozenberg, D. C. Quigley, and W. K. Melville, "Laboratory study of polarized scattering by surface waves at grazing incidence: Part 1 – Wind waves," *IEEE Transactions on Geoscience and Remote Sensing*, vol. 33, no. 4, pp. 1037–1046, 1995.

[37] A. D. Rozenberg, D. C. Quigley, and W. K. Melville, "Laboratory study of polarized scattering by surface waves at grazing incidence: The influence of long waves," *IEEE Transactions on Geoscience and Remote Sensing*, vol. 34, no. 6, pp. 1331–1342, 1996.

[38] N. L. Hicks, N. Knable, J. J. Kovaly, G. S. Newell, J. P. Ruina, and C. W. Sherwin, "The spectrum of X-band radiation backscattered from the sea surface," *Journal of Geophysical Research*, vol. 65, no. 3, pp. 825–837, 1960.

[39] L. Rosenberg, D. J. Crisp, and N. J. Stacy, "Characterisation of low-PRF X-band sea-clutter Doppler spectra," *International Radar Conference*, 2008, pp. 100–105.

[40] L. Rosenberg, "The effect of temporal correlation with K and KK-distributed sea-clutter," *IEEE Radar Conference*, 2012, pp. 303–308.

[41] L. Rosenberg, S. Watts, S. Bocquet, and M. Ritchie, "Characterisation of the Ingara HGA dataset," *International Radar Conference*, 2015, pp. 0027–0032.

[42] S. Watts, L. Rosenberg, and M. Ritchie, "Characterising the Doppler spectra of high grazing angle sea clutter," *International Radar Conference*, 2014, pp. 1–6.

[43] L. Rosenberg, "Parametric modelling of sea clutter Doppler spectra," *IEEE Transactions on Geoscience and Remote Sensing*, 2021.

[44] R. J. A. Tough and K. D. Ward, "The generation of correlated K-distributed noise," Defence Research Establishment UK, Tech. Rep. DRA/CIS/CBC3/WP94001/2.0, 1994.

Chapter 5

Sea clutter simulation

5.1 Introduction

Realistic simulation of radar sea clutter is extremely important to stimulate radar processors during development and testing, generate realistic displays in radar trainers and to evaluate radar detection algorithms. A simulated signal must reproduce as faithfully as possible the statistical characteristics that are present in real data, including the mean backscatter, amplitude statistics, short-term temporal correlation (including that represented by the Doppler spectra) and any spatial or longer term temporal variations. It must also reflect the chosen radar parameters, collection geometry and model the effect of platform motion if the data is being collected by an airborne radar system.

In this chapter, the simulation of sea clutter is restricted to representations using compound Gaussian models. As discussed in Chapter 2, compound models have a speckle component with an exponential distribution for intensity, and a texture component that is approximately constant over the tens of milliseconds typically associated with a radar dwell. When using pulse-to-pulse frequency agility with sufficiently large frequency steps, the speckle fluctuation will be approximately random and for a fixed frequency radar, will fluctuate according to a short-term temporal correlation, usually modelled by a Doppler spectrum. The texture will also have correlations that influence both the spatial and longer term temporal fluctuations of the sea clutter. Section 5.2 describes techniques for simulating coherent clutter samples with different statistical distributions, correlated texture and a uniform Doppler spectrum (i.e. constant centre frequency and width).

There are many statistical methods for generating sea clutter with correlated samples, but only a few that model the range and/or time-varying nature of the Doppler spectrum. Some notable work includes Greco *et al.* [1] who proposed an autoregressive scheme for modelling for the texture variation observed in the IPIX radar data and Davidson [2] who presented a method for recreating the coherent properties of a specific sea clutter data set. Unfortunately, it is not easy to utilise these techniques for other radar systems and collection geometries. The work by Watts [3] describes a more general method for simulating sea clutter by representing the time and/or range evolution of the Doppler spectra. It was originally presented as a single Gaussian model and then extended to the bimodal case in [4,5] with parameters characterised over a wide range of geometries and sea conditions. Section 5.3 describes two approaches to simulation using this model. These include a frequency domain implementation for a fixed set of parameters and a time domain approach that allows the evolution of parameters. Details about the simulation of multiple phase centres and a practical example for a scanning radar are also presented.

McDonald and Cerutti-Maori [6] have attempted to increase the realism by simulating a tilted wave spectrum. This is achieved with a spatially and temporally limited clutter (STLC) model which links the two domains together. Good agreement was demonstrated between simulated and real data collected with a Fraunhofer FHR's multichannel PAMIR radar. The model has a stronger physical basis than the evolving Doppler spectrum model but is not able to accurately recreate data from the up- and down-swell directions [7]. Building on these models, a different representation of the coherent spectrum has been proposed by Bocquet [8]. The compound G2 model includes separate spatial correlations for both components of the bimodal spectrum, with the intensity of each component represented by an independent gamma variate with its own scale and shape parameter. The overall local clutter power or texture is then distributed as the sum of these two gamma variates. In Section 5.4, these two coherent simulation techniques are outlined with details on how they can be implemented to simulate sea clutter. The accuracy of the evolving Doppler spectrum, the STLC and G2 models are then compared with a run from the Ingara X-band data set.

5.2 Uniform Doppler spectrum

The simplest model of Doppler spectra for sea clutter has a constant mean Doppler centre frequency and spectrum width (typically defined as the standard deviation of a Gaussian-shaped spectrum). This is referred to as a 'uniform Doppler spectrum'. If an accurate spectrum model is not required, then simulated data with a specified amplitude distribution can be generated by first creating a sequence of uniform random variables between 0 and 1, and then transforming it appropriately using an inverse CDF function [9]. However, for a compound Gaussian model, it is often desirable to generate the speckle and texture components separately to allow for different correlation scales to be incorporated. As described in Chapter 2, the speckle will be represented by a complex Gaussian process, with an autocorrelation function (ACF) that is determined by an inverse Fourier transform of the desired Doppler spectrum. The Gaussian distributed in-phase and quadrature returns, x, will have an envelope with a Rayleigh distribution and the intensity or power of the returns will have an exponential distribution. The mean power or texture, τ, has a PDF chosen to achieve the desired compound model. For example, a K-distribution will have a gamma distributed texture, while a Pareto distribution will have an inverse gamma. In the amplitude domain, the resultant random variable is given by

$$y = \sqrt{\tau}x. \tag{5.1}$$

To generate a uniform Doppler spectrum with a constant centre frequency and spectral width, the texture, τ, is treated as a real stationary non-Gaussian process and the speckle contains correlated in-phase and quadrature components. There are a number of implementation approaches described in [10, Chapter 4], including a recursive relationship and a method based on Fourier synthesis (FS). If the speckle covariance, $\mathbf{R}$, is circulant with the autocorrelation defined along the top row, then another approach is to use a single value decomposition of the covariance, $\mathbf{R} = \mathbf{U}\boldsymbol{\Lambda}\mathbf{V}^H$. The correlated random vector, $\mathbf{x}_c$, is then given by

$$\mathbf{x}_c = \mathbf{U}\sqrt{\boldsymbol{\Lambda}}\mathbf{x} \tag{5.2}$$

where $\mathbf{x}$ are realisations of $N \times 1$ complex normal random vectors, $\mathcal{CN}(0, 1) = \mathcal{N}(0, 1/2) + j\mathcal{N}(0, 1/2)$.

More generally, a spherically invariant random vector (SIRV) is defined such that the output is a random process whose PDF is uniquely determined by the specification of a mean vector, a covariance matrix and a first-order PDF. A spherically invariant random process (SIRP) is then defined as a random process such that every random vector obtained by sampling the process is an SIRV [11,12]. The SIRP can then be written as an analogue to the compound Gaussian model,

$$P_y(\mathbf{y}) = \int_0^{\infty} \frac{1}{(\pi\tau)^N |\mathbf{R}|} \exp\left(-\frac{\mathbf{y}^H\mathbf{R}^{-1}\mathbf{y}}{\tau}\right) P_\tau(\tau)d\tau. \tag{5.3}$$

where $\mathbf{y} = [y_1 \ldots, y_N]$. While the FS and SIRP methods generally achieve a high level of accuracy, only a few random stochastic processes can be linearly connected to the Gaussian process. The generation of correlated gamma random fields using SIRP theory is examined in [13,14]. In these papers, the gamma random process is expressed as a sum of Gaussian processes when the gamma shape parameter ν has special values. The method of Conte *et al.* [13] shows that the gamma distribution with shape values in the range of $0 < \nu \leq 2$ closely resembles a compound Gaussian process and hence the correlated gamma variables can be simulated with a bi-dimensional SIRP. However, it is not clear how to map the correlation property of a gamma process to the correlation property of a Gaussian process. On the other hand, the method of Armstrong and Griffiths [14] considers a special case where the correlation follows a geometric progression (i.e. the correlation coefficients form a geometric sequence) for ν having values of $\nu = 0.5n$ where n is a natural number. An alternative method which can be used for generating random sequences from any distribution is known as the memoryless non-linear transform (MNLT) [10,15].

5.2.1 Memoryless non-linear transform

The MNLT is used to transform a set of random Gaussian distributed variables to a different PDF. Samples are first generated from a zero mean, unit variance Gaussian process and for each sample value p, the cumulative distribution of the Gaussian process is equated to a cumulative distribution with the desired PDF, P_{dist} evaluated at q. For successive random samples, the resulting values of q will then have the required PDF. This can be written as

$$\int_q^\infty P_{\text{dist}}(q')dq' = \frac{1}{\sqrt{2\pi}} \int_p^\infty \exp\left(-\frac{p'^2}{2}\right) dp'$$
$$= \frac{1}{2}\text{erfc}\left(\frac{p}{\sqrt{2}}\right) \tag{5.4}$$

where erfc(·) is the complementary error function. If the complementary quantile function, $Q_c(\cdot)$, is defined so that

$$\int_{Q_c(\kappa)}^\infty P_{\text{dist}}(q)dq = \kappa, \tag{5.5}$$

then the required value q can be given by

$$q(p) = Q_c\left(\frac{1}{2}\text{erfc}\left(\frac{p}{\sqrt{2}}\right)\right). \tag{5.6}$$

For example, if we want to generate a gamma distribution with

$$P_{\text{dist}}(p) = \frac{1}{\Gamma(\nu)} p^{\nu-1} \exp(-p), \tag{5.7}$$

then

$$\int_{Q_c(\kappa)}^\infty P_{\text{dist}}(q)dq = \frac{\Gamma(\nu, Q_c(\kappa))}{\Gamma(\nu)} = \frac{1}{2}\text{erfc}\left(\frac{p}{\sqrt{2}}\right) \tag{5.8}$$

and the gamma random variable is recovered by solving this relationship and evaluating the quantile function. The quantile function for both the gamma and inverse gamma distributions is the inverse incomplete gamma function, which is available in recent versions of MATLAB® [16] and Mathematica [17].

If correlated Gaussian samples are transformed using the MNLT, the resulting samples will have the desired PDF but their ACF will have changed. Within some limitations discussed later, it is possible to determine the ACF that the Gaussian samples should have in order to achieve the required ACF. This method was first described in [15] and is further illustrated in [10]. If the ACF of the transformed variable, y, is given by $\langle y(0)y(t)\rangle$, it can be shown that the desired normalised ACF, $\rho_G(t)$, of the Gaussian samples is related by

$$\langle y(0)y(t)\rangle = \frac{1}{2\pi}\sum_{n=0}^{\infty}\frac{\rho_G^n(t)}{2^n n!}\times\left(\int_{-\infty}^{\infty} e^{-x^2/2}H_p\left(\frac{x}{\sqrt{2}}\right)Q_c\left(0.5\text{erfc}\left(\frac{x}{\sqrt{2}}\right)\right)dx\right)^2 \tag{5.9}$$

where $H_p(x)$ is a Hermite polynomial. This expression looks complicated, but typically only a few terms of the infinite summations are required and it can readily be processed using numerical computing methods. The MNLT method has the advantage that it can reproduce samples with negative values of the correlation coefficient.

Consider the generation of correlated gamma distributed samples. If $\langle y\rangle = 1$, then $\langle y^2\rangle = 1 + 1/\nu$ and if the required correlation coefficient is $\rho(t)$, then $\langle y(0)y(t)\rangle = 1 + \rho(t)/\nu$ and (5.9) can be solved to give

$$\langle y(0)y(t)\rangle = f(0) + f(1)\rho_G(t) + f(2)\rho_G^2(t) + \cdots . \tag{5.10}$$

This result has to be inverted for each value of the lag, t, to find the required values of $\rho_G(t)$. It is then a simple matter to generate Gaussian samples with the required ACF, as discussed earlier.

Figure 5.1 shows examples of the mapping between the desired correlation coefficient of gamma distributed variables, $\rho(t)$, and the correlation coefficient of the

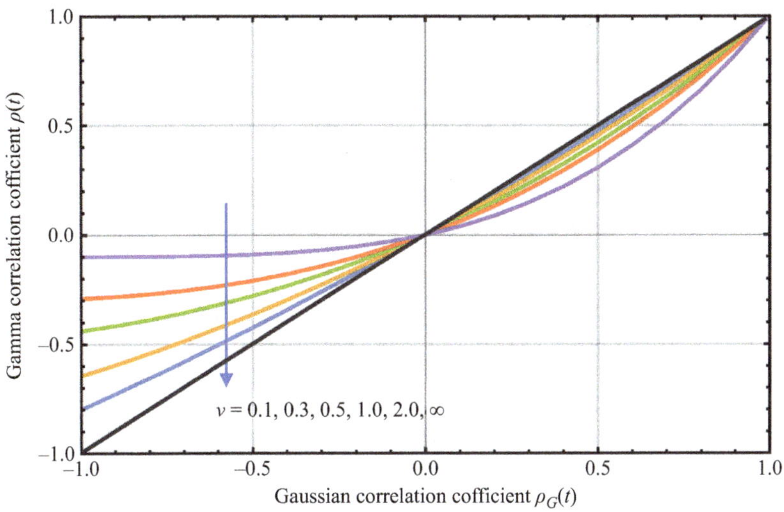

Figure 5.1 Mapping between the correlation coefficient of gamma distributed variables, $\rho(t)$, and the correlation coefficient of the Gaussian variables $\rho_G(t)$, for various values of shape parameter, ν [18].

Gaussian variables $\rho_G(t)$, prior to application of the MNLT. For $\nu = \infty$, the mapping is $1 : 1$, but as ν reduces it is not possible to map all negative values of $\rho(t)$ onto $\rho_G(t)$. For example with $\nu = 0.1$, only $\rho(t) > -0.1$ can be modelled.

The texture is correlated along both time and range. This is most apparent for a ground-based maritime radar which continually stares or scans over the same spatial region for a period of seconds. In an airborne scenario, the effect of the time-varying texture is harder to observe due to the aircraft motion and the revisit time of the scanning radar. To model the texture, the MNLT approach can be used [10], with the texture being gamma or inverse gamma distributed to achieve a K or Pareto distribution, respectively.

A model for the ACF has been developed in [19] based on a number of sources [20]. It is defined in the range and slow-time, $\rho(R,t)$, by the weighted sum of two components representing the fluctuations of the wind-driven sea and the swell:

$$\rho(R,t) = (1 - a_{\text{tex}})\exp\left(-R^2/R_L^2(\boldsymbol{\Omega}_p) - t^2/T_d^2\right) + a_{\text{tex}}\exp\left(-\sqrt{(R/(0.44R_e))^2 + (t/(0.44t_e))^2}\right)\cos\left(|k_{sr}|R + 2\pi t/T_s\right) \tag{5.11}$$

where a_{tex} is the weighting parameter, T_d is the temporal decorrelation time and R_L is the spatial decorrelation length obtained from the parameter set $\boldsymbol{\Omega}_p \equiv \{\theta_{\text{plat}}, \psi, \theta_{\text{sw}}, U, H_{1/3}\}$. This set represents the variables used to characterise the model parameters: θ_{plat} is the radar-look direction at an angle relative to the platform motion, ψ is the angle between the radar-look direction and the wind, θ_{sw} is the angle between the wind and swell, U is the wind speed and $H_{1/3}$ is the significant wave height. In the second term of (5.11), the decorrelation length and time in the swell component of the ACF are proportional to the spatial and temporal extent (R_e, t_e) of the illuminated region, which determines the wave number and frequency resolution in the radar image. The parameter, $k_{sr} = k_s \cos\phi \cos\theta_{\text{sw}}$, is the component of the swell wave vector parallel to the radar beam for a grazing angle, ϕ. The swell wave number $k_s = (2\pi/T_s)^2/g$, assuming deep water with no surface current, where g is the gravitational acceleration and T_s is the swell period. A gamma realisation with a shape parameter of 3 is shown in Figure 5.2 with a swell moving through the scene with waves lasting for several seconds. The model parameters used are $R_d = 3.2$ m, $T_d = 1.4$ s, $T_s = 5$ s, $\phi_{\text{deg}} = 30°$, $\theta_{\text{sw,deg}} = 0°$, $R_e = 360$ m, $t_e = 20$ s and $a_{\text{tex}} = 0.7$.

The ACF model in (5.11) is only suitable if the 'sea' component is not resolved by the radar, so that the Gaussian model is a satisfactory representation in both the range/time domain and wave number/frequency domains, and the swell is monochromatic, so that the single cosine term is a satisfactory representation. Modelling of a more general wave spectrum needs to be done in the wave number/frequency domain, as each component in the spectrum propagates at a different phase velocity, according to the dispersion relation. A representation of the time-varying sea surface can then be obtained via a 2D (range/time) or 3D (range/cross-range/time) inverse Fourier transform. This approach has been used for computer graphics imagery and marine engineering applications [21] and for physical modelling of radar backscatter [22,23].

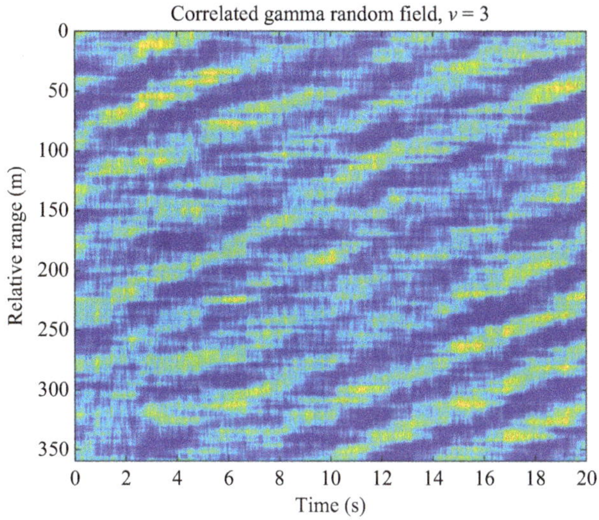

Figure 5.2 Two-dimensional correlated gamma texture with $\nu = 3$ *[19].* *© Commonwealth of Australia 2020*

Alternatively, an MNLT can be applied to the time-varying Gaussian random surface realisation to produce the appropriate radar backscatter statistics [24]. However, further research is needed to properly relate this approach to the wind and sea conditions and geometry. The dispersion relation from linear hydrodynamics is satisfactory if the wave steepness is low and the spatial and temporal extent of the simulation is not too large; otherwise, non-linear hydrodynamics is needed to create a realistic representation of the sea surface [25].

5.3 Evolving Doppler spectrum

The evolving Doppler spectrum model was first presented by Watts [3,10] for modelling the clutter spectra observed by a ground-based radar system. It describes a method for simulating clutter by representing the time and/or range evolution of the Doppler spectra. It was presented in Chapter 2 and provides the basis for the simulation methods described in this section.

5.3.1 Frequency domain simulation

The frequency domain simulation approach assumes that both the texture and model parameters are constant and hence this approach is only suitable when considering a limited geometric region. To generate clutter samples directly in the frequency domain, the average power spectrum for the nth frequency bin at a given range is modelled as a Gaussian function,

$$G(n, \tau, \boldsymbol{\Omega}_{\mathrm{p}}) = \frac{\tau}{\sqrt{2\pi}w(\boldsymbol{\Omega}_{\mathrm{p}})} \exp\left(-\frac{(f_{\mathrm{r}}n/N - m_{\mathrm{f}}(\tau, \boldsymbol{\Omega}_{\mathrm{p}}))^2}{2w^2(\boldsymbol{\Omega}_{\mathrm{p}})}\right) \tag{5.12}$$

where the frequency bins, $n = -N/2, \ldots, N/2 - 1$ for a coherent processing interval (CPI) of length N, f_{r} is the pulse repetition frequency (PRF) and τ is the speckle mean power or texture. The spectral width, $w(\boldsymbol{\Omega}_{\mathrm{p}})$, of the clutter is a Gaussian random variable with mean value $\sqrt{m_{\mathrm{w}}^2 + \sigma_{\mathrm{CPI}}^2}$ and standard deviation σ_{w}. If is a window function applied in the time domain, then $Q_{\mathrm{w}}(\cdot)$ is defined as the power spectral density of the window function. The effect of the CPI length can then be determined by calculating the normalised second-order moment,

$$\sigma_{\mathrm{CPI}} = \left[\frac{\int_{-f_{\mathrm{r}}/2}^{f_{\mathrm{r}}/2} Q_{\mathrm{w}}(f) f^2 df}{\int_{-f_{\mathrm{r}}/2}^{f_{\mathrm{r}}/2} Q_{\mathrm{w}}(f) df}\right]^{1/2}. \tag{5.13}$$

The mean Doppler frequency is related to the normalised mean speckle power, $\bar{\tau} = \tau/\langle\tau\rangle$ with

$$m_{\mathrm{f}}(\tau, \boldsymbol{\Omega}_{\mathrm{p}}) = A_{\mathrm{c}} + B_{\mathrm{c}}\bar{\tau} + f_{\mathrm{lin}} \tag{5.14}$$

where A_{c} and B_{c} relate the centre point to the texture and the term f_{lin} was added to the model in [5] and captures the fluctuation around the model fit. It is described by a Gaussian random variable with zero mean and standard deviation, σ_{lin}. Alternatively, the clutter representation in (5.12) can be replaced with the bimodal spectrum presented in Chapter 2 where the mean Doppler offset, $m_{\mathrm{f}}(\tau, \boldsymbol{\Omega}_{\mathrm{p}})$, is replaced with its equivalent bimodal version. This Gaussian mixture model uses a ratio, $\alpha \leq 1$, to determine how much of each Gaussian component is present,

$$G_{\mathrm{bi}}(n, \tau, \boldsymbol{\Omega}_{\mathrm{p}}) = (1 - \alpha)G(n, \tau, w_1, \boldsymbol{\Omega}_{\mathrm{p}}) + \alpha G(n, \tau, w_2, \boldsymbol{\Omega}_{\mathrm{p}}) \tag{5.15}$$

where $w_1(\boldsymbol{\Omega}_{\mathrm{p}})$ and $w_2(\boldsymbol{\Omega}_{\mathrm{p}})$ are the Doppler spreads for each component and the centre frequencies are determined by

$$m_{\mathrm{f}_1}(\tau) = \begin{cases} A_{\mathrm{c}} + B_{\mathrm{c}}\bar{\tau} + f_{\mathrm{bi}}, & \bar{\tau} \leq \tau_{\mathrm{bi}} \\ A_{\mathrm{c}} + B_{\mathrm{c}}\tau_{\mathrm{bi}} + f_{\mathrm{bi}}, & \bar{\tau} > \tau_{\mathrm{bi}} \end{cases} \tag{5.16}$$

$$m_{\mathrm{f}_2}(\tau) = A_{\mathrm{c}} + B_{\mathrm{c}}\bar{\tau} + f_{\mathrm{bi}}$$

where f_{bi} is a zero mean Gaussian random variable with standard deviation, σ_{bi}. This model for the Doppler spectrum has the effect of broadening the spectrum if $\bar{\tau}$ exceeds the threshold, τ_{bi}. Also, to ensure a single Gaussian model, when $\bar{\tau} \leq \tau_{\mathrm{bi}}$, the weighting factor, $\alpha = 0$. The mean and variance of the overall bimodal spectrum can be expressed as

$$\begin{aligned} m_f(\tau) &= \alpha m_{f_1}(\tau) + (1-\alpha) m_{f_2}(\tau), \\ w^2(\boldsymbol{\Omega}_p) &= \alpha w_1^2 + (1-\alpha) w_2^2 \\ &\quad + \alpha(1-\alpha)\left(m_{f_1}^2(\tau) + m_{f_2}^2(\tau) - 2m_{f_1}(\tau)m_{f_2}(\tau)\right) \end{aligned} \tag{5.17}$$

which can be simplified by defining $w(\boldsymbol{\Omega}_p) = w_1(\boldsymbol{\Omega}_p) = w_2(\boldsymbol{\Omega}_p)$. On an airborne platform, the underlying Doppler spectrum will also be broadened due to the platform motion and the effect of sampling by a finite CPI. Across the antenna beampattern, there are Doppler shifts given by $f = 2v_p \sin\theta/\lambda$ where λ is the radar wavelength and v_p is the platform velocity. The azimuth two-way antenna beampattern is described by $A_{az}(\theta)$ where θ is the azimuth angle relative to the antenna pointing direction and is maximum when $\theta = 0$. This can be discretised as $A_{az}(n)$ and included in the convolution,

$$G_{az}(n, \tau, \boldsymbol{\Omega}_p) = A_{az}(n) \times G(n, \tau, \boldsymbol{\Omega}_p). \tag{5.18}$$

If $A_{az}(n)$ is assumed to have a Gaussian shape, then the standard deviation of the spectrum will be increased to $\sqrt{w(\boldsymbol{\Omega}_p)^2 + \sigma_{plat}^2}$ with

$$\sigma_{plat} = \frac{v_p \theta_{3dB} |\sin\theta_{plat}|}{\sqrt{2\ln 2}\lambda} \tag{5.19}$$

where θ_{3dB} is the two-way azimuth 3 dB beamwidth. The coherent spectra can then be generated as

$$C(n) = \sqrt{G_{az}(n, \tau, \boldsymbol{\Omega}_p)}x(n) \tag{5.20}$$

where the speckle component comprises random samples from a complex Gaussian distribution, $\mathcal{CN}(0, 1)$ with zero mean and unity variance. Thermal noise can then be added to the frequency domain signal,

$$\mathbf{H}_{evolving}(f) = \mathbf{C} + \boldsymbol{\xi} \tag{5.21}$$

where $\boldsymbol{\xi} = \mathcal{CN}(0, p_n/N)$ are realisations of complex normal random variables. Finally, if a time domain representation is desired, the complex time domain output can obtained by an inverse Fourier transform, $\mathbf{h}_{evolving} = \mathcal{F}^{-1}\left(\mathbf{H}_{evolving}\right)$. The simulation steps for the frequency domain implementation include the following:

- Define the correlation coefficient of the intensity as a function of both range and time.
- Create a realisation of the texture, τ, with unity mean using the MNLT and the desired correlation function. This could be gamma or inverse gamma distributed to achieve a compound K or Pareto distribution, respectively.

- Generate a Gaussian distributed realisation of the width, w, with mean $\sqrt{m_\mathrm{w}^2 + \sigma_\mathrm{CPI}^2}$ and standard deviation, σ_w.
- Generate the Doppler spectrum model in (5.12) or (5.15) across N frequency bins and convolve it with the frequency modulation due to the azimuth two-way beampattern, $A_\mathrm{az}(\cdot)$.
- Generate complex Gaussian realisations for the speckle with zero mean and unity variance.
- Form the frequency domain clutter signal according to (5.20).
- Thermal noise can be added as complex Gaussian realisations with zero mean and variance, p_n/N.
- Complex time domain samples can then be obtained by an inverse Fourier transform of $\mathbf{H}_\mathrm{evolving}$.

5.3.2 Time domain simulation

A different approach is required to simulate coherent time domain data with model parameters that can be updated over a longer time period [3,10]. The implementation is based on a convolution of the desired clutter impulse response $g_\mathrm{az}(\cdot)$ and the compound Gaussian signal model, $x\sqrt{\tau}$. The time domain output of the clutter component comprises the filter at pulse index m,

$$c(m) = \sum_{l=-L_\mathrm{O}/2}^{L_\mathrm{O}/2} g_\mathrm{az}(l, \tau, \mathbf{\Omega}_\mathrm{p}) x(m+l) \sqrt{\tau(m+l)} \tag{5.22}$$

where L_O is the filter order and the sampling interval is $1/f_\mathrm{r}$. The discretised clutter impulse response for the unimodal model, $g_\mathrm{az}(\cdot)$, is found by taking the Fourier transform of the spectrum in (5.18),

$$g_\mathrm{az}(l, \tau, \mathbf{\Omega}_\mathrm{p}) = a_\mathrm{az}\left(\frac{l}{f_\mathrm{r}}\right) \exp\left(-j\frac{m_\mathrm{f}(\tau, \mathbf{\Omega}_\mathrm{p}) 2\pi l}{f_\mathrm{r}} - \left(\frac{2\pi w(\mathbf{\Omega}_\mathrm{p}) l}{f_\mathrm{r}}\right)^2\right) \tag{5.23}$$

where $a_\mathrm{az}(\cdot)$ is the azimuth two-way beampattern in the time domain. Alternatively, the bimodal Doppler spectrum from (5.15) could be used. The total received signal of size $N \times 1$ is then given by

$$\mathbf{h}_\mathrm{evolving} = \mathbf{c} + \boldsymbol{\xi}. \tag{5.24}$$

This signal represents the return for a single pulse and hence the technique can be implemented by repeating (5.24) for each pulse in the simulation (i.e. as a sliding window). If the filter weights change slowly enough over time, the resulting data will then have an appropriate time-varying spectrum. For best fidelity, it may be appropriate to either increase the PRF and later down-sample as required, or increase the length of the filter, L_O. A possible guideline is that $wL_\mathrm{O}/f_\mathrm{r} \approx 1$.

The steps required to generate complex bursts of data include two stages. Prior to the sliding window:

- Define the correlation coefficient of the intensity as a function of both range and time.
- Create a realisation of the texture, τ, with unity mean using the MNLT and the desired correlation function. This could be gamma or inverse gamma distributed to achieve a compound K or Pareto distribution, respectively.
- Generate realisations of the speckle random variable, x, and the spectrum standard deviation, w, for each desired pulse and range bin. The speckle has a complex normal distribution, $\mathcal{CN}(0, 1)$, while the spectrum standard deviation initially has a normal distribution with zero mean and unit variance, $\mathcal{N}(0, 1)$, which is later scaled to the desired mean and standard deviation within the simulation loop.

Then at each pulse through the simulation,

- Determine the model parameters and extract the correct blocks from the pre-calculated random variables, x and w. The standard deviation random variables are now offset by m_{w} and scaled by σ_{w}.
- Form the clutter impulse response using either the unimodal or bimodal representations.
- Generate complex time domain signals using (5.24) with the desired clutter and noise power levels.

5.3.3 Multiple phase centres

Modelling the simultaneous returns from multiple phase centres requires the representation of a clutter patch across azimuth [19,26,27]. This is shown in Figure 5.3 for two antenna phase centres separated by d, and at a distance R_0, where the centre of each sub-patch subtends an angle θ_i from the centre of the array and $i = 1, \ldots, I$. In general $I \gg 1$ within the antenna main beam and it is important that the sub-patches also extend over the antenna sidelobes if a moving platform is to be modelled. Either the frequency or time domain implementations can be used to simulate multiple channels. Both require the total intensity of the returns from all clutter sub-patches to be scaled by the two-way antenna beam pattern, $\tau_i = A_{\mathrm{az}}(\theta_i)\tau$, with their sum equal to the local mean intensity,

$$\tau = \sum_{i=1}^{I} \tau_i. \tag{5.25}$$

In the frequency domain, the coherent spectra are generated as

$$C(n, \tau_i, \mathbf{\Omega}_{\mathrm{p}}) = \sqrt{G(n, \tau_i, \mathbf{\Omega}_{\mathrm{p}})}x(n) \tag{5.26}$$

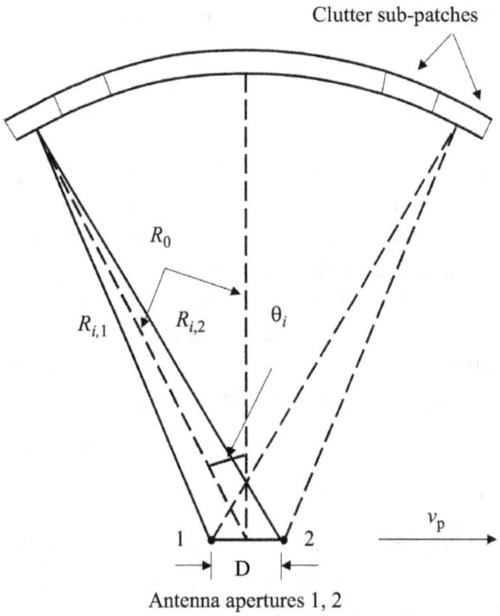

Figure 5.3 Clutter sub-patches observed by two phase centres [19]. © Commonwealth of Australia 2020

with complex time domain samples for each sub-patch given by $c(m, \tau_i, \mathbf{\Omega}_p) = \mathcal{F}^{-1}\left(C(n, \tau_i, \mathbf{\Omega}_p)\right)$. In the time domain, the complex time domain samples are determined by

$$c(m, \tau_i, \mathbf{\Omega}_p) = \sum_{l=-L_0/2}^{L_0/2} g_{az}(l, \tau_i, \mathbf{\Omega}_p) x(m+l)\sqrt{\tau_i(m+l)}. \tag{5.27}$$

For either the frequency or time domain case, the total return in a single aperture is given by

$$c(m, \tau, \mathbf{\Omega}_p) = \sum_{i=1}^{I} c(m, \tau_i, \mathbf{\Omega}_p). \tag{5.28}$$

This model assumes that the normalised power spectral densities of the clutter returns are the same in each clutter sub-patch, but independent of each other (i.e. with independent speckle components). However, this is just for convenience and each clutter sub-patch could have a different normalised power spectral density. For example, the clutter returns observed through the antenna sidelobes may have a power spectral density that varies with look direction. Also, the areas of the sub-patches and their angular spacing do not need to be equal. In fact, an uneven spacing may help one to avoid artefacts in any angle dependent processing of the returns. Discrete spikes can be added to the returns by modelling point responses within the received beam or as distributed returns across two or more clutter sub-patches.

Consider a linear array receiving a signal in the far field and pointing normal to the array (broadside). The return from the lth receive channel when $l = 1, \ldots, L$ is then given by

$$c_l(m, \boldsymbol{\Omega}_\mathrm{p}) = \sum_{i=1}^{I} c(m, \tau_i, \boldsymbol{\Omega}_\mathrm{p}) \exp\left(j\varphi_{i,l}\right) \exp\left(j\frac{4\pi v_\mathrm{p} n}{\lambda f_\mathrm{r}} \sin\theta_i\right) \tag{5.29}$$

where the spatial phase shift, φ, is given by

$$\varphi_{i,l} = \frac{2\pi}{\lambda}\left(R_{i,l} - R_{i,1}\right) \approx \frac{2\pi(l-1)d\sin\theta_i}{\lambda}. \tag{5.30}$$

The second phase term in (5.29) is required to account for the antennas moving with an along-track velocity, v_p, that causes the phase from each clutter sub-patch to increment or decrement from pulse-to-pulse due to the change in platform position. This implementation is clearly more computationally intensive than the single channel model due to the extra summation. However, the range of sub-patches which contribute to the sum in (5.29) can be reduced by only including those where $A(\theta_i) > \epsilon_\mathrm{lim}$, where $\epsilon_\mathrm{lim} \approx 10^{-3}$. Finally, thermal noise can be added as complex Gaussian samples to the time domain returns in each channel, according to the clutter-to-noise ratio (CNR) required.

The simulation steps for the multichannel implementation are similar to the previous two. First, create realisations of the mean clutter intensity, speckle and width. Then divide the antenna beam into I angular sectors and calculate the mean intensity of the return in each clutter sub-patch according to (5.25). For the frequency domain implementation, generate a complex spectrum for each clutter sub-patch using either the unimodal or bimodal models with the appropriate weighted mean intensity and random values of speckle as shown in (5.26). These complex spectra are then transformed into the time domain, giving N values of a complex time domain series $s(n, \tau_i)$ for each sub-patch. For the time domain implementation, first determine the model parameters for each pulse and extract the correct blocks from the pre-calculated random variables, x and $w(\boldsymbol{\Omega}_\mathrm{p})$. The standard deviation random variables are now offset by m_w and scaled by σ_w. The clutter impulse response can then be generated using either the unimodal or bimodal representations. The final three steps are then common to both and include adding any discrete spikes before applying the appropriate phase weighting to the returns from each sub-patch and summing the result as shown in (5.29). The result is then normalised to have unity mean power and thermal noise is added independently to each channel to achieve the desired CNR.

5.3.4 Scanning radar scenario

For scanning radars, there may be both coherent pulse-to-pulse integration within a dwell and non-coherent scan-to-scan integration. The pulse-to-pulse integration provides a coherent gain when detecting targets, while the scan-to-scan integration allows the operator to control false alarms, manage fading targets, improve the quality

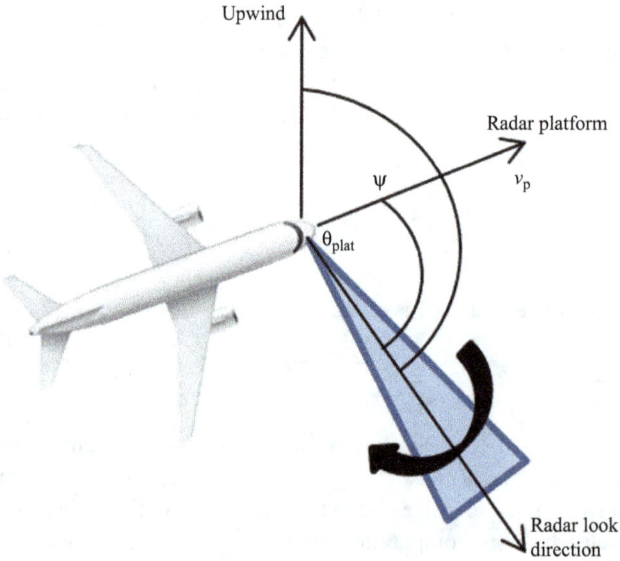

Figure 5.4 Scanning airborne radar scenario [19]. © Commonwealth of Australia 2020

of the displayed output to the radar operator and reduce the impact of persistent spikes. Consider the airborne scanning radar in Figure 5.4, where the radar platform is travelling with velocity, v_p, and the radar-look direction is at an angle, θ_{plat}, relative to the platform motion. The angle between the radar-look direction and the wind is given by ψ. As the radar scans with a rate, $\dot{\theta}_{az}$ rad/s, there are $N_p = \text{Floor}\left(2\pi f_r/\dot{\theta}_{az}\right)$ pulses in a given scan. Also, if the two-way azimuth 3 dB beamwidth is θ_{3dB}, then a point target will be in the beam for $T_{targ} = \theta_{3dB}/\dot{\theta}_{az}$ seconds or $N_{targ} = \text{Floor}\left(T_{targ} f_r\right)$ pulses. This limit determines the 'dwell' or the number of pulses which can be used for within-beam processing.

For radars that include pulse-to-pulse integration, the beamshape loss needs to be considered. It results from the integration of pulses that vary in power according to the two-way beampattern [28]. However, if there is no pulse-to-pulse integration, there will be an antenna pattern loss when the radar beam does not point directly at a target [29]. As the scan rate increases, both the beamshape loss and the antenna pattern loss increase. If these losses are a problem, one solution would be to increase the PRF, subject to maintaining unambiguous range performance. To extend the unambiguous range for non-coherent radars, range ambiguities can be resolved through frequency agility or by slightly changing the PRF after each pulse (also known as pulse jittering). Also, for fast scanning radars, there is a shorter time to perform real time signal processing and there will be challenges in the stabilisation of the antenna gimbal, while for slow scanning radars, the biggest constraint is the reduced revisit time in each azimuth-look direction.

As the radar scans in azimuth, the statistics of the clutter patch will gradually change, and the parameter values for the clutter simulation will require updating. However, over a small azimuth region, the model parameters do not change significantly and hence the frequency domain implementation of the evolving Doppler spectrum model from Section 5.3.1 can be used. However, if wider azimuth regions are required, then the more complicated time domain implementation must be used from Section 5.3.2. For simulating a continuously scanning radar, a full scan of data is required and the 2D texture simulation is not feasible. Instead, a smaller number of texture realisations, N_{tex}, can be generated at points around the scan with the correct statistical distribution and range correlation. These are determined from the user-defined input parameters, $\boldsymbol{\Omega}_{\text{p}}$, with the empirical models given in Chapter 4. The small number of positions ensures that the texture does not vary significantly over the integration time of the filter. If K is the number of range samples and N_{tex} is the number of texture realisations, then the size of the texture is $N_{\text{tex}} \times K$. After this processing step is completed, the texture is then interpolated to the correct time scale, $1/f_r$, with final size, $N \times K$.

A point target or discrete clutter spike within the beam will also have its spectrum broadened due to azimuth scanning of the antenna, even from a stationary platform. For a dwell that matches the full beamwidth of the antenna, the spectrum will be broadened by an additional component which can be modelled as a Gaussian function with standard deviation,

$$\sigma_{\text{scan}} = \frac{\sqrt{2 \ln 2}\dot{\theta}_{\text{scan}}}{\theta_{3\text{dB}}}. \tag{5.31}$$

In the bimodal version of the clutter model, the sea-spike broadening might only be applicable to the higher Doppler frequency component. In that case, the Doppler spread for the first component, $w_1(\boldsymbol{\Omega}_{\text{p}}) = w(\boldsymbol{\Omega}_{\text{p}})$, and the second component is

$$w_2(\boldsymbol{\Omega}_{\text{p}}) = \sqrt{w_1^2(\boldsymbol{\Omega}_{\text{p}}) + \sigma_{\text{scan}}^2}. \tag{5.32}$$

If the processing dwell is significantly less than an antenna beamwidth, this effect may be ignored.

5.3.5 Simulation examples

In this section, both frequency and time domain implementations are demonstrated with key simulated parameters measured to verify the simulation accuracy. The following example considers a high-resolution radar mode suitable for detecting small targets with the radar scanning at 20 rpm and a PRF of 1,500 Hz. The azimuth 3 dB beamwidth is 3° and the centre frequency and bandwidth are 10 GHz and 200 MHz, respectively. The radar is scanning in a clockwise direction with the radar platform travelling North at 100 m/s with the wind and swell from the West. The bimodal spectrum is used with a K-distributed amplitude and the filter order, $L_0 = 32$. Other radar, environmental and implementation parameters are given in Table 5.1.

Table 5.1 Simulated data characteristics

Parameter	Value
Radar scan rate	20 rpm
Aircraft speed	100 m/s
Azimuth 3 dB beamwidth (two-way)	3°
Centre frequency	10 GHz
Bandwidth	200 MHz
Pulse repetition frequency	1 500 Hz
No. of pulses a target will be in the beam	37
Grazing angle	30°
Sea state	3
Radar heading	0°
Wind and swell direction	270°
Amplitude PDF model	K-distribution
Spectral model	Bimodal

Inserting realistic target models into a simulated radar scene can be quite involved. For the purpose of verifying detection algorithms, point targets are commonly used with appropriate target power, fluctuation, point spread function (PSF) and spectrum. For example, the target radar cross section, A, can be modelled with a Swerling fluctuation and the target Doppler frequency can be modelled with a sinusoidal function. For a target located at azimuth θ_0 (or slow-time $t_0 = \theta_0/\dot{\theta}_{\text{scan}}$), frequency f_0 and range, R_0, the target model is given by

$$S(t, R) = A\sqrt{S_t(t - t_0)S_R(R - R_0, f_0)} \exp(j2\pi t f_0) \tag{5.33}$$

where the slow-time PSF can be formed using a Gaussian approximation for the beampattern,

$$S_t\left(t = \frac{\theta}{\dot{\theta}_{\text{scan}}}\right) = \exp\left(-\frac{4\ln 2\theta^2}{\theta_{3\text{dB}}^2}\right). \tag{5.34}$$

As a result, if there are more pulses illuminating a target, the extent of the PSF in the Doppler domain will be smaller. For the range or fast-time dimension, the extent of the PSF is determined by the radar bandwidth, B, and pulse width, T_p. It can be modelled using a slice of the ambiguity function for a linear frequency modulated waveform at the target Doppler frequency [29],

$$S_R\left(R = \frac{c_0 t_f}{2}, f_0\right) = \left|\left(1 - |t_f|/T_p\right) \frac{\sin\left(\pi T_p(\mu_0 t_f + f_0)\left(1 - |t_f|/T_p\right)\right)}{\pi T_p(\mu_0 t_f + f_0)\left(1 - |t_f|/T_p\right)}\right|^2 \tag{5.35}$$

where t_f is fast-time, c_0 is the speed of light and $\mu_0 = B/T_p$.

Figures 5.5 and 5.6 show a comparison of the frequency and single-channel time domain simulation outputs. The first figure shows the forward looking case which corresponds to cross-wind ($\theta_{\text{plat,deg}} = 0°$, $\psi_{\text{deg}} = 270°$) and the second figure shows the side looking case which is in the up-wind direction ($\theta_{\text{plat,deg}} = 90°$, $\psi_{\text{deg}} = 0°$). A synthetic target has been injected at range $R_0 = 187$ m, frequency $f_0 = -300$ Hz and an azimuth, $\theta_{0,\text{deg}}$, which is 2° more than the radar-look direction. As expected, the forward looking case has a narrower spectrum, while the side looking case is broadened by the convolution between the azimuth beampattern and the Doppler spectrum of the clutter.

To verify the accuracy of the time domain simulation, a block of data has been simulated for an entire scan period. For every 10° in azimuth, a window of pulses has then been extracted from the data corresponding to a 20° azimuth region. The spatial correlation length, shape and the spectral width are then measured and compared to the desired value. Figure 5.7 shows this comparison for both time domain simulation cases, where $I = 50$ clutter patches have been used for the simulation. While there is some fluctuation in the estimates, in general there is a good match between the measured and desired values.

To test the multichannel aspects of the simulation, a linear array of $L = 4$ channels has been simulated with $d = \lambda/2$ spacing between the phase centres. A side looking scenario is considered and the Fourier power spectra, $G_{\text{FP}}(\cdot)$ and optimal (minimum variance) spectra, $G_{\text{MV}}(\cdot)$, calculated for different steering angles by

$$G_{\text{FP}}(\theta,f) = \mathbf{s}^H(\theta,f)\mathbf{R}\mathbf{s}(\theta,f), \tag{5.36}$$

$$G_{\text{MV}}(\theta,f) = \frac{1}{\mathbf{s}^H(\theta,f)\mathbf{R}^{-1}\mathbf{s}(\theta,f)} \tag{5.37}$$

where the space–time steering vector is the Kronecker product,

$$\mathbf{s}(\theta,f) = \mathbf{s}_\theta(\theta) \otimes \mathbf{s}_t(f). \tag{5.38}$$

The spatial steering vector given by

$$\mathbf{s}_\theta(\theta) = \left[1, \exp\left(-j2\pi\frac{d}{\lambda}\sin\theta\right), \ldots, \exp\left(-j2\pi\frac{d}{\lambda}(L-1)\sin\theta\right)\right] \tag{5.39}$$

and the temporal steering vector by

$$\mathbf{s}_t(f) = \frac{1}{\sqrt{N}}[1, \exp(-j2\pi f T_r), \ldots, \exp(-j2\pi f(N-1)T_r)]^T \tag{5.40}$$

where T_r is the pulse repetition interval. The covariance matrix, $\mathbf{R}$, can be estimated using a sample covariance matrix estimate for the kth range bin and then averaging

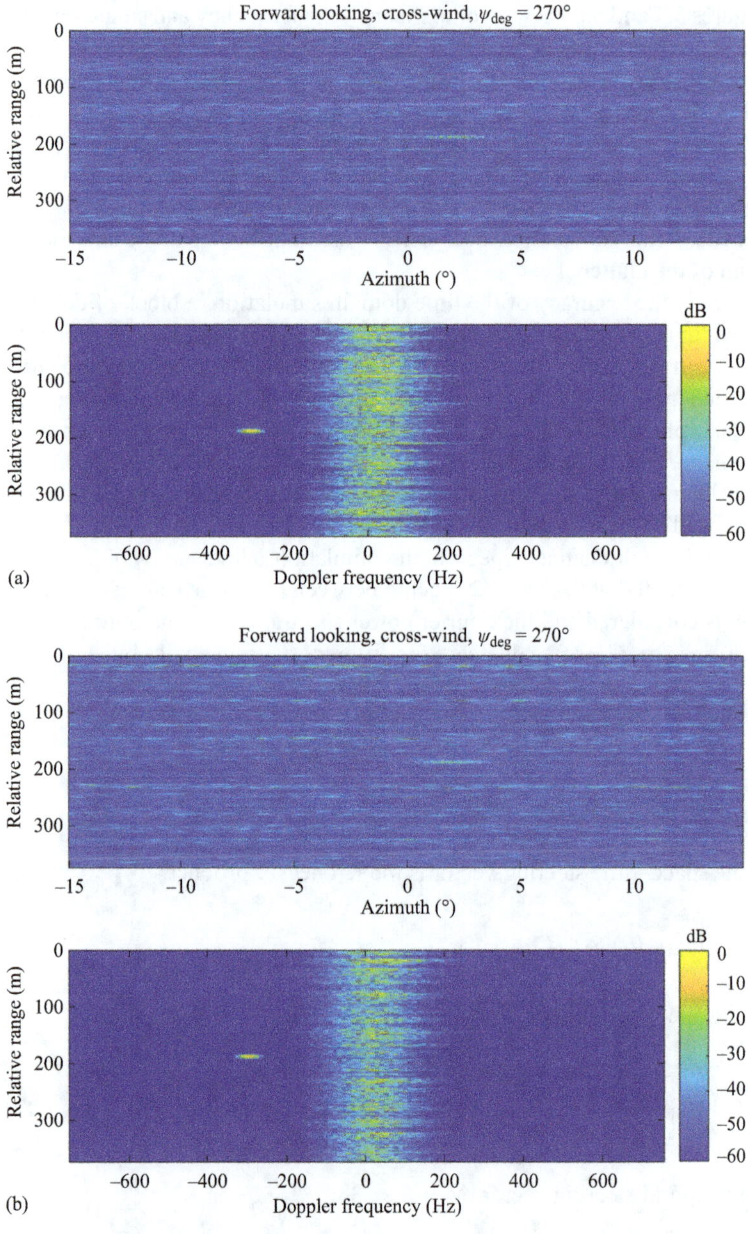

Figure 5.5 Forward looking examples with a target at range $R_0 = 187$ m, frequency $f_0 = -300$ Hz and azimuth, $\theta_{0,\mathrm{deg}} = 2°$: (a) frequency domain simulation and (b) time domain simulation [19].

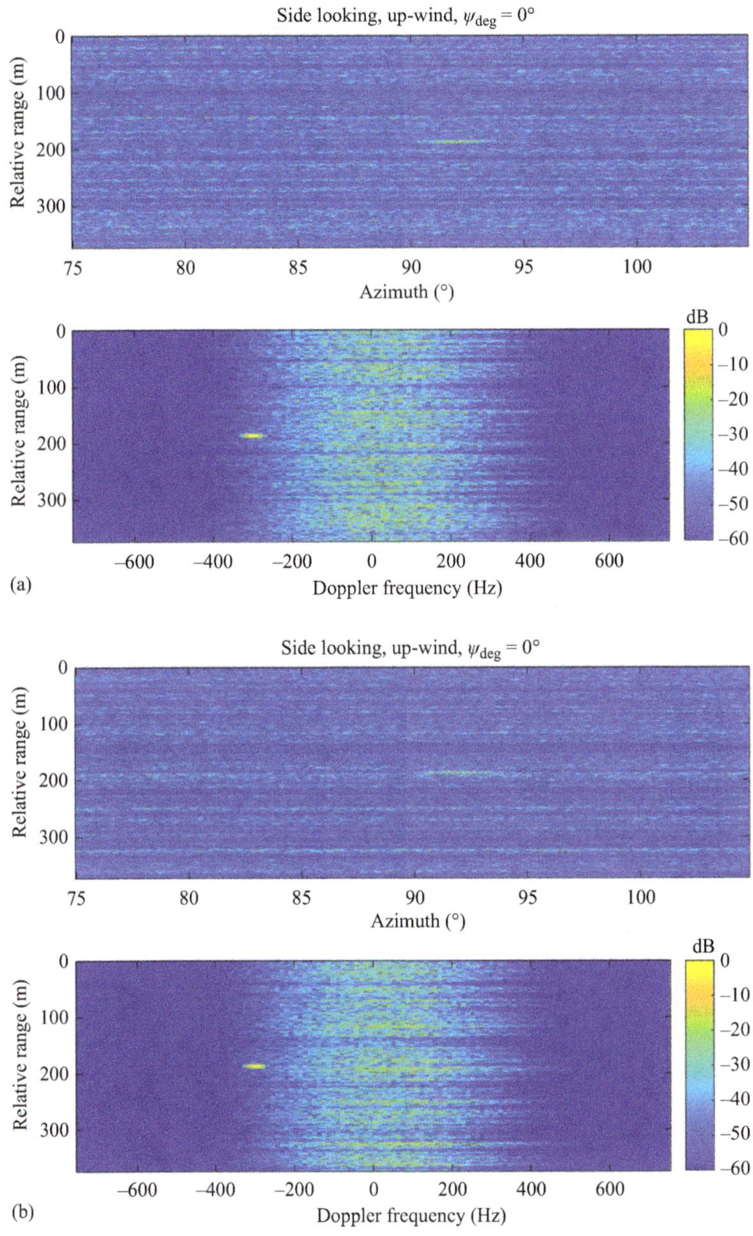

Figure 5.6 Side looking examples with a target at range $R_0 = 187$ m, frequency $f_0 = -300$ Hz and azimuth, $\theta_{0,\text{deg}} = 92°$: (a) frequency domain simulation and (b) time domain simulation [19].

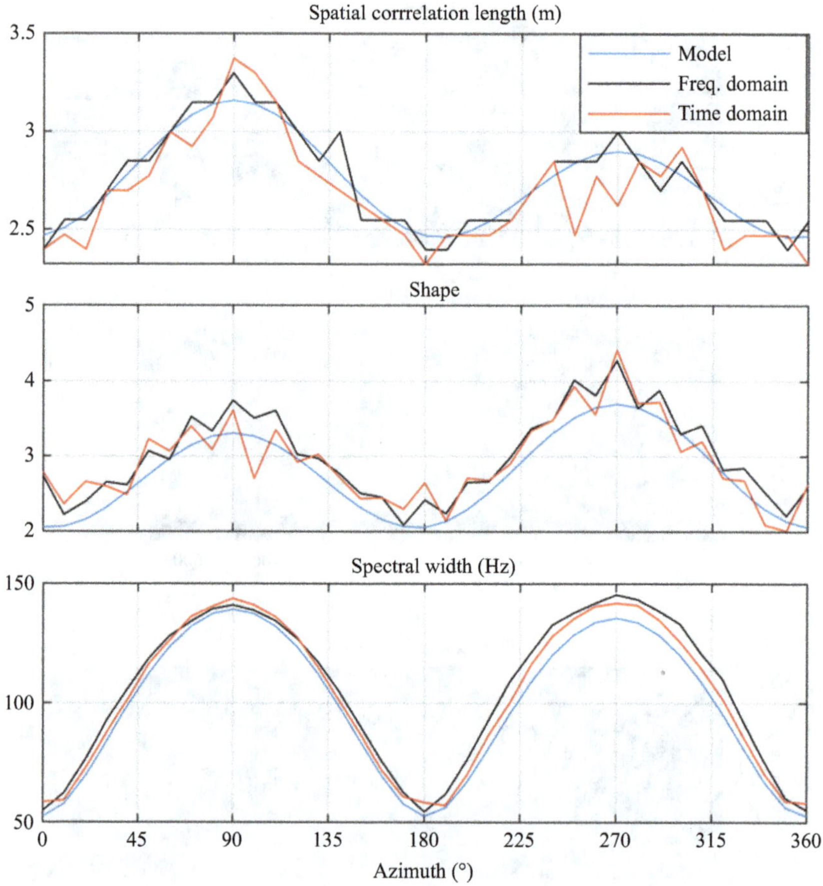

Figure 5.7 Simulation validation: (—) desired model values, (—) frequency domain, (—) time domain [19]. © Commonwealth of Australia 2020

over K range bins, excluding the cell under test. This space–time data vector, $\mathbf{h}$, is formed by stacking the L spatial elements of the input data for each pulse and then stacking N pulses:

$$\widehat{\mathbf{R}}_{\mathrm{SCM}}(k) = \frac{1}{K} \sum_{\substack{n=k-K/2,\\ n\neq k}}^{k+K/2} \mathbf{h}_n \mathbf{h}_n^H. \tag{5.41}$$

Using a CPI of $N = 16$ pulses, Figure 5.8 shows the Fourier and the optimal (minimum variance) power spectra for the target range bin with the sample covariance matrix estimate formed by averaging over the neighbouring $K = 2NL$ range bins. As expected, the optimal power spectrum is much narrower with a diagonal slope indicating the effect of platform motion over the measurement time, and the target is localised at the desired position.

5.4 Other simulation algorithms

In this section, two other sea clutter models are presented with details of how they can be applied for simulation. These include the STLC model in Section 5.4.1 and the compound G2 model in Section 5.4.2. The accuracies of these algorithms are then compared with the evolving Doppler spectrum model in Section 5.4.3 using an example from the Ingara X-band data set. Verification of each model is achieved by comparing a number of statistical measures over a range of swell angles and polarisations.

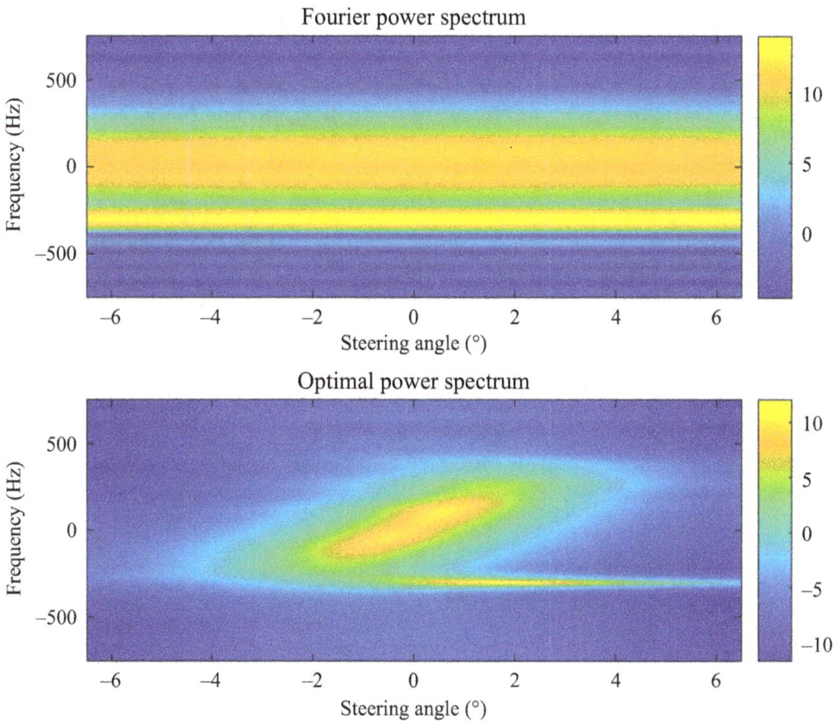

Figure 5.8 Fourier and optimal power spectra for clutter with a target located at $f_0 = -300$ Hz and relative azimuth, $\theta_{0,\text{deg}} = 2°$ [19].

5.4.1 Spatially and temporally limited clutter model

The STLC model is a bimodal formulation [30,31] that represents both the Bragg and fast scattering components, G_{Br} and G_{F}, respectively. It is an extension of the uniform model [31], where each component is represented by a Gaussian spectral response with the mean clutter spectrum given by

$$G_{\mathrm{uni}}(f) = \alpha G_{\mathrm{Br}}(f) + (1 - \alpha) G_{\mathrm{F}}(f) \tag{5.42}$$

where each Gaussian spectrum is represented by

$$G(f) = \frac{1}{\sqrt{2\pi} w} \exp\left(-\frac{(f - m_{\mathrm{f}})^2}{2w^2}\right) \tag{5.43}$$

with mean m_{f} and width w. The amplitude distribution for each component is then gamma distributed with the final compound distribution given as the sum of two K-distributed random variables, otherwise referred to as the K+K-distribution.

The STLC model improves realism by relating the spatial and temporal domains to achieve a tilted wave spectrum. The Bragg component is associated with scattering off the persistent waves with tilted wave-like features across both range and Doppler, while the fast component models the response from the discrete spikes without range spreading or significant Doppler spread. Also due to the representation of the spatial domain, it is straightforward to include multiple phase centres in the model.

To achieve the wave-like patterns observed in some Doppler spectra, the Bragg scattering component requires a model of the underlying wave pattern of the sea. An example is shown in the top plot of Figure 5.9 as a function of range and azimuth with the swell spacing or gravity wavelength given by λ_{G}. The azimuth angle, θ, is defined as the angular change between the radar position and scene centre over the collection time. The wave pattern can greatly affect the result of the STLC Doppler spectrum and is dependent on the wavelength and the range and azimuth extent of the data being characterised. For this example, the wave pattern is defined over 75 m in range and 3.1° in azimuth.

The Bragg component, $\mathbf{c}_{\mathrm{Br}}$, is modelled as a complex Gaussian random variable of size $N \times 1$ with probability

$$P_x(\mathbf{c}_{\mathrm{Br}}) = \frac{1}{(\pi \tau_{\mathrm{Br}})^N |\mathbf{R}_{\mathrm{Br}}|} \exp\left(-\mathbf{c}_{\mathrm{Br}}^H (\tau_{\mathrm{Br}} \mathbf{R}_{\mathrm{Br}})^{-1} \mathbf{c}_{\mathrm{Br}}\right) \tag{5.44}$$

where τ_{Br} is the mean power of the Bragg component and $\mathbf{R}_{\mathrm{Br}}$ is the covariance. The first step in determining the covariance is to consider each range bin separately and define the location of Bragg scattering events at the minima of each wave slice. These events then have an associated Bragg spectrum $G_{\mathrm{Br}}(f)$ located at sequential positions in Doppler and truncated to a width of N_σ times the Bragg spectral width.

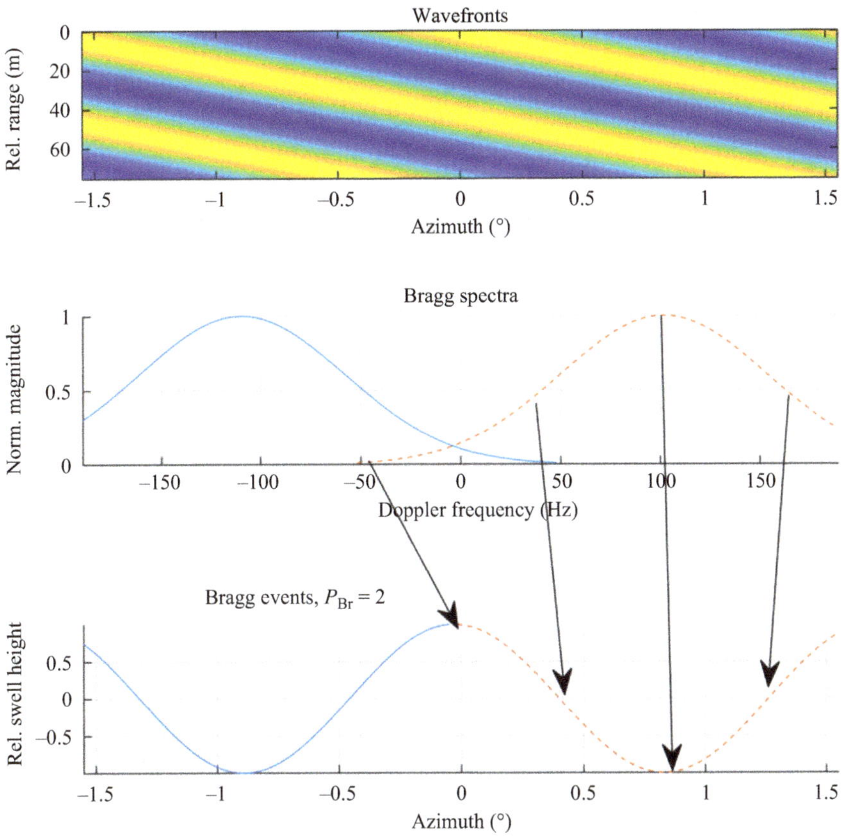

Figure 5.9 Wavefront matching example with wavefronts shown in the top plot, the associated Doppler spectrum in the middle plot and the mapping for a Bragg event shown in the bottom plot. The blue (—) and red (- -) lines indicate different Bragg events [7].

This is shown for the first range bin in the second plot of Figure 5.9 with the mapping between Doppler frequency and azimuth angle are shown by the arrows. The relationship between Doppler frequency and azimuth angle is given by $f = 2v_p \sin\theta/\lambda$. The covariance, $\mathbf{R}_{Br}$, is then the summation of P_{Br} events observed over azimuth in a single range cell,

$$\mathbf{R}_{Br} = \sum_{p=1}^{P_{Br}} \mathbf{R}_{Br}(p) \tag{5.45}$$

where the covariance of each event is given by

$$\mathbf{R}_{\mathrm{Br}}(p) = \sum_{n=1}^{N_{\mathrm{wave}}} \mathbf{R}_{\mathrm{t}}(p,n) \otimes \mathbf{R}_{\theta}(p,n) \tag{5.46}$$

where N_{wave} is the number of points mapped between Doppler and azimuth (i.e. the arrows shown in Figure 5.9). The spatial component is given by

$$\mathbf{R}_{\theta} = A_{\mathrm{az}}(f(p,n))\mathbf{s}_{\theta}^{H}\mathbf{s}_{\theta} \tag{5.47}$$

and the temporal component by

$$\mathbf{R}_{\mathrm{t}}(p,n) = \mathbf{s}_{\mathrm{t}}(f(p,n))\mathbf{s}_{\mathrm{t}}^{H}(f(p,n)). \tag{5.48}$$

For each Bragg event, the associated Doppler frequencies are defined relative to the wave minima shown in the bottom plot of Figure 5.9.

Similar to the uniform model, the fast scattering mechanism is modelled as a K-distribution, except its existence is contingent on a second random variable with a Bernoulli distribution. This random variable is referred to as the probability of occurrence, $P_{\mathrm{occ}}(\theta)$. At each range bin, the received fast scatterer vector of size $N \times 1$ is given by the sum of realisations for each fast scattering event, $\mathbf{c}_{\mathrm{F}}(\theta)$. These each have probability

$$P_y(\mathbf{c}_{\mathrm{F}}(\theta)) = P_{\mathrm{occ}}(\theta) \int_0^{\infty} \frac{1}{(\pi\tau_{\mathrm{F}})^N |\mathbf{R}_{\mathrm{F}}(\theta)|} \exp\left(-\mathbf{c}_{\mathrm{F}}^H(\tau_{\mathrm{F}}\mathbf{R}_{\mathrm{F}}(\theta))^{-1}\mathbf{c}_{\mathrm{F}}\right) P_{\tau}(\tau_{\mathrm{F}}) d\tau_{\mathrm{F}} \tag{5.49}$$

where $P_{\tau}(\tau_{\mathrm{F}})$ is modelled as a gamma distribution with shape parameter ν_{occ}. The covariance matrix is defined for a single azimuth location as $\mathbf{R}_{\mathrm{F}}(\theta) = \mathbf{R}_T \otimes \mathbf{R}_{\theta}$, with the temporal component defined by

$$\mathbf{R}_T = \int_f G_{\mathrm{F}}(f)\mathbf{s}_{\mathrm{t}}(f)\mathbf{s}_{\mathrm{t}}^H(f)df \tag{5.50}$$

and the spatial component was defined in (5.47).

Simulation of the Bragg component includes the following steps:

- Create a realisation of the underlying wave pattern over a defined range and azimuth extent. The swell spacing, λ_{G}, can be determined by measuring the dominant oscillation in the up-swell spatial correlation.

- Determine the Doppler mapping for each event.
- Form the corresponding covariance matrix $\mathbf{R}_{\mathrm{Br}}$.
- Generate correlated complex Gaussian realisations, $\mathbf{c}_{\mathrm{Br}}$, to model the fast scattering events with mean power determined by τ_{Br}.

Generation of the fast scattering component then involves the following steps:

- Generate realisations of the probability of occurrence across a range and angle grid.
- Form the spatial–temporal covariance matrix $\mathbf{R}_{\mathrm{F}}(\theta)$ for each event location.
- Generate correlated complex Gaussian realisations, $\mathbf{x}_{\mathrm{F}}$, to model the fast scattering events.
- Generate gamma realisations, τ_{F}, for each location to model the texture and multiply with the correlated complex Gaussian realisations, $\mathbf{c}_{\mathrm{F}}(\theta) = \sqrt{\tau_{\mathrm{F}}}\mathbf{x}_{\mathrm{F}}$.
- For each range bin, sum the realisations across azimuth, $\mathbf{c}_{\mathrm{F}} = \sum_{\theta} \mathbf{c}_{\mathrm{F}}(\theta)$.

The complete STLC time domain signal of size $N \times 1$ is then given by the coherent sum of the two components with additive Gaussian noise, $\boldsymbol{\xi}$,

$$\mathbf{h}_{\mathrm{STLC}} = \mathbf{c}_{\mathrm{Br}} + \mathbf{c}_{\mathrm{F}} + \boldsymbol{\xi}. \tag{5.51}$$

Note that the Doppler broadening due to platform motion is accounted for through the azimuth antenna beampattern in the spatial covariance.

5.4.2 Compound G2 model

The compound G2 model was presented in [32] and further extended in [8]. It comprises features from both the STLC and evolving Doppler models. For example, the texture is modelled as the sum of Bragg and fast components, $\tau = \tau_{\mathrm{Br}} + \tau_{\mathrm{F}}$, with each represented by a correlated gamma random variable. The overall texture is then distributed as the sum of the independent gamma variates,

$$P_{\tau}(\tau) = \frac{b_{\mathrm{Br}}^{\nu_{\mathrm{Br}}} b_{\mathrm{F}}^{\nu_{\mathrm{F}}} \tau^{\nu_{\mathrm{Br}}+\nu_{\mathrm{F}}-1}}{\Gamma(\nu_{\mathrm{Br}} + \nu_{\mathrm{F}})} e^{-b_{\mathrm{Br}}\tau} {}_1F_1(\nu_{\mathrm{F}}, \nu_{\mathrm{Br}} + \nu_{\mathrm{F}}; (b_{\mathrm{Br}} - b_{\mathrm{F}})\tau) \tag{5.52}$$

where $\nu_{\mathrm{Br}}, \nu_{\mathrm{F}}$ and $b_{\mathrm{Br}}, b_{\mathrm{F}}$ are the shape and scale parameters for the Bragg and fast components, respectively, and ${}_1F_1(\cdot)$ is the confluent hypergeometric function. The compound Gaussian form is the same as the coherent K+K-distribution and is referred to as the compound G2 model,

$$P_z(z) = \int_0^{\infty} \frac{1}{\tau + p_{\mathrm{n}}} \exp\left(-\frac{z}{\tau + p_{\mathrm{n}}}\right) P_{\tau}(\tau) d\tau. \tag{5.53}$$

There is no analytic solution for this integral and it must be evaluated using numerical integration. For the spatial correlation, a variation of the model presented in [33] is used

$$\rho_{\mathrm{spat}}(R) = \alpha \exp\left(-\frac{R}{R_{\mathrm{L}}}\right) + (1-\alpha)\exp\left(-\frac{R}{R_{\mathrm{G}}}\right)\cos\left(\frac{2\pi R}{\lambda_G}\right) \tag{5.54}$$

where R_{G} is the correlation length of the gravity wave component, λ_G is the gravity wavelength, R_{L} is the short-range correlation length for the spatial fluctuations and the Bragg component weighting is

$$\alpha = \frac{\langle \tau_{\mathrm{Br}} \rangle}{\langle \tau \rangle} = \left(1 + \frac{\nu_{\mathrm{F}} b_{\mathrm{Br}}}{\nu_{\mathrm{Br}} b_{\mathrm{F}}}\right)^{-1}. \tag{5.55}$$

The overall model for the Doppler spectrum in each range bin is then represented as the sum of two Gaussian components,

$$G_{\mathrm{G2}}(f) = \tau_{\mathrm{Br}} G_{\mathrm{Br}}(f) + \tau_{\mathrm{F}} G_{\mathrm{F}}(f). \tag{5.56}$$

The key difference between the compound G2 model and the evolving Doppler model is that the Bragg spectral component is treated as asymmetric and the fast spectral component as symmetric. Similarly to the evolving Doppler spectrum model, there is a correlation between the Doppler frequency and intensity, except that it only applies to the Bragg component in the compound G2 model. Some of the asymmetry in the Bragg component is due to the Doppler-intensity correlation, while some is due to the fast component being centred away from zero Doppler. Changes in the overall spectrum width arise from independent fluctuations in the centre frequencies of the two components. This is in contrast to the evolving Doppler model where fluctuations in the overall spectrum width are modelled separately. Also in common to both previous models is the Gaussian distribution used to represent the spectral components in each range cell. Finally, in the compound G2 model, special attention is paid to the relative phase difference between the Bragg and fast components and between the Bragg centre frequency and its intensity.

In each range cell, the Bragg component is represented by a Gaussian spectral shape with width, u_{Br}, and a gamma distributed intensity with shape, ν_{Br}. The centre frequency of the Bragg component is allowed to fluctuate with its value in each range cell drawn from a Gaussian distribution with mean, f_{Br}, and standard deviation, $\sigma_{f_{\mathrm{Br}}}$. The mean is related linearly to the normalised intensity, $\bar{\tau}_{\mathrm{Br}}$, with parameters A_{c} and B_{c},

$$f_{\mathrm{Br}} = A_{\mathrm{c}} + B_{\mathrm{c}} \bar{\tau}_{\mathrm{Br}}. \tag{5.57}$$

Averaged over many range cells, the Bragg component is asymmetric in shape with the degree of asymmetry depending on B_c and ν_{Br}. The component centroid is

$$\bar{f}_{Br} = A_c + B_c\left(1 + 1/\nu_{Br}\right) \tag{5.58}$$

with an overall width

$$w_{Br} = \sqrt{u_{Br}^2 + \sigma_{f_{Br}}^2 + B^2\left(1 + 1/\nu_{Br}\right)/\nu_{Br}}. \tag{5.59}$$

A correlation in the form (5.54) is also applied to the fluctuations in centre frequency over range.

In each range cell, the fast component is represented by a Gaussian spectral shape with width u_F, a gamma distributed intensity with shape ν_F and a centre frequency normally distributed with mean f_F and standard deviation σ_{f_F}. There is no correlation between the fast component centre frequency and intensity, but there is correlation present along range in the centre frequency which can be modelled by an exponential decay with the correlation length, R_C. Averaged over many range cells, the fast component is Gaussian in shape, with centre frequency $\bar{f}_F$ and standard deviation,

$$w_F = \sqrt{u_F^2 + \sigma_{f_F}^2}. \tag{5.60}$$

Note that if there are no fluctuations in centre frequency, the uniform model for the component is obtained with $w_F = u_F$.

The steps for simulating the Bragg component of the compound G2 model include the following:

- Generate a sequence of correlated normal variates using the model in (5.54).
- For each range cell, use the MNLT to create a sequence of gamma variates, τ_{Br}, with shape parameter, ν_{Br}, representing the Bragg component intensity in each range cell.
- Generate a sequence of correlated normal variates for the Bragg centre point, with a correlation in range given by the second term in (5.54). In each range cell, the Bragg component is centred at $A_c + B_c\bar{\tau}_{Br}$ with standard deviation $\sigma_{f_{Br}}$.
- Apply a circular shift to the sequence of centre points, to reproduce the phase difference between the Bragg component centre frequency and intensity.

The steps for simulating the fast component are as follows:

- Take the same sequence of correlated normal variates used for the Bragg component and apply a different MNLT to create a sequence of gamma variates, τ_F, with shape parameter, ν_F, representing the fast component intensity in each range cell.

- Apply a circular shift to the sequence, τ_F, to reproduce the phase difference between the intensities of the two components.
- Generate another sequence of correlated normal variates for the fast component centre point with a negative exponential correlation in range, mean, $\bar{f}_F$, and standard deviation, σ_{f_F}.

These can then be combined to form the final signal as follows:

- Form the overall spectrum in each range cell as the weighted sum of two Gaussian components with widths u_{Br}, u_F and intensities τ_{Br} and τ_F, respectively. The centre frequency of each Gaussian is obtained from the sequences of random variates described earlier.
- In order to model the spreading due to radar platform motion, convolve the spectrum with the two-way azimuth beampattern. A circular convolution is used to reproduce aliasing of the Doppler spectrum.
- Generate a complex Gaussian distributed realisation for the speckle, $\mathbf{x}$, with unit mean and form the compound Gaussian clutter signal.
- Generate a complex Gaussian distributed realisation for the thermal noise, $\boldsymbol{\xi}$, with variance, p_n/N. The final signal model of size $N \times 1$ is then

$$\mathbf{H}_{G2} = \sqrt{\mathbf{G}_{G2}}\mathbf{x} + \boldsymbol{\xi} \tag{5.61}$$

- The time domain signal is then obtained by an inverse Fourier transform of (5.61).

5.4.3 Algorithm comparison

To compare these three algorithms, a dual-polarised run from the Ingara X-band data set has been used which covers 31°–37° in grazing or 600 m in range. The wind and swell directions are aligned and the sea comes from a Douglas sea state 3–4 with a wind speed of 10.2 m/s and wave height of 1.2 m. A CPI of $N = 64$ pulses is used with a −55 dB Dolph–Chebyshev window applied prior to any measurement of the Doppler domain. The Doppler spectrum has also been centred due to an unknown offset during the data collection and all model fits and simulations are based on the observed Doppler spectrum which includes broadening due the radar platform motion. To evaluate the two models, the Ingara parameters are estimated for both the horizontal transmit and receive (HH) and vertical transmit and receive (VV) polarisations and for three swell angles, $\theta_{sw} = 0°, 45°$ and $90°$. Table 5.2 summarises the key parameters for each model, with the exception of the spatial correlation parameters for the compound G2 model, which are given in Table 5.3.

Details on how to estimate the parameters for the evolving Doppler spectrum model were originally given in [4,34]. For the K-distribution shape, the $z \log z$ estimator has been used [35], while the spatial decorrelation length is measured at the point where the spatial correlation decays to $1/e$. To determine the Doppler spectrum

Table 5.2 Estimated model parameters for each model

Polarisation	**HH**			**VV**		
Swell angle (°)	**0**	**45**	**90**	**0**	**45**	**90**
Evolving Doppler model						
Bragg weighting, α	0.1	0.2	0.8	1	0	0.9
K-distribution shape, ν	2.7	3.6	1.4	29	10	21
Spatial decorrelation length (m), R_L	1.1	1.4	0.8	0.7	0.8	0.6
Mean bimodal fit mean (Hz), A_c	−39	−28	−8	−15	−12	7
Mean bimodal fit slope (Hz), B_c	30	24	7	17	26	−3
Bimodal threshold (Hz), τ_{bi}	3	3	9	3.7	0.7	4
Std. of fluct. around mean (Hz), w_{bi}	28	26	28	22	20	17
Spectrum width mean (Hz), m_w	56	54	56	52	50	51
Spectrum width std. (Hz), σ_w	19	19	23	15	13	12
STLC model						
Bragg weighting, α	0.93	0.82	0.38	0.93	0.61	0.29
Bragg centre freq. (Hz), f_{Br}	−9	−7	−47	−9	1	44
Bragg width (Hz), w_{Br}	63	59	30	63	42	49
Fast centre freq. (Hz), f_F	89	43	43	81	32	−20
Fast width (Hz), w_F	78	70	57	71	88	66
Prob. of fast occ. ($\times 10^{-3}$), P_{occ}	8.8	8.8	8.5	7.0	8.4	7.3
Fast shape, ν_{occ}	11	13	10	37	37	36
Compound G2 model						
Bragg weighting, α	0.72	0.67	0.78	0.77	0.91	0.83
Bragg shape parameter, ν_{Br}	5.0	5.3	5.6	9.3	9.5	9.5
Bragg intensity phase shift (°)	130	4	–	−3	107	–
Bragg centre freq. mean (Hz), f_{Br}	−19	−16	−2	−11	0.7	6
Bragg centre freq. corr. (Hz), B_c	65	46	4	30	21	4
Bragg width (Hz), w_{Br}	56	55	58	54	58	58
Bragg uniform width (Hz), u_{Br}	44	49	55	51	55	55
Bragg centre freq. phase shift (°)	98	−20	–	−35	94	–
Fast shape parameter, ν_F	0.16	0.24	0.079	0.65	0.10	0.34
Fast intensity phase shift (°)	218	275	–	174	275	–
Fast mean centre freq. (Hz), f_F	62	27	19	30	10	−4
Fast width (Hz), w_F	59	75	67	71	56	42
Fast uniform width (Hz), u_F	30	38	32	36	29	30

Table 5.3 Estimated spatial correlation parameters for the compound G2 model

Swell angle (°)	**0**	**45**	**90**
Gravity wavelength (m), λ_G	30	25	–
Gravity correlation length (m), R_G	300	300	–
HH short range correlation lengths (m), R_L	2.9	2.9	2.7
VV short range correlation lengths (m), R_L	3.0	5.0	5.0

parameters, the mean and variance of the spectrum needs to be measured at each range bin. For the nth Doppler and kth range bin, these can be found from the first two moments of the Doppler spectrum, $H(n,k)$ [36],

$$\widehat{m}_{\mathrm{f}}(k) = \frac{1}{h_{\mathrm{n}}}\sum_{n=1}^{N}\left(\frac{nf_{\mathrm{r}}}{N}-\frac{f_{\mathrm{r}}}{2}\right)H(n,k), \tag{5.62}$$

$$\widehat{w}_{\mathrm{sp}}^{2} = \frac{1}{h_{\mathrm{n}}}\left(\sum_{n=1}^{N}\left(\frac{nf_{\mathrm{r}}}{N}-\frac{f_{\mathrm{r}}}{2}-\widehat{m}_{\mathrm{f}}(k)\right)^{2}H(n,k)-p_{\mathrm{n}}\frac{f_{\mathrm{r}}^{3}}{12}\right) \tag{5.63}$$

where w_{sp} is the width of the total Doppler spectrum, N is the CPI length, f_{r} is the PRF, $h_{\mathrm{n}} = \left(\sum_{n=1}^{N} H(n,k) - p_{\mathrm{n}} f_{\mathrm{r}}\right)$ is the normalising factor and the last term in (5.63) is a correction for the thermal noise contribution to the spectral width. To get an accurate measurement of underlying clutter spectra width, it is important to account for broadening due to motion of the aircraft and the CPI length. If the broadening factor is modelled as a Gaussian function, then its standard deviation is given by (5.19) and the underlying spectral width of the clutter can then be determined by

$$\widehat{w}^{2}(k) = \widehat{w}_{\mathrm{sp}}^{2}(k) - \sigma_{\mathrm{plat}}^{2} - \sigma_{\mathrm{CPI}}^{2}. \tag{5.64}$$

The values of $\alpha, A_{\mathrm{c}}, B_{\mathrm{c}}$ and τ_{bi} are then determined by a joint least-squares minimisation between the data and relationships in (5.17). The variance of the residual centre point, σ_{bi}, can then be determined by subtracting the model fit in (5.16) with the measured centre point values.

For the STLC model, a collection time of 2 s is used which results in the wave pattern shown in the top part of Figure 5.9. The uniform model parameters, $\alpha, f_{\mathrm{Br}}, w_{\mathrm{Br}}, f_{\mathrm{F}}$ and w_{F} are determined using a least-squares fit of the model in (5.42) with an equivalent relationship for the spectrum variance. For the Ingara dual-polarised data set, the uniform model is slightly extended using the formulation in [30] where the spectral components have a common centre frequency and width. This technique for fitting the model was developed in [32] and does not require direct use of the effective shape parameters. An example model fit for the HH polarisation at $\theta_{\mathrm{sw}} = 45°$ is shown in Figure 5.10. The probability of occurrence of the discrete spikes, P_{occ}, can be determined using the method described in [37], where the intensity data is thresholded at 5 standard deviations above the mean. Figure 5.11 shows an example of the discrete spike distribution for the up-swell direction. A gamma model with shape ν_{occ} has then been determined using a maximum likelihood estimator and overlayed on the figure. However, this does not appear to properly model the data distribution as suggested in [6]. The swell spacing or wavelength in range is determined by fitting the model (5.54) to the spatial correlation in the up-swell direction. For the Ingara dual-polarised data set, the wavelength is estimated as $\lambda_{\mathrm{G}} = 30$ m. Finally, the number of points used to map between the Bragg events and the Bragg spectrum, $N_{\mathrm{wave}} = 30$ and the truncated width of the Bragg component is $N_{\sigma} = 3$ times the standard deviation of the Bragg component estimated by the uniform fit.

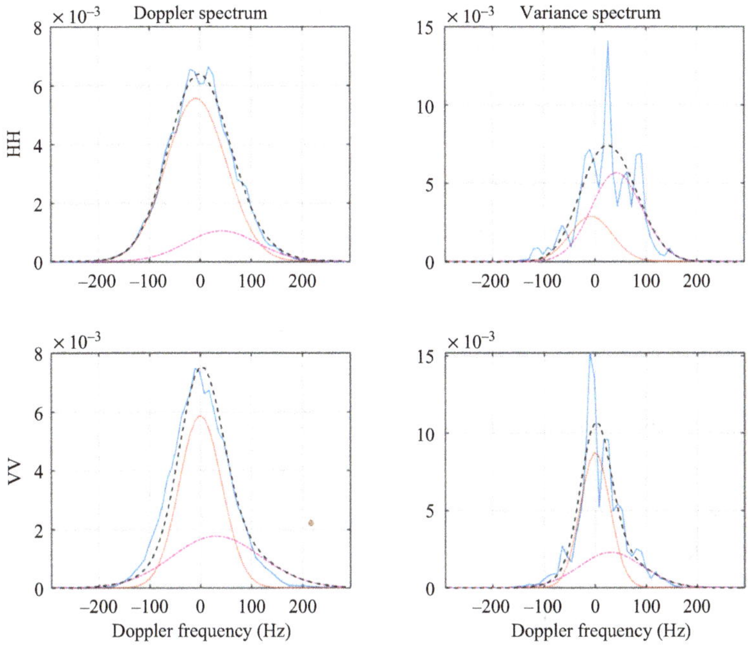

Figure 5.10 Uniform fit example for swell angle $\theta_{sw} = 45°$. (—) Ingara data, (···) Bragg component, (-·-) fast component, (- -) complete model fit [7]. © Commonwealth of Australia 2019

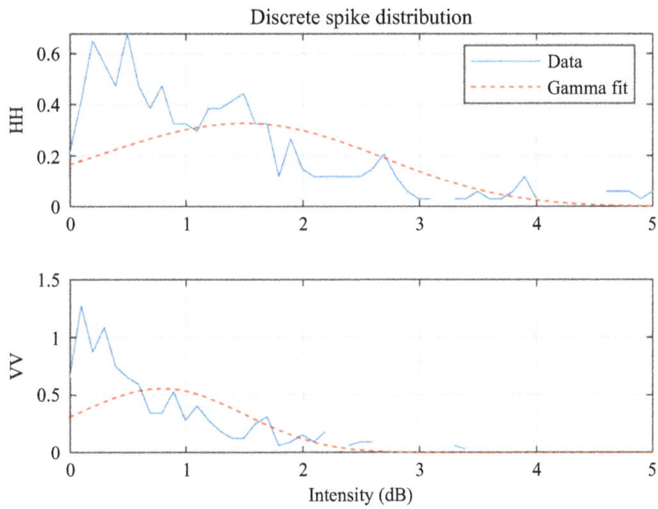

Figure 5.11 Discrete spike distribution example for swell angle $\theta_{sw} = 45°$. (—) Data PDF, (-·-) gamma model fit [7]. © Commonwealth of Australia 2019

For the compound G2 model, 13,050 spectra were generated using the 64 pulse CPI. The distribution of clutter power, or texture, was obtained from the intensities of these individual spectra and modelled with the G2 PDF in (5.52). The shape parameters ν_{Br}, ν_F and the Bragg component weighting, α, were estimated from this distribution using the constrained maximum likelihood estimator [8]. The component centre frequencies, f_{Br}, f_F, widths, w_{Br}, w_F, and uniform widths, u_{Br}, u_F, were estimated by simultaneous fitting of the mean Doppler and variance spectra to the spectrum model, as described in [8]. In these fits, the shape parameters and Bragg weighting were fixed to the values estimated from the texture distribution. The Doppler-intensity correlation parameter, B_c, and the phase shifts for the component intensities and slow centre frequency were estimated from cyclic averaging of the Doppler spectra over the gravity wavelength. The spatial correlation was estimated using the model in (5.54) for the three swell directions considered here, 0°, 45° and 90°, with the exception of the 90° cross-swell case, where the cosine term does not apply. The results for each swell angle are shown in Table 5.3. The shorter gravity wavelength of 25 m that is estimated for the 45° case is likely the result of a complicated interplay between the Bragg and fast components when the radar beam is not aligned with the waves.

This section now considers realisations of each model using a CPI of 64 pulses and 600 range bins. To assess the model over the chosen swell directions and polarisations, data has been simulated for each case and with three metrics compared with the original Ingara dual-polarised data. The first is the Bhattacharyya distance (BD) described in Appendix A, which determines the similarity between the amplitude distributions of the real and simulated data. The second metric is the spatial decorrelation length and the third determines the accuracy of the frequency domain representation by measuring the root mean square (RMS) error between the real and estimated K-distribution shapes in the endo-clutter region of the frequency domain. The Doppler spectrum plots are shown in Figures 5.12–5.14 for swell directions, $\theta_{sw} = 0°$, 45° and 90° with the measured metrics in Table 5.4.

The evolving Doppler spectrum model has produced a good representation of the data, although it does not reproduce the tilted range/Doppler pattern seen in the data. The BDs are low and the K-distribution shape and spatial decorrelation lengths are also quite close for each result. For the STLC model, the up-swell results, $\theta_{sw} = 0°$, have produced strong Bragg components at each wave location with a broad fast component. These results also have an extremely large BD with low spatial decorrelation and shape values that are very different to the data. For the $\theta_{sw} = 45°$ results, the STLC model has produced a tilted wave structure in the spectrum and the BD is low indicating a good match with the amplitude distribution. The result with $\theta_{sw} = 90°$ shows artificial looking edges on the spectra which are more pronounced for the VV polarisation. To determine the exact wave structure which corresponds to the real data is extremely difficult as the sea clutter will likely contain a more complicated wave structure than can be represented by the STLC model. For the compound G2 model, the BD is low and there is a better match visually with the tilted wave spectrum observed in the data. However, when looking at the spatial decorrelation lengths and shape parameter, there is more variation than the evolving model. This mismatch is slightly worse for the VV polarisation.

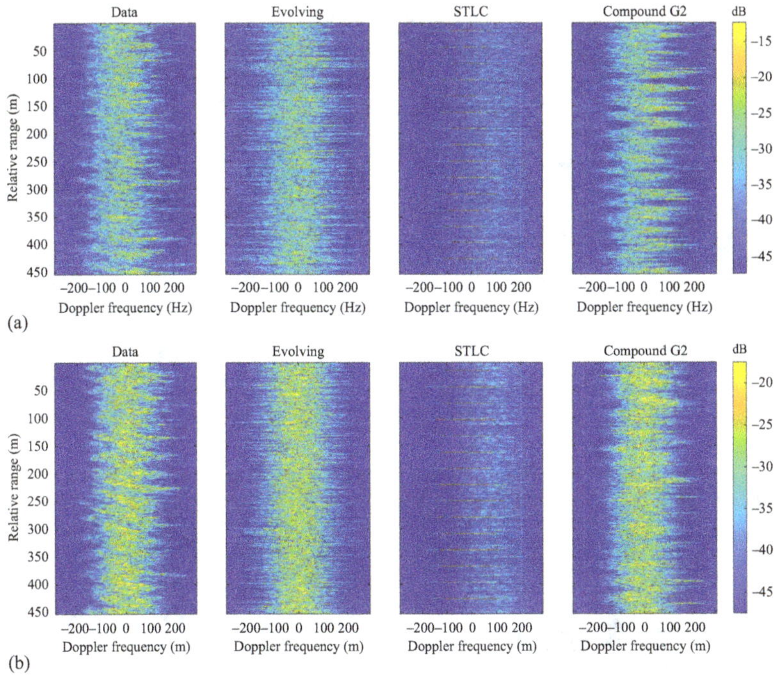

Figure 5.12 Doppler spectrum comparison for swell angle $\theta_{sw} = 0°$: (a) HH and (b) VV [7]. © Commonwealth of Australia 2019

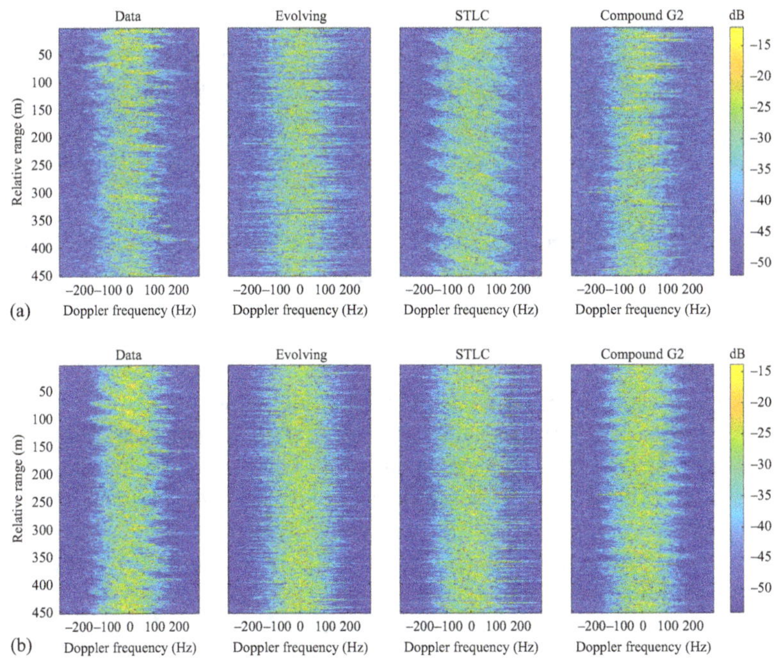

Figure 5.13 Doppler spectrum comparison for swell angle $\theta_{sw} = 45°$: (a) HH and (b) VV [7]. © Commonwealth of Australia 2019

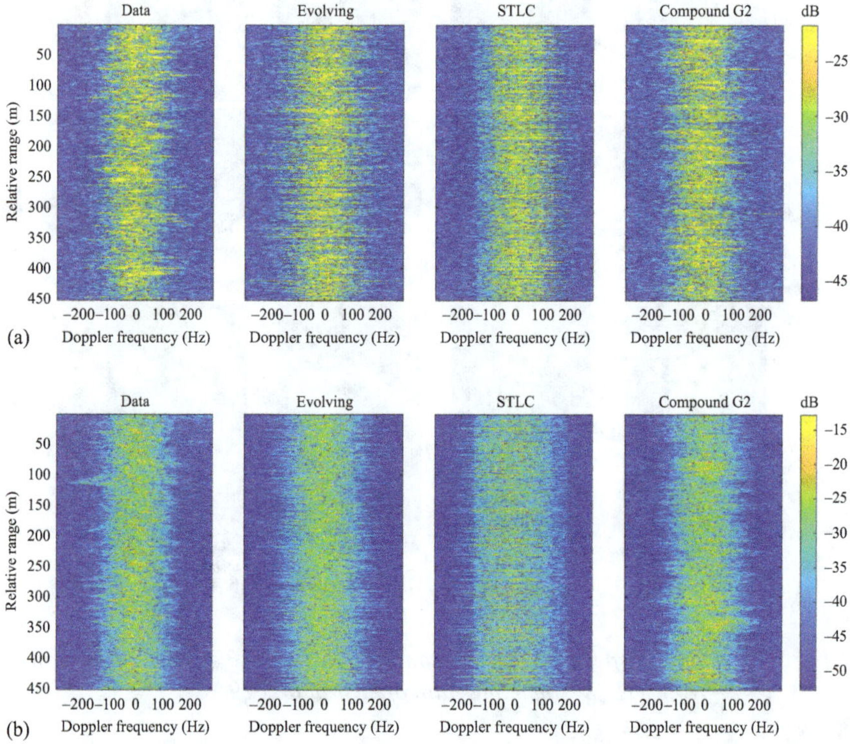

Figure 5.14 Doppler spectrum comparison for swell angle $\theta_{sw} = 90°$: (a) HH and (b) VV [7]. © Commonwealth of Australia 2019

The final results in Figures 5.15–5.17 show the CNR, the coefficient of variation and the K-distribution shape estimates over frequency. Note that the coefficient of variation is determined by the ratio of the standard deviation to the mean of the intensity, $\sigma_z / \langle z \rangle$. For the evolving Doppler model, the majority of estimates are extremely close, although there are some differences in the CNR for the negative Doppler region with $\theta_{sw} = 0°$ and some slightly higher shape values observed for $\theta_{sw} = 0°$ and $\theta_{sw} = 90°$. For the STLC model, the only results that match the data are for $\theta_{sw} = 45°$ with a slightly better match for the VV polarisation. For the other results, there are some frequency regions that match, but not across the entire spectrum. For the compound G2 model, the all three parameters show consistent results for each swell direction. In terms of the RMS shape errors, the compound G2 model has the lowest errors for both polarisations. For the other two models, the evolving Doppler model has lower errors for the HH polarisation, while the STLC model has lower errors for VV.

Table 5.4 Real and simulated parameter comparison

Polarisation	HH			VV		
Swell angle (°)	**0**	**45**	**90**	**0**	**45**	**90**
Data						
Clutter to noise ratio (dB)	16.4	13.6	10.7	23.1	19.7	17.2
K-distribution shape	4.4	3.6	2.6	31.5	10.3	57.7
Spatial decorrelation length (m)	1.1	1.4	0.8	0.7	0.8	0.6
Evolving Doppler model						
K-distribution shape	4.4	4.0	3.0	37.0	13.4	70.3
Spatial decorrelation length (m)	0.9	1.4	0.7	0.4	0.5	0.4
Bhattacharyya distance (dB)	–23.5	–23.6	–23.6	–24.5	–23.8	–24.6
K-distribution shape (freq.) RMS error	1.8	2.3	1.4	6.8	6.2	16.9
STLC model						
K-distribution shape	0.04	12.8	1.2	0.07	3.0	1.0
Spatial decorrelation length (m)	0.3	0.5	0.4	0.3	0.4	0.4
Bhattacharyya distance (dB)	–3.4	–22.3	–20.6	–2.2	–21.2	–13.9
K-distribution shape (freq.) RMS error	2.4	5.0	1.8	2.7	2.1	7.9
Compound G2 model						
K-distribution shape	2.2	3.0	2.3	6.7	7.9	5.4
Spatial decorrelation length (m)	1.5	2.1	1.1	3.8	1.3	3.2
Bhattacharyya distance (dB)	–22.0	–23.6	–23.9	–23.0	–23.7	–22.7
K-distribution shape (freq.) RMS error	0.9	1.2	0.4	1.8	3.4	6.6

5.4.4 Summary

In this section, two models for representing sea clutter were presented. Through the analysis here, the evolving Doppler spectrum and the compound G2 model have consistently produced good results, with the STLC model producing a realistic looking tilted wave spectrum for $\theta_{sw} = 45°$. However, for the up- and down-swell cases, the spectrum did not appear realistic and has poor estimates across frequency. Also, for the STLC results, the overall shape and spatial decorrelation estimates are wrong, indicating that the wave pattern of the data is not being modelled correctly. The compound G2 model produced realistic looking spectrum with the closest match to real data across the frequency domain. Unfortunately, the implementation of this model is difficult due to a large number of parameters to estimate. The evolving Doppler spectrum model produced reasonable results and required a much smaller number of parameters to implement. In terms of the algorithm run times, both the evolving Doppler spectrum and compound G2 models are simple and quick to run, while the STLC model takes an order of magnitude longer due to the effort required

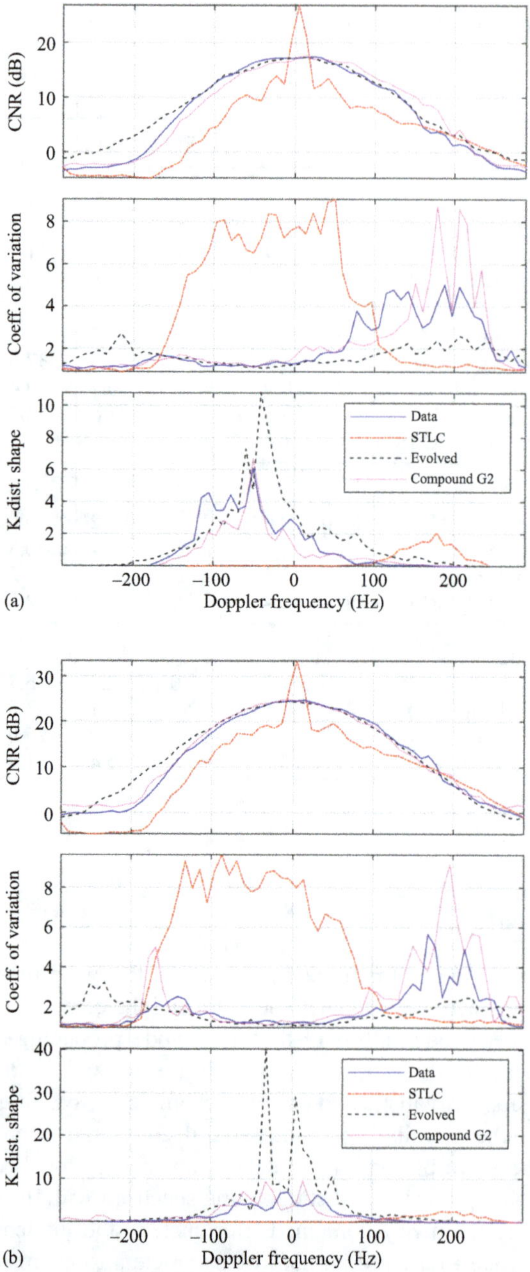

Figure 5.15 *Model comparison for $\theta_{sw} = 0°$ against frequency showing the CNR, the coefficient of variation and the K-distribution shape parameter: (a) HH, $\theta_{sw} = 0°$ and (b) VV, $\theta_{sw} = 0°$. (—) Ingara data, (-·-) STLC model, (- -) evolving Doppler model, (· · ·) compound G2 model [7].*

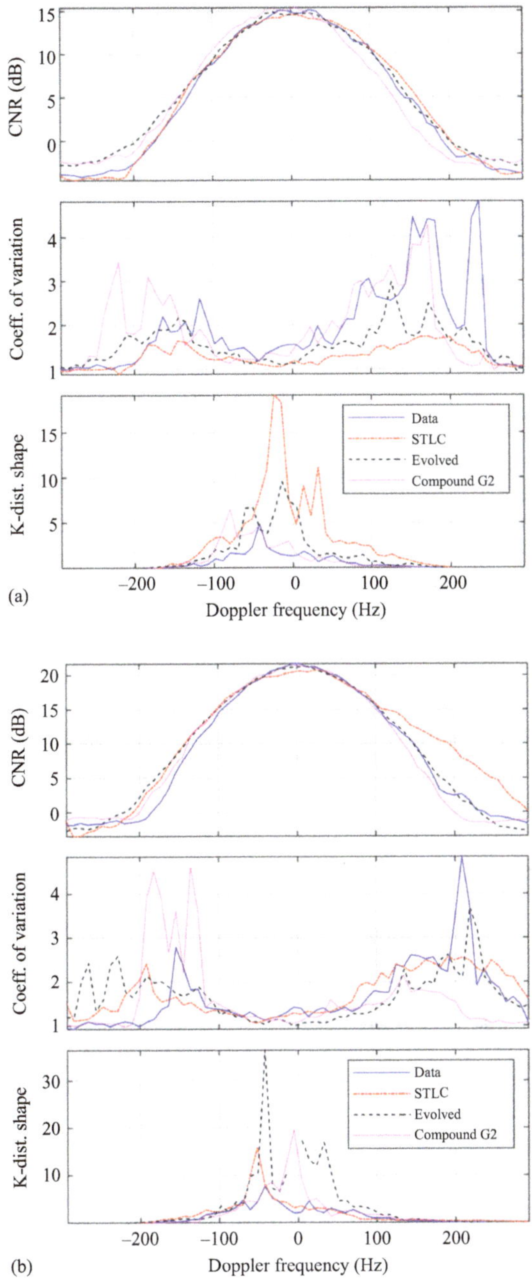

Figure 5.16 Model comparison for $\theta_{sw} = 45°$ against frequency showing the CNR, the coefficient of variation and the K-distribution shape parameter: (a) HH, $\theta_{sw} = 45°$ and (b) VV, $\theta_{sw} = 45°$. (—) Ingara data, (-·-) STLC model, (- -) evolving Doppler model, (···) compound G2 model [7]. *© Commonwealth of Australia 2019*

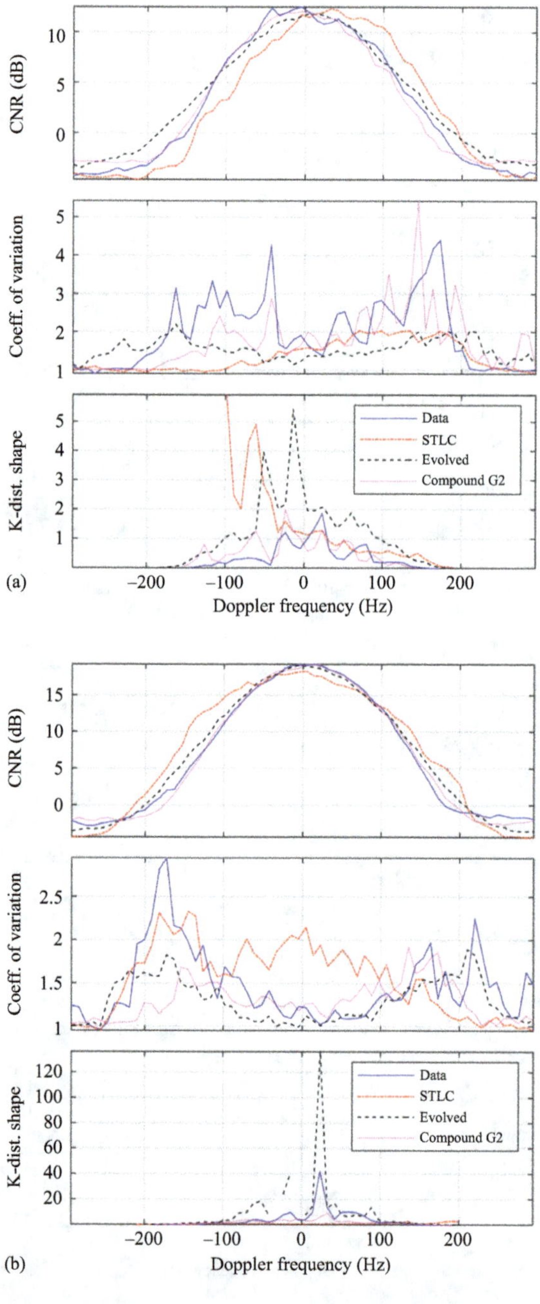

Figure 5.17 Model comparison for $\theta_{sw} = 90°$ against frequency showing the CNR, the coefficient of variation and the K-distribution shape parameter: (a) HH, $\theta_{sw} = 90°$ and (b) VV, $\theta_{sw} = 90°$. (—) Ingara data, (-·-) STLC model, (- -) evolving Doppler model, (· · ·) compound G2 model [7].

to correctly model the wavefronts. For these models to be useful in designing new detection schemes, they need parametric models to relate each model parameter to the environment and collection geometry. Currently, only the evolving Doppler spectrum model has such models as presented in Chapter 4.

References

[1] M. Greco, F. Bordoni, and F. Gini, "X-band sea-clutter nonstationary: influence of long waves," *IEEE Journal of Oceanic Engineering*, vol. 29, no. 2, pp. 269–293, 2004.

[2] G. Davidson, "Simulation of coherent sea clutter," *IET Radar Sonar and Navigation*, vol. 4, no. 2, pp. 168–177, 2010.

[3] S. Watts, "Modeling and simulation of coherent sea clutter," *IEEE Transactions on Aerospace and Electronic Systems*, vol. 48, no. 4, pp. 3303–3317, 2012.

[4] S. Watts, L. Rosenberg, S. Bocquet, and M. Ritchie, "The Doppler spectra of medium grazing angle sea clutter; part 1: Characterisation," *IET Radar Sonar and Navigation*, vol. 10, no. 1, pp. 24–31, 2016.

[5] S. Watts, L. Rosenberg, S. Bocquet, and M. Ritchie, "The Doppler spectra of medium grazing angle sea clutter; part 2: Exploiting the models," *IET Radar Sonar and Navigation*, vol. 10, no. 1, pp. 32–42, 2016.

[6] M. McDonald and D. Cerutti-Maori, "Multi-phase centre coherent radar sea clutter modelling and simulation," *IET Radar Sonar and Navigation*, vol. 11, no. 9, pp. 1359–1366, 2017.

[7] L. Rosenberg and S. Bocquet, "Comparison of bi-modal coherent sea clutter simulation techniques," *IET Radar Sonar and Navigation*, vol. 13, no. 9, pp. 1519–1529, 2019.

[8] S. Bocquet, "Compound G2 model of coherent radar sea clutter," *IET Radar Sonar and Navigation*, vol. 13, no. 9, pp. 1508–1518, 2019.

[9] L. Devroye, *Non-uniform random variable generation*. Springer-Verlag, Berlin, 1986.

[10] K. D. Ward, R. J. A. Tough, and S. Watts, *Sea clutter: Scattering, the K-distribution and radar performance*, 2nd ed. The Institute of Engineering Technology, London, 2013.

[11] M. Rangaswamy, D. Weiner, and A. Ozturk, "Non-Gaussian random vector identification using spherically invariant random processes," *IEEE Transactions of Aerospace and Electronic Systems*, vol. 29, pp. 111–123, 1993.

[12] I. Antipov, "Simulation of sea clutter returns," Defence Science Technology Organisation, Technical Report DSTO-TR-0679, Jun. 1998.

[13] E. Conte, M.-D. Bisceglie, M. Lops, and G. Ricci, "Simulation of correlated random fields with gamma-distributed amplitude for SAR applications," *IEEE Geoscience and Remote Sensing Symposium*, 1991, pp. 2397–2400.

[14] B. C. Armstrong and H. D. Griffiths, "Modelling spatially correlated K-distributed clutter," *Electronic Letters*, vol. 27, no. 15, pp. 1355–1356, 1991.

[15] R. J. A. Tough and K. D. Ward, "The correlation properties of gamma and other non-Gaussian processes generated by memoryless nonlinear transformation," *Journal of Applied Physics*, vol. 32, pp. 3075–3084, 1999.

[16] MATLAB, *9.10 (R2021a)*. The MathWorks Inc., Natick, MA, 2021.

[17] Wolfram Research, Inc., *Mathematica*. Wolfram Research, Inc., Champaign, IL, 2021.

[18] L. Rosenberg, S. Watts, and M. Greco, "Modelling the statistics of microwave radar sea clutter," *IEEE Aerospace and Electronic Systems Magazine*, vol. 34, no. 10, pp. 44–75, 2019.

[19] L. Rosenberg, S. Watts, and S. Bocquet, "Scanning radar simulation in the maritime environment," *IEEE Radar Conference*, 2020, pp. 1–6.

[20] C. J. Oliver, "Representation of radar sea clutter," *Proceedings of the IEE*, vol. 135, no. F, pp. 497–500, 1988.

[21] S. Park and J. Park, "Realistic simulation of mixed sea using multiple spectrum-based wave systems," *Simulation: Transactions of the Society for Modeling and Simulation*, vol. 96, no. 3, pp. 281–296, 2020.

[22] E. A. Caponi, B. M. Lake, and H. C. Yuen, "Hydrodynamic effects in low-grazing angle backscattering from the ocean," *IEEE Transactions on Antennas and Propagation*, vol. 47, no. 2, pp. 354–363, 1999.

[23] J. T. Johnson, R. J. Burkholder, J. V. Toporkov, D. R. Lyzenga, and W. J. Plant, "A numerical study of the retrieval of sea surface height profiles from low grazing angle radar data," *IEEE Transactions on Geoscience and Remote Sensing*, vol. 47, no. 6, pp. 1641–1650, 2009.

[24] S. Zhang, J. Li, Y. Li, P. Wei, and M. Zhang, "Fast realization of 3-D space-time correlation sea clutter of large-scale sea scene based on FPGA: From EM model to statistical model," *IEEE Journal of Selected Topics in Applied Earth Observations and Remote Sensing*, vol. 14, pp. 567–576, 2021.

[25] R. Stuhlmeier and M. Stiassnie, "Deterministic wave forecasting with the Zakharov equation," *Journal of Fluid Mechanics*, vol. 913, pp. 1–21, 2021.

[26] S. Kemkemian, L. Lupinski, J. Degurse, V. Corretja, R. Cottron, and S. Watts, "Performance assessment of multi-channel radars using simulated sea clutter," *International Radar Conference, Radar 2015*, 2015, pp. 1015–1020.

[27] L. Rosenberg and S. Watts, "Coherent simulation of sea-clutter for a scanning radar," *NATO SET-239 Workshop on Maritime Radar Surveillance from Medium and High Grazing Angle Platforms*, 2016, pp. 1–10.

[28] M. I. Skolnik, *Introduction to radar systems*, 3rd ed. McGraw-Hill, New York, 2001.

[29] B. R. Mahafza, *Radar systems analysis and design using Matlab*, 2nd ed. Chapman and Hall, Boca Raton, FL, 2005.

[30] L. Rosenberg, "Characterisation of high grazing angle X-band sea-clutter Doppler spectra," *IEEE Transactions on Aerospace and Electronic Systems*, vol. 50, no. 1, pp. 406–417, 2014.

[31] M. K. McDonald and D. Cerutti-Maori, "Coherent radar processing in sea clutter environments part 1: Modelling and partially adaptive STAP performance," *IEEE Transactions on Aerospace and Electronic Systems*, vol. 52, no. 4, pp. 1797–1817, 2016.

[32] S. Bocquet, "Two component statistical model of coherent radar sea clutter," *International Radar Conference*, 2018, pp. 1–6.

[33] S. Bocquet, L. Rosenberg, and S. Watts, "Simulation of coherent sea clutter with inverse gamma texture," *International Radar Conference, Radar 2014*, Oct. 2014, pp. 1–6.

[34] L. Rosenberg, S. Watts, S. Bocquet, and M. Ritchie, "Characterisation of the Ingara HGA dataset," *International Radar Conference*, 2015, pp. 0027–0032.

[35] S. Bocquet, "Parameter estimation for Pareto and K distributed clutter with noise," *IET Radar Sonar and Navigation*, vol. 9, no. 1, pp. 104–113, 2015.

[36] S. Watts, L. Rosenberg, and M. Ritchie, "Characterising the Doppler spectra of high grazing angle sea clutter," *International Radar Conference*, 2014, pp. 1–6.

[37] L. Rosenberg, "Sea-spike detection in high grazing angle X-band sea-clutter," *IEEE Transactions on Geoscience and Remote Sensing*, vol. 51, no. 8, pp. 4556–4562, 2013.

Chapter 6

Performance prediction modelling

6.1 Introduction

One of the important uses of clutter models is their application to the prediction of a radar's detection performance in different conditions. Such modelling may be required at various stages in the design cycle of a radar (see Chapter 1). For example, it may be necessary to make accurate performance predictions at the proposal stage of a new design, for the purposes of marketing and to make comparisons with competing systems. As a radar design develops, alternative solutions may be proposed and performance prediction modelling will play an important role in comparing their relative merits. When a radar is to be delivered to a customer, it will usually be necessary to demonstrate its performance. As discussed in [1], measuring the detection performance of a maritime radar is extremely difficult and modelling will inevitably play an important role, allowing trial results to be interpreted in the light of predicted performance in the prevailing conditions. Finally, modelling may be used to provide guidance on the appropriate configuration of the radar in different operational scenarios.

This chapter describes how sea clutter models are used in the prediction of radar detection performance. For these results to have value, the radar analyst must have confidence in their accuracy. If an analytic modelling approach is taken, considerable attention to detail is required to understand and model the radar operation over a wide range of environmental conditions and viewing geometries. These include the grazing angle, the wind speed and direction and the sea swell and direction. The models must also be able to accurately represent the effects of the radar waveforms and processing on the received signals. The key characteristics include [1]

- mean backscatter,
- amplitude statistics,
- temporal pulse-to-pulse correlation/Doppler spectrum,
- spatial correlation,
- discrete clutter spikes,
- carrier frequency,
- polarisation,
- spatial resolution,
- effects of pulse-to-pulse frequency agility.

The premise of target detection is to first select an acceptable probability of false alarm, P_{fa} (i.e. false detection), and determine the corresponding threshold, γ, based on analysis of target-free data. Then using the received data, any response above the threshold is declared a potential target. The detection decision process is based on the null and alternative hypotheses:

$$\begin{aligned} H_0 &: \mathbf{h} = \mathbf{c} + \boldsymbol{\xi} \\ H_1 &: \mathbf{h} = \mathbf{c} + \boldsymbol{\xi} + \mathbf{s} \end{aligned} \tag{6.1}$$

where $\mathbf{c}$, $\boldsymbol{\xi}$ and $\mathbf{s}$ are the clutter, noise and target signal vectors, respectively, representing N successive coherent returns for a pulsed radar. Figure 6.1 shows a graphical representation where two histograms are shown that correspond to these hypotheses. The test statistic, ζ, could be the raw magnitude of the returns or the processed output after signal processing. If a detector can adapt to the changing clutter conditions and update the threshold accordingly to maintain the P_{fa}, then the detector can be considered to have a constant false alarm rate (CFAR). The P_{fa} and the probability of detection, P_{d}, can be defined as

$$P_{\mathrm{fa}}(\gamma) = \mathrm{Prob}(\zeta > \gamma | H_0) = \int_{\gamma}^{\infty} P_{\zeta}(\zeta) d\zeta, \tag{6.2}$$

$$P_{\mathrm{d}}(\gamma) = \mathrm{Prob}(\zeta > \gamma | H_1) = \int_{\gamma}^{\infty} P_{\zeta|s}(\zeta | s) d\zeta \tag{6.3}$$

where $P_{\zeta}(\zeta)$ represents the clutter-plus-noise (interference) probability density function (PDF), and $P_{\zeta|s}(\zeta|s)$ is the target-plus-interference PDF. Also shown in Figure 6.1 is P_{m}, which denotes the probability of a missed detection.

It is convenient to consider coherent and non-coherent processing separately. Coherent processing retains the phase information of the received signal, which must

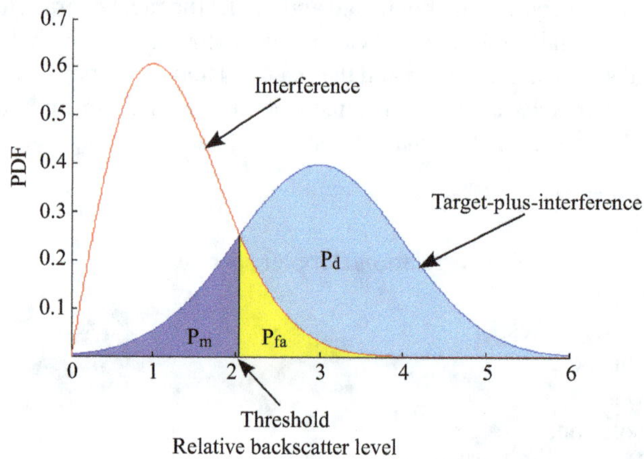

Figure 6.1 Target detection regions

be processed at some intermediate frequency or more usually at baseband using in-phase and quadrature samples. There will always be a point when the signal is demodulated, usually by forming its amplitude or intensity. In older equipment, this would have been achieved by a diode detector applied at an intermediate frequency in the receiver, but now is included in the digital processing of the in-phase and quadrature samples. Further signal processing may then be undertaken until a detection decision is made by comparing the signal return with a detection threshold. Performance prediction modelling can be broadly divided between analytic and Monte Carlo methods. Analytic methods use mathematical models to estimate the received signal levels and their statistical properties at various stages of the radar processing in order to predict probabilities of detection and false alarm. Monte Carlo methods may use real or simulated data as inputs to actual or modelled representations of the radar processing, accumulating probabilities of detection and false alarm over many repeated runs. Monte Carlo methods may be needed when the signal processing cannot be represented by analytic models.

The prediction of radar performance starts with the radar range equation and its use for signals in sea clutter is described in Section 6.2. At the point where a detection decision is to be made, predictions of the detection threshold required for a desired P_{fa} and associated probability of detection, P_{d}, will be required. Section 6.3 provides an overview of the mathematical methods used for non-coherent processing, including the effects of pulse-to-pulse integration. Section 6.4 then extends the discussion to coherent processing methods. Finally, Section 6.5 looks at the application of these methods to the assessment of end-to-end radar performance prediction.

6.2 Radar range equation

The radar range equation provides a means for estimating the power received from targets and clutter in the radar scene, following the transmission of a single radar pulse. In its simplest form, it can be written as

$$p_{\text{rec}} = \frac{p_{\text{t}} G_{\text{t}}}{4\pi R^2} \cdot \frac{\sigma F_{\text{p}}^4}{4\pi R^2} \cdot \frac{G_{\text{r}} \lambda^2}{4\pi} \cdot \frac{1}{L_{\text{RF}} L_{\text{a}} L_{\text{sp}}}. \tag{6.4}$$

The first term in the expression is the power density at a range, R, for a transmitter with peak power p_{t} and an antenna with directive transmit gain, G_{t}. The second term represents the power density following reflection from a target at range, R, with a radar cross-section (RCS), σ, and propagation factor, F_{p}, which represents the variation due to anomalous propagation (ducting) and multipath scattering (see Chapter 2). The third term represents the power at the receive antenna with a directive gain, G_{r}, for a radar wavelength, λ. The final term captures the three key losses in the radar. These include the RF losses due to the radome and transmit and receive hardware, L_{RF}, losses due to two-way propagation through the atmosphere, L_{a}, and the signal processing losses, L_{sp}. For a monostatic radar with a single antenna it will usually be the case that $G_{\text{r}} = G_{\text{t}}$.

The thermal noise power in the receiver is usually the dominant source of noise for this class of radar,

$$p_{\mathrm{n}} = k_{\mathrm{B}} T_0 F_{\mathrm{n}} B \tag{6.5}$$

where k_{B} is Boltzmann's constant, T_0 is the ambient reference temperature and F_{n} is the receiver noise figure. For operational systems, many of the variables in the radar equation will be measured in the factory with slight deviations from their original specification. Other parameters such as the noise figure may vary slightly due to changes in temperature. To account for these changes, some radars will maintain a real-time internal calibration of parameters such as the thermal noise power to better adjust the detection thresholds and gain levels.

To assess the performance of subsequent signal processing, it is necessary to estimate the peak power levels following matched filtering, p_{mf}. This is obtained from the received power, p_{rec}, and the pulse compression gain, μ_{c}, so that

$$p_{\mathrm{mf}} = p_{\mathrm{rec}} \mu_{\mathrm{c}} \tag{6.6}$$

where $\mu_{\mathrm{c}} = \alpha_{\mathrm{mf}} BT$ and B is the pulse bandwidth, T is the uncompressed pulse width and $\alpha_{\mathrm{mf}} < 1$ is a factor to account for mismatch losses in the matched filter.

To determine the target received power in (6.4), σ represents the target RCS and $p_{\mathrm{rec}}/p_{\mathrm{n}}$ determines the target signal-to-noise ratio. Alternatively, when determining the clutter power, the clutter RCS is often written as $\sigma = \sigma^0 A_{\mathrm{cl}}$, where σ^0 is the normalised RCS per unit area (reflectivity or mean backscatter coefficient), A_{cl} is the clutter patch area illuminated by a pulse and the clutter-to-noise ratio (CNR) is given by $p_{\mathrm{rec}}/p_{\mathrm{n}}$. The resolution cell area for a pulsed radar is defined by the azimuth antenna beam shape and the compressed pulse width, $T_{\mathrm{p}} = T/\mu_{\mathrm{c}}$, which accounts for the pulse compression gain. A simplified illustration is given in Figure 6.2 with the area approximately given by

$$A_{\mathrm{cl}} \approx \alpha_{\mathrm{bp}} R \theta_{3\mathrm{dB}} c_0 (T_{\mathrm{p}}/2) \sec \phi \tag{6.7}$$

where ϕ is the local grazing angle, $\theta_{3\mathrm{dB}}$ is the two-way azimuth 3 dB antenna beamwidth and c_0 is the speed of light. The factor α_{bp} in (6.7) can be used to account for the actual antenna beamshape compared with an idealised rectangular beam. As discussed in [1], $\alpha_{\mathrm{bp}} = 1$ for a rectangular beam and $\alpha_{\mathrm{bp}} = 0.75$ for a Gaussian-shaped beam. Although this calculation is adequate to estimate the clutter mean power (especially given the variability in the value of σ^0), care must be taken when assessing the clutter contribution in an airborne coherent radar. For example, the energy in the clutter Doppler spectrum outside the main beam will have much lower power, but potentially a much wider Doppler spread. This will be determined by the antenna sidelobes which can contribute Doppler frequencies up to $\pm 2v_{\mathrm{p}}/\lambda$ where v_{p} is the platform speed.

Calculating the loss terms in the radar equation requires significant attention to detail. For example, determining the RF losses, L_{RF}, requires careful measurement of

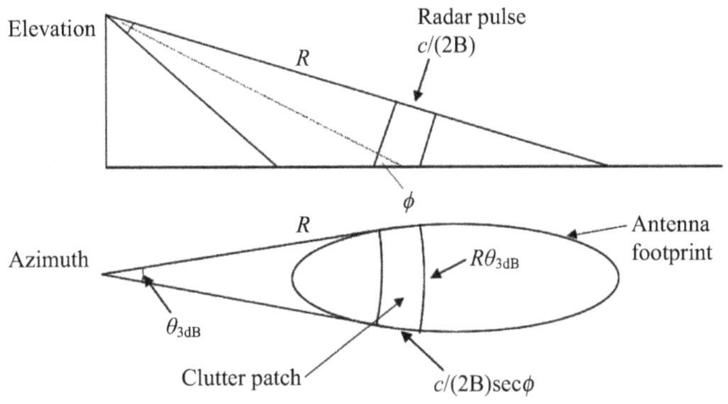

Figure 6.2 Clutter patch for a pulsed radar [1]

the radar system to determine waveguide and cable losses, radome losses and anything else in the RF transmit and receive chains. The atmospheric losses, L_a, will depend on the aircraft altitude and whether there is also rain, snow or fog [2–4]. Further detail on the propagation loss is given in Appendix C. Signal-processing losses, L_{sp}, are discussed in detail in [4] and can include items such as quantisation losses, beamshape losses, cusping losses (sampling) and matched filter losses. These may all be considered 'losses' compared to the ideal case assumed in the radar power budget and subsequent signal processing analysis. When undertaking analytic modelling, there are other types of loss that relate to the signal processing techniques being used. It may be convenient to include these in the overall estimates of target signal-to-clutter ratio and signal-to-noise ratio derived from the radar range equation. For example, if Doppler filters are formed using coherent processing, there may be windowing losses, as well as cusping losses for point targets with a narrow spectrum. When a cell-averaging CFAR (CA-CFAR) filter is used to set a detection threshold (see Chapter 7), it may be convenient to include a mean loss term to account for differences between the fluctuating threshold actually set and an ideal fixed threshold. However, when undertaking Monte Carlo modelling using simulated data, these types of loss are fully accounted for if the signal processing steps are also simulated in full.

6.3 Non-coherent processing

There are many non-coherent processing techniques that are used for target detection in the maritime environment, with some described in Chapter 7. A typical processing chain is shown in Figure 6.3 using a square law detector (providing signal intensity) for the sampled signal. This example shows analogue in-phase, I and quadrature, Q signals, produced by an analogue phase sensitive detector followed by a number of signal processing stages. Often, the signal may be sampled and digitised at an intermediate frequency, with the in-phase and quadrature samples produced as part of the signal processing. As alternative to the square law detector, a logarithmic or linear detector (providing signal amplitude) may be used to reduce the dynamic range

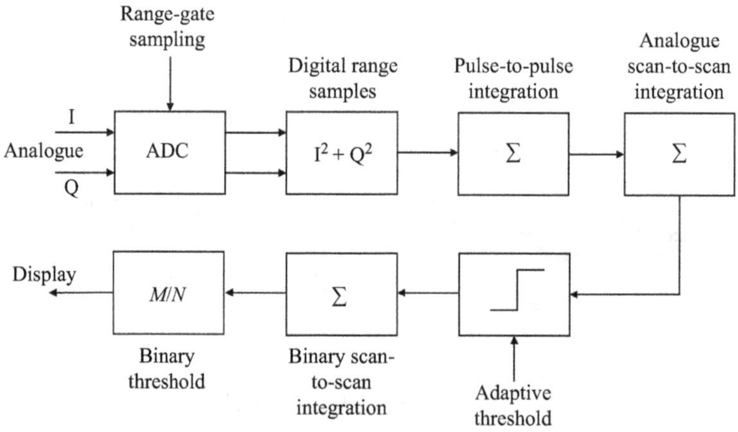

Figure 6.3 Non-coherent signal processing with a square-law detector, followed by pulse-to-pulse and scan-to-scan integration

of the signals, although the logarithmic approach has a detection loss of 1–2 dB relative to a square law detector [1, Ch. 12].

The first processing stage within each range cell is usually pulse-to-pulse integration. For a scanning radar, the total number of pulses that can usefully be integrated is determined by

$$N = \mathrm{Floor}\left(\frac{\theta_{3\mathrm{dB}} f_{\mathrm{r}}}{\dot{\theta}_{\mathrm{az}}}\right) \tag{6.8}$$

where f_r is the pulse repetition frequency (PRF) and $\dot{\theta}_{az}$ is the azimuth antenna scan rate. Ideally the integrator would be weighted to match the amplitude modulation imparted by the antenna beamshape, but this is often costly to implement in a practical radar and a uniform weighting is typically used. Performance prediction is based on N equal amplitude returns with a 'beamshape loss' used to account for the loss of gain at the beam edges. Pulse-to-pulse integration may be followed by further analogue integration over successive antenna scans. On a moving platform, this will require good ground stabilisation and moving targets may need to be tracked. As discussed later, scan-to-scan integration may also be undertaken by a binary integrator, operating after the first detection threshold.

The calculation of probability of detection, P_d, and probability of false alarm, P_{fa}, following the first detection threshold requires models of the interference and target-plus-interference PDFs and their correlations from-pulse-to pulse and scan-to-scan. As discussed in Chapter 2, sea clutter is best described using one of the compound Gaussian models. These have a speckle component with Gaussian statistics that allow for the accurate addition of thermal noise as well as modelling of the pulse-to-pulse correlations of the clutter and target components. In general, clutter returns will be partially correlated from pulse-to-pulse and scan-to-scan and modelling the radar performance is more difficult than in noise alone.

Using the compound Gaussian model, it is often assumed that the local underlying clutter intensity (texture) is constant over an antenna beam dwell. This will not be precisely correct for an azimuth scanning radar [5] but is usually a good approximation for most applications. The clutter speckle component will be fluctuating from pulse-to-pulse in accordance with its Doppler spectrum (or equivalently its autocorrelation function (ACF)). Typical decorrelation times are about 10 ms or less and must be compared with the antenna beam dwell time of the radar to assess its impact on pulse-to-pulse integration. If the radar dwell time for a burst of pulses is short relative to the clutter decorrelation time, then the clutter can be considered fully correlated for a fixed radar frequency. Also, for most practical applications with a fixed frequency radar and a typical sea clutter Doppler spectrum, it is unlikely that the clutter would ever be totally decorrelated. An alternative is to use pulse-to-pulse frequency agility, where the clutter speckle components of successive pulses can be considered statistically independent as long as the frequency step from pulse-to-pulse exceeds the pulse bandwidth, B [6]. If the total frequency agile bandwidth over a beam dwell is B_{a}, then the maximum number of independent samples in a dwell due to the frequency agility is $\sim B_{\mathrm{a}}/B$.

Recall from Chapter 2 that following a square law detector, the PDF of the clutter-plus-noise intensity, z, is given by

$$P_z\left(z|\tau,p_{\mathrm{n}}\right) = \frac{1}{\tau+p_{\mathrm{n}}}\exp\left(-\frac{z}{\tau+p_{\mathrm{n}}}\right) \tag{6.9}$$

where τ is the mean local clutter intensity (texture) and p_{n} is the mean noise power. If the speckle component of the radar is independent from pulse-to-pulse (i.e. by using frequency agility) and the local clutter texture, τ, remains constant over a beam dwell, then the PDF following integration of N pulses is given by

$$P_Z\left(Z|\tau,p_{\mathrm{n}}\right) = \frac{Z^{N-1}}{\left(\tau+p_{\mathrm{n}}\right)^N\Gamma(N)}\exp\left(-\frac{Z}{\tau+p_{\mathrm{n}}}\right) \tag{6.10}$$

where

$$Z = \sum_{n=1}^{N} z_n. \tag{6.11}$$

Further integration may be undertaken using scan-to-scan processing to reduce the sea clutter fluctuation. While the speckle will be fully decorrelated over a scan interval, the texture is likely to have some residual correlation depending on the scan interval and sea conditions. Following pulse-to-pulse and scan-to-scan integration as shown in Figure 6.3, a detection threshold is set with a binary decision representing the presence or absence of a target. The detection threshold will need to be adjusted according to the signal statistics to achieve the desired P_{fa}. In general, the radar will attempt to estimate (directly or indirectly) the local clutter and thermal noise statistics in order to set an appropriate threshold. The output from the detector may then be followed by scan-to-scan binary integration [7]. This can be undertaken jointly or as an alternative to analogue scan-to-scan integration. The M/N binary detector declares a detection when there are at least M_b detections out of N_b previous scans. Scan-to-scan integration is discussed further in Chapter 7.

This section provides details of the P_fa and P_d calculations for a target present in clutter and noise (interference). Initially, the clutter speckle is assumed to be decorrelated from pulse-to-pulse, such as the case in a frequency agile radar. It is also assumed that target returns are either constant or fully decorrelated from pulse-to-pulse over an integration period. Section 6.3.1 considers a constant target present in Gaussian interference covering a 'local area'. This is known as a Marcum model and the solution for P_d was first described in [8]. In most radar scenarios, the target RCS will fluctuate between pulses or from scan-to-scan and Section 6.3.2 describes a method for representing this fluctuation in terms of generalised Weinstock fluctuations [9] and Swerling models [10]. Section 6.3.3 then looks at the extension to compound clutter.

There will often be some level of correlation present from pulse-to-pulse for both the target and the speckle component of the clutter. For example, if the radar has a fixed frequency with a pulse repetition interval that is shorter than the decorrelation time of the clutter speckle, then successive speckle samples will be partially correlated. Also, the target returns are commonly assumed to be either totally correlated or uncorrelated and fluctuate around a mean level with an exponential distribution (Swerling cases 1 and 2) or chi-square distribution (cases 3 and 4). This is not necessarily true and Section 6.3.4 presents a study of methods which can be used to account for these partial correlations.

6.3.1 Constant target

The first model represents a target with a non-fluctuating or constant RCS present in Gaussian interference and is known as a Marcum model or Swerling case 0. In the intensity domain, the PDF for a single pulse is given by a Rician distribution [1],

$$P_{z|A}(z|A,\tau,p_\mathrm{n}) = \frac{1}{\tau+p_\mathrm{n}}\exp\left(-\frac{z+A^2}{\tau+p_\mathrm{n}}\right)I_0\left(\frac{2|A|\sqrt{z}}{\tau+p_\mathrm{n}}\right) \tag{6.12}$$

where the target complex amplitude is given by A and $I_0(\cdot)$ is a modified Bessel function of the first kind, order zero. For the following derivations, it is useful to define the local received signal and the target power as being normalised to the interference power

$$\mu = \frac{1}{\tau+p_\mathrm{n}}\sum_{n=1}^{N} z_n, \tag{6.13}$$

$$\mathcal{S}_0 = \frac{N\langle A^2\rangle}{\tau+p_\mathrm{n}}. \tag{6.14}$$

The expectation denoted by $\langle\cdot\rangle$ generalises the term so it can be applied to the fluctuating targets in the following section. With these definitions, the PDF describing the interference is given by

$$P_\mu(\mu) = \frac{\mu^{N-1}}{\Gamma(N)}\exp(-\mu) \tag{6.15}$$

with the P_{fa} determined by

$$\begin{aligned} P_{\text{fa}}(Y_0) = P_{Y_0}(Y_0|N) &= \int_{Y_0}^{\infty} P_\mu(\mu)d\mu \\ &= \frac{\Gamma(N, Y_0)}{\Gamma(N)} \end{aligned} \tag{6.16}$$

where $\Gamma(N, Y_0)$ is the upper incomplete gamma function, $\Gamma(N) = (N-1)!$ and Y_0 is the normalised threshold, related to the 'true' threshold γ by

$$Y_0 = \frac{\gamma}{\tau + p_{\text{n}}}. \tag{6.17}$$

For the total received signal, the PDF becomes a multi-look Rician distribution,

$$\begin{aligned} P_{\mu|\mathcal{S}_0}(\mu|\mathcal{S}_0, N) &= \left(\frac{\mu}{\mathcal{S}_0}\right)^{(N-1)/2} \exp\left(-(\mu+\mathcal{S}_0)\right) I_{N-1}\left(2\sqrt{\mu\mathcal{S}_0}\right) \\ &= \sum_{k=0}^{\infty} \frac{e^{-\mu}\mu^{N+k-1}}{(N+k-1)!}\frac{e^{-\mathcal{S}_0}\mathcal{S}_0^k}{k!} \end{aligned} \tag{6.18}$$

and the probability of detection for a constant target can be written in terms of the normalised threshold Y_0 as

$$\begin{aligned} P_{\text{d}}(Y_0) = P_{Y_0|\mathcal{S}_0}(Y_0|\mathcal{S}_0, N) &= \int_{Y_0}^{\infty} P_{\mu|\mathcal{S}_0}(\mu|\mathcal{S}_0, N)d\mu \\ &= \sum_{k=0}^{N-1} e^{-Y_0}\frac{Y_0^k}{k!} + \sum_{k=N}^{\infty} e^{-Y_0}\frac{Y_0^k}{k!}\left(1 - \sum_{n=0}^{k-N}\frac{e^{-\mathcal{S}_0}\mathcal{S}_0^n}{n!}\right). \end{aligned} \tag{6.19}$$

The series in (6.19) is difficult to evaluate, but effective methods were devised by Shnidman [11,12]. It can also be expressed in terms of the generalised Marcum Q-function [11]:

$$P_{Y_0|\mathcal{S}_0}(Y_0|\mathcal{S}_0, N) = Q_N\left(\sqrt{2N\mathcal{S}_0}, \sqrt{2Y_0}\right). \tag{6.20}$$

For most radar targets of interest however, the target reflection will not be constant. The following section looks at a generalisation of this model.

6.3.2 Fluctuating target

A target will typically fluctuate from pulse-to-pulse and scan-to-scan. Shnidman [13] describes four different target fluctuation models, including chi-square, log–normal, Weibull and Rician distributions. The chi-square distribution is the most popular as it can be used to represent the well-known Swerling models. With this model, the target fluctuation with parameter K_t is given by

$$P_{s_0|\mathcal{S}_0}(s_0|\mathcal{S}_0, K_t) = \frac{s_0^{K_t-1}}{\Gamma(K_t)} \left(\frac{K_t}{\mathcal{S}_0}\right)^{K_t} \exp\left(-\frac{K_t s_0}{\mathcal{S}_0}\right) \tag{6.21}$$

where

$$s_0 = \frac{1}{\tau + p_n} \sum_{n=1}^{N} A_n^2 \tag{6.22}$$

is the sum of target powers, A_n^2, normalised by the noise and local speckle power. The probability of detection for a fluctuating target is obtained by substituting s_0 for $\mathcal{S}_0$ in the constant target result in (6.19), multiplying by the PDF in (6.21) and integrating over the random variable s_0,

$$P_d(Y_0) = P_{Y_0|\mathcal{S}_0}(Y_0|\mathcal{S}_0, N, K_t) = \int_0^\infty P_{s_0|\mathcal{S}_0}(s_0|\mathcal{S}_0, K_t) P_{Y_0|\mathcal{S}_0}(Y_0|s_0, N) ds_0. \tag{6.23}$$

A general solution is given in [13],

$$\begin{aligned} P_d(Y_0) = P_{Y_0|\mathcal{S}_0}(Y_0|\mathcal{S}_0, N, K_t) &= \sum_{n=0}^{N-1} e^{-Y_0} \frac{Y_0^n}{n!} + \sum_{n=N}^{\infty} e^{-Y_0} \frac{Y_0^n}{n!} \\ &\times \left(1 - \sum_{k=0}^{n-N} \frac{\Gamma(K_t + k)}{k!\Gamma(K_t)} \left(\frac{1}{1 + \mathcal{S}_0/K_t}\right)^{K_t} \left(\frac{\mathcal{S}_0/K_t}{1 + \mathcal{S}_0/K_t}\right)^k\right) \end{aligned} \tag{6.24}$$

where the P_{fa} is identical to the non-fluctuating target case. This series is also difficult to evaluate, but an effective method is given by Shnidman [13].

There are also a number of special cases depending on the choice of the parameter K_t. For example, when $K_t \to \infty$ this equation reverts back to the constant target case from the previous section. Other studies have looked at the region where $0 \leq K_t \leq 1$, where the models are known as Weinstock fluctuations [9]. These represent slowly fluctuating targets and have been shown to accurately model the fluctuation from objects that are cylindrical in shape. Four other special cases of this model are when $K_t = 1, N, 2$ and $2N$. These are known as the Swerling models 1–4 and were originally

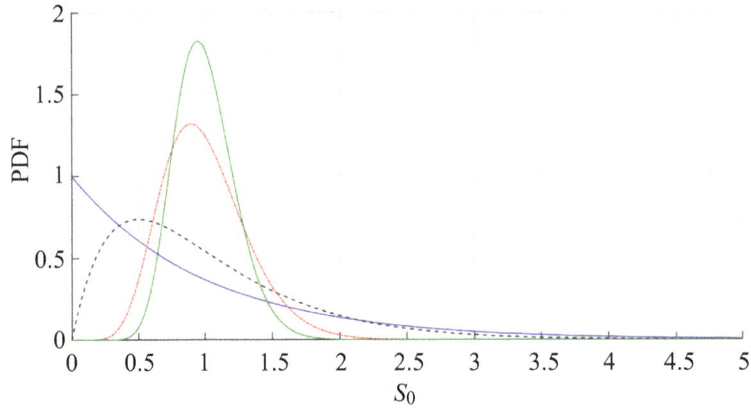

Figure 6.4 Swerling model comparison for N=10 and with an SIR of $\mathcal{S}_0 = 0$ dB, (—) case 1, (-.-) case 2, (- -) case 3, (—) case 4. Adapted from [17].

proposed in [10] and detailed in the books [14,15]. These models are able to capture the majority of target RCS fluctuations after either pulse-to-pulse or scan-to-scan processing and are popular due to their physical interpretation [16]. The Swerling 1 fluctuation can be used to represent targets with many independent scatterers such as ships and other physically large targets. The return is correlated from pulse-to-pulse with an exponential distribution but remains uncorrelated from scan-to-scan. The Swerling 2 model represents a target with many scatterers, but the fluctuation is now uncorrelated from pulse-to-pulse with each pulse having an exponential distribution. As a result, it should primarily be used to represent the returns from frequency agile radars. Swerling 3 and 4 target models have the same pulse-to-pulse and scan-to-scan fluctuations as the Swerling 1 and 2 models, respectively. They are represented by a chi-square distribution and represent targets such as buoys or small yachts with one dominant plus other small scatterers or a single non-uniform scatterer subject to modest changes in orientation [15]. The Swerling 4 model is rarely used in maritime radar. Examples of the four Swerling PDF models from (6.21) are shown in Figure 6.4 for $N = 10$ and a signal-to-interference ratio (SIR) of $\mathcal{S}_0 = 0$ dB. Cases 1 and 3 are clearly more spread out due to the larger number of scatterers being observed over the slower observation time. Conversely, cases 2 and 4 have narrower distributions due to the faster change of RCS.

6.3.3 Compound clutter

Thus far the mean clutter power τ has been treated as a constant. In the compound model, τ is a random variable with a PDF, $P_\tau(\tau)$. The global probability of detection is obtained by integrating the local P_d over this PDF. This has the effect of spreading out the clutter power distribution with the resulting detection threshold set higher to achieve the same P_{fa} [16,18]. With this extension, the signal and threshold are now normalised by the global average interference power, so that the parameters $\mathcal{S}_0$ and Y_0 are replaced by

$$\mathcal{S} = \frac{N\langle A^2\rangle}{\langle\tau\rangle + p_n}, \tag{6.25}$$

$$Y = \frac{\gamma}{\langle\tau\rangle + p_n} \tag{6.26}$$

where $\langle\tau\rangle$ and $\langle A^2\rangle$ are the mean clutter and target powers, respectively. These can both be rewritten in terms of the short scale definitions,

$$\mathcal{S}_0 = \frac{\mathcal{S}(\langle\tau\rangle + p_n)}{\tau + p_n}, \tag{6.27}$$

$$Y_0 = \frac{Y(\langle\tau\rangle + p_n)}{\tau + p_n}. \tag{6.28}$$

The P_d is determined by multiplying by the PDF, $P_\tau(\tau)$, and then integrating over τ,

$$P_d(Y) = P_{Y|\mathcal{S}}(Y|\mathcal{S}, N, K_t) = \int_0^\infty P_{Y_0|\mathcal{S}_0}\left(\frac{Y(\langle\tau\rangle + p_n)}{\tau + p_n} | \frac{\mathcal{S}(\langle\tau\rangle + p_n)}{\tau + p_n}, N, K_t\right) P_\tau(\tau)d\tau. \tag{6.29}$$

There is no documented analytic solution to this integral, and numerical integration must be used for evaluation as described in [19, Section 2.3]. Different target fluctuations can be modelled by changing the value of K_t as in the previous section.

The solution for P_d can also be expressed in terms of Y, $\mathcal{S}$ and the CNR, $\mathcal{C}$. If the integration variable is redefined as $\bar{\tau} = \tau/\langle\tau\rangle$,

$$P_d(Y) = P_{Y|\mathcal{S}}(Y|\mathcal{S}, N, K_t, \mathcal{C}) = \int_0^\infty P_{Y_0|\mathcal{S}_0}\left(\frac{Y(\mathcal{C}+1)}{\bar{\tau}\mathcal{C}+1} | \frac{\mathcal{S}(\mathcal{C}+1)}{\bar{\tau}\mathcal{C}+1}, N, K_t\right) P_\tau(\bar{\tau})d\bar{\tau} \tag{6.30}$$

where $P_\tau(\bar{\tau}) = P_\tau(\tau/\langle\tau\rangle)/\langle\tau\rangle$ and $d\tau = \langle\tau\rangle d\bar{\tau}$. The probability of false alarm can be calculated similarly,

$$P_{fa}(Y) = P_Y(Y|N, \mathcal{C}) = \frac{1}{\Gamma(N)} \int_0^\infty \Gamma\left(N, \frac{Y(\mathcal{C}+1)}{\bar{\tau}\mathcal{C}+1}\right) P_\tau(\bar{\tau})d\bar{\tau} \tag{6.31}$$

where again numerical integration can be used for evaluation.

6.3.4 Correlated returns

The previous calculations have assumed that the speckle component of the clutter is fully decorrelated from pulse-to-pulse, either through the clutter's own internal motion or by using pulse-to-pulse frequency agility. At the other extreme, if the speckle returns are approximately constant over a dwell, then the clutter can be modelled as a Swerling 1 target embedded in thermal noise and the mean clutter intensity will be given by the local clutter texture [20]. More commonly, there will be correlation between successive samples such as a radar with a fixed frequency from pulse-to-pulse and a pulse repetition interval that is shorter than the decorrelation time of the clutter speckle. Such partial correlation may also be observed in targets. For example, using an X-band continuous wave radar system, it was observed in [21] that radar echoes from a number of aircraft were partially correlated from pulse-to-pulse and the traditional Swerling models were not appropriate.

There are a number of approaches in the literature which can account for correlation of the target and speckle. Ideally, they should be able to model the ACF through its representation of the received signal's Doppler spectrum. The simplest method was presented by Farina *et al.* [22] and is based on an improvement factor that can be used to modify the required SIR in the P_{d} calculation. However, it does not consider the impact of different amplitude statistics or target correlation and assumes that the Doppler spectrum is modelled with a linear combination of Gaussian, Lorentzian and Voigtian components.

Another popular approach is to use an effective number of independent looks $1 \leq N_{\mathrm{eff}} \leq N$ to account for correlation, where N_{eff} does not need to be an integer and ranges from 1 for complete correlation up to N for completely uncorrelated signals. Barton [23] used this approach for modelling target correlation with N_{eff} determined by the correlation time relative to the total observation time. Kanter [24] derived the effective number of looks from this ratio in terms of the pulse-to-pulse correlation of the data samples, $0 \leq \tilde{\rho} \leq 1$,

$$N_{\mathrm{eff}} = \mathrm{Min}\left[N, 1 - (N-1)\ln(\tilde{\rho})\right]. \tag{6.32}$$

where Min[·] indicates the minimum value. This correlation value can also be related to decorrelation time, T_{d}, using a Gaussian model where n is the temporal lag:

$$\rho_n = \exp\left(-\frac{n^2}{f_{\mathrm{r}}^2 T_{\mathrm{d}}^2}\right) = \tilde{\rho}^{n^2}. \tag{6.33}$$

Another method for determining the effective number of looks is the ratio of the squared mean to the variance for correlated gamma random variables. The result was originally obtained by Kotz and Neumann [25] and later derived independently by Kanter [24]. If the values of the ACF are represented by ρ_n, the effective number of looks is given by

$$N_{\mathrm{eff}} = \frac{N^2}{N + 2\sum_{n=1}^{N-1}(N-n)\rho_n^2}. \tag{6.34}$$

This expression has a better statistical justification and is more general since all the correlation lags are used. For comparison, Figure 6.5 shows the number of effective looks for each case using the Gaussian ACF. Although both offer similar results, the second expression has a smoother response as the correlation increases. The approximation can be made more accurate if the number of effective looks is calculated from the ratio of squared mean to variance for the clutter-plus-noise, rather than the clutter alone [17]. This amounts to replacing N_{eff} with

$$\tilde{N}_{\text{eff}} = N_{\text{eff}} N \frac{C^2 + 1}{NC^2 + N_{\text{eff}}}. \tag{6.35}$$

The effective number of looks approximation was used to account for correlated speckle by Ward, Tough and Watts (WTW) [1,27]. Their approach is useful as it treats the speckle, thermal noise and target separately. However, it becomes computationally intensive as the CNR gets higher and fails completely when there is only clutter present. This led to a modification in [28] that evaluated (6.29) using an effective number of looks for the speckle. Thermal noise has also been included in this formulation, with the caveat that it will be correlated with the speckle and is therefore not as accurate at low CNR. A trade-off was therefore proposed, where the WTW method is used for low CNR and the modification used for larger CNR.

While the use of an effective number of looks is attractive, it can be quite inaccurate in the distribution tail, meaning that the detection threshold can be incorrect by several decibels. Another approach is to use moment generating functions (MGF) to describe the probability of detection and false alarm relationships. If the MGF is

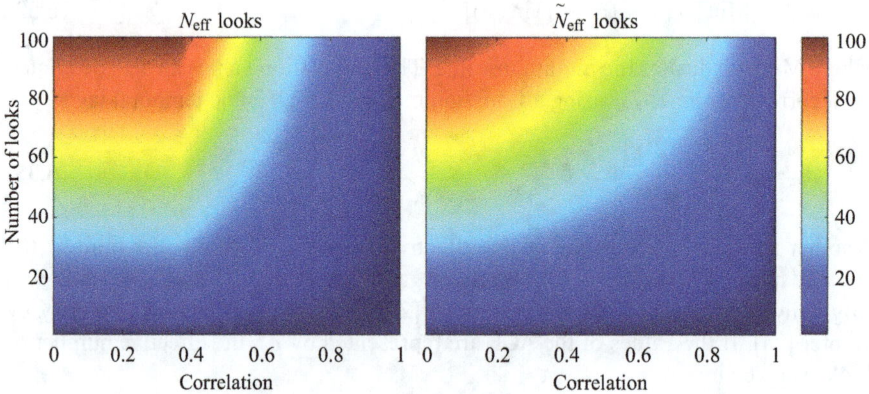

Figure 6.5 Effective number of looks comparison, left: N_{eff}, right: $\tilde{N}_{eff}$ [26].

$\mathcal{M}(\mathrm{s})$, then the Laplace transforms of the cumulative distribution function (CDF), $F_{-}(\cdot)$, and its complement, $F_{+}(\cdot)$ (which is equivalent to P_{fa}), can be expressed as

$$\begin{aligned}\mathcal{L}\left(F_{-}(z)\right) &= \mathcal{M}(\mathrm{s})/\mathrm{s},\\ \mathcal{L}\left(F_{+}(z)\right) &= \left(1-\mathcal{M}(\mathrm{s})\right)/\mathrm{s}.\end{aligned} \tag{6.36}$$

Determining the P_{fa} then involves inverting the second equation in the absence of targets, while P_{d} is determined by inverting the first equation when targets are present. Note that this formulation only models the speckle component of the clutter and a further integration is required for compound models using the texture distribution $P_{\tau}(\tau)$ as discussed in Section 6.3.3.

In the literature, many authors have used this approach to determine P_{fa} and P_{d} for different combinations of partially correlated targets and/or speckle. For example, Kanter [24] derived an exact solution for correlated targets with an exponential fluctuation in uncorrelated speckle and Weiner [29] did the same for targets with a chi-squared fluctuation. Hou and Morinaga [30] and Bocquet *et al.* [31] have both looked at the detection of exponential fluctuating targets with correlated speckle. Further work by Zuk *et al.* [32] then proposed a solution for uncorrelated speckle with correlated targets that can have either exponential or chi-squared fluctuations. Zuk [33] then further generalised the approach to cover both correlated speckle and correlated targets with exponential or chi-squared fluctuations. The biggest contribution in this work was to reformulate the problem in terms of voltages with phase terms in order to accurately model the effect of two correlated sources.

For the case of a single correlated source (speckle or target), the MGF is given by

$$\begin{aligned}\mathcal{M}(\mathrm{s}) &= \frac{1}{\det\left(\mathbf{I}_N+\mathbf{R}_{\mathrm{m}}\mathrm{s}\right)}\\ &= \frac{1}{\prod_{n=1}^{N}\left(1+\lambda_n\mathrm{s}\right)}\end{aligned} \tag{6.37}$$

where $\det(\cdot)$ is the determinant, $\mathbf{I}_N$ is the identity matrix of size $N\times N$ and $\mathbf{R}_{\mathrm{m}}$ is the covariance matrix comprising the target, clutter and noise with eigenvalues $\lambda_n, n=1,\ldots,N$. The covariance can be written as

$$\mathbf{R}_{\mathrm{m}} = \mathbf{I}_N + \mathcal{C}\mathbf{R} + \mathcal{S}\mathbf{R}_{\mathrm{s}} \tag{6.38}$$

where $\mathbf{R}$ is the speckle covariance matrix, $\mathcal{S}$ is the signal-to-noise ratio and $\mathbf{R}_{\mathrm{s}}$ is the target covariance matrix. Using this MGF, Bocquet *et al.* [31] studied a number of solutions for inverting (6.36), including the residue series [30], gamma series [34–36] and contour integration [37]. The residue series is the most straightforward method for inverting the MGF [30]. While it gives an exact result, the series is numerically unstable for high correlation and large N [24,37]. For the gamma series, the distribution function is expressed as an infinite sum of gamma distributions, with the shape parameter incremented by one in each successive term of the series until the desired accuracy is attained [38]. When applied to this problem, it was found to have a slow convergence, requiring a large number of terms in the series to achieve the required accuracy [31].

Numerical quadrature along a contour, $\mathbb{C}$, in the complex plane is the only method that avoids the numerical instability of the residue series and the slow convergence of the gamma series. It is advantageous to choose a contour that passes through a saddlepoint on the real line [37]. Convergence is most rapid along the path of steepest descent, but numerical root finding is usually needed to find this path, so it can be preferable to use a simple approximation to the steepest descent path such as a parabola. The CDF and CCDF integrals can be expressed in the form

$$F_{\pm}(Z) = \frac{1}{2\pi i} \int_{\mathbb{C}-i\infty}^{\mathbb{C}+i\infty} e^{\Phi(\mathrm{s}|Z)} d\mathrm{s} \tag{6.39}$$

where Z is the integrated return and the ‘phase’ function,

$$\Phi(\mathrm{s}|Z) = \mathrm{s}Z - \sum_{n=1}^{N} \ln(1 + \lambda_n \mathrm{s}) - \ln(\mp \mathrm{s}). \tag{6.40}$$

The contour for the CDF, F_- and CCDF, F_+ are chosen to pass through the positive saddlepoint s_0^+ and the negative saddlepoint s_0^-, respectively, and Newton’s method can then be used to find the saddlepoints [37]. Once these are determined, Helstrom proposed a parabolic contour to determine the CCDF with the integral (6.39) given by

$$F_{\pm}(Z) = \frac{1}{\pi} \mathrm{Re}\left(\int_0^{\infty} e^{\Phi(\mathrm{s}|Z)} (1 - i\chi_{\mathrm{c}} y_{\mathrm{s}})\, dy_{\mathrm{s}} \right) \tag{6.41}$$

where $s = \mathrm{s}_0^{\mp} + \chi_{\mathrm{c}} y_{\mathrm{s}}^2/2 + j y_{\mathrm{s}}$ and $\mathrm{Re}(\cdot)$ is the real component. The parabola with the curvature of steepest descent, χ_{c}, is defined at the saddlepoints, $\mathrm{s}_0^{\mp}$,

$$\chi_{\mathrm{c}} = \frac{\Phi'''\left(\mathrm{s}_0^{\mp}\right) + 2(\mathrm{s}_0^{\mp})^{-3}}{3\left(\Phi''\left(\mathrm{s}_0^{\mp}\right) - (\mathrm{s}_0^{\mp})^{-2}\right)} \tag{6.42}$$

and the kth derivative of the phase function is determined for $k \geq 2$

$$\Phi^{(k)}(\mathrm{s}) = (-1)^k (k-1)! \left(\mathrm{s}^{-k} + \sum_{n=1}^{N} \frac{\lambda_n^k}{(1 + \lambda_n \mathrm{s})^k} \right). \tag{6.43}$$

The integrand in (6.41) is complex because the parabolic contour is only an approximation to the path of steepest descent and can fail if N is large and the correlation is high. In these cases a straight line path from the saddlepoint can be used [37]. An alternative approach is to just evaluate the integral at the saddlepoint [37,39],

$$F_{\pm}(Z) = \left(2\pi\Phi''\left(s_0^{\mp}\right)\right)^{-1/2}\exp\left(\Phi\left(s_0^{\mp}|Z\right)\right)(1+R_{\mathrm{res}}) \tag{6.44}$$

with the basic saddlepoint approximation obtained by setting the residual component, R_{res}, to zero. This approximation is satisfactory in the tails of the distribution but becomes inaccurate close to the mean. It can be improved by using an asymptotic expansion for R_{res} [39].

The extension to correlated targets and speckle was given by Zuk [33] who defined a 'first principles' solution for the MGF. Unfortunately, there is no closed form solution and no easy way to invert the Laplace transform in (6.36). An alternative 'effective' model was therefore proposed that leads to a number of convenient solutions. For this case, the MGF is given by

$$\mathcal{M}(\mathrm{s}) = \frac{\left(\det\left(\mathbf{I}_N + (\mathrm{s}/N)(p_{\mathrm{n}}^2 + \langle\tau\rangle^2\,\mathbf{R})\right)\right)^{K_{\mathrm{t}}-1}}{\left(\det\left(\mathbf{I}_N + (\mathrm{s}/N)(p_{\mathrm{n}}^2 + \langle\tau\rangle^2\,\mathbf{R} + \langle A^2\rangle^2\,\mathbf{R}_{\mathrm{s}}/K_{\mathrm{t}})\right)\right)^{K_{\mathrm{t}}}}. \tag{6.45}$$

Two approximations for solving the inverse problem with correlated speckle and target components include the Dalle Mese Giuli (DMG) and diagonal approaches. The DMG approximation is significantly faster than the full effective model, while the diagonal approximation is simpler to implement as it does not require repeated calculation of the eigenvalues of the full covariance matrix when evaluating the compound integral in (6.29).

Table 6.1 summarises the advantages and disadvantages of the different approaches discussed in this section. Contour integration is the best of the exact methods, while the preferred approximation is the DMG method.

Table 6.1 Summary of calculation methods for correlated clutter and/or targets

Method	**Advantages**	**Disadvantages**
Improvement factor	Fast, simple	Restricted in use
Ward, Tough and Watts	Fast for low C	Poor tail approximation, complicated
Residue series	Fast, simple	Fails for large N with high $\tilde{\rho}$, low C
Gamma series	Exact, robust	Slow for large N and high $C, \tilde{\rho}$
Contour integral	Exact	Convergence problems for large $N, \tilde{\rho}$
Effective model	Accurate	Slower to run
DMG approximation	Extremely fast	Complicated
Diagonal approximation	Very accurate	Slower to run

6.4 Coherent processing

Coherent signal processing describes the techniques that retain phase information in the received signal. The assessment of coherent processing requires knowledge of the target and clutter spectra in the cell under test (CUT), as well as their amplitude statistics. Coherent processing will usually be followed by a non-linear processing stage and the techniques described for non-coherent processing in Section 6.3 can be used to predict the final detection performance.

There are many different detection schemes that have been proposed for use with compound Gaussian clutter [40]. Simple Doppler filtering with an independently controlled threshold in each Doppler bin is one approach that is commonly used. This may seem appealing as it makes no assumptions about the spectrum characteristics, but is not necessarily the best approach for controlling false alarms at the edges of the spectrum [41]. If the clutter has Gaussian or compound Gaussian statistics, the optimum approach is to whiten the interference in the CUT [40,42]. However, this requires knowledge of the interference covariance matrix which is in general unknown. Many schemes have been proposed for estimating the covariance matrix from surrounding range cells [43–45]. However, the assessment of their performance is usually predicated on the clutter being a spherically invariant random process (SIRP) with the texture and speckle being mutually independent processes and hence the normalised covariance matrix of the speckle can be estimated by averaging over surrounding cells. This assumption is often invalid, with observations of real clutter showing that the mean Doppler shift can be correlated with the local texture intensity and the normalised covariance matrix fluctuates randomly from one cell to the next. As a result, the ability to estimate a covariance matrix from surrounding cells is then dependent on the how the spectral characteristics change over range.

Traditional coherent detection techniques for detecting targets in sea clutter are described in Chapter 7. It is not the intention here to explore all the different coherent detection techniques that have been proposed, but to indicate how clutter models can be used to assess their performance. Section 6.4.1 describes how models of the Doppler spectra can be used in the prediction of performance. This typically requires the amplitude statistics of the signals in each Doppler bin to be estimated with consideration for how they vary over range or time. Section 6.4.2 then describes how analytic models can be used to assess the performance of adaptive pre- or post-Doppler whitening filters.

6.4.1 Doppler filtering

For the relatively simple case of a pulse-Doppler processor applied to a burst of pulses, the radar range equation combined with the evolving Doppler spectrum model can be used to predict the CNR, the clutter amplitude statistics and the target SIR in each frequency bin. The procedure for calculating P_d and P_{fa} following amplitude detection at the output of the filter is then the same as described in Section 6.3 for single pulse detection.

The evolving Doppler spectrum model was described in Chapter 2 and is given by

$$G(f) = \frac{\tau}{\sqrt{2\pi}w} \exp\left(-\frac{(f - m_f(\tau))^2}{2w^2}\right) \tag{6.46}$$

where $G(f)$ is the power spectral density for Doppler frequency, f, and the texture, τ, has a gamma distributed PDF if the clutter has K-distribution. The mean Doppler frequency is related to the normalised mean speckle power, $\bar{\tau} = \tau / \langle\tau\rangle$ with

$$m_f(\tau) = A_c + B_c\bar{\tau} + f_{lin} \tag{6.47}$$

where the term f_{lin} was added to the model in [46] and captures the fluctuation around the model fit. It is described by a Gaussian random variable with zero mean and standard deviation, σ_{lin}. The final term in (6.46) is the spectrum width, w (standard deviation of the Gaussian-shaped spectrum), which has a PDF, $P_w(w)$, with mean value m_w and standard deviation σ_w.

The expression in (6.46) represents the mean clutter power spectral density. As the texture varies over range and time, the mean of the spectrum will change according to (6.47) and the spectrum width will fluctuate randomly. If the mean clutter spectrum was constant over time and only modulated in power by τ, then the statistics of the clutter component in each Doppler bin observed over time or range would be the same as those for τ. Of course, added noise must also be modelled and will dominate for Doppler shifts well outside the clutter spectrum. The effect of the range-varying spectrum width and mean Doppler will be to change the statistics of the clutter intensity in each Doppler bin when observed over range, especially near the edges of the spectra, where the clutter statistics become spikier [1, Ch. 2]. This effect can be directly estimated by determining the moments of the clutter intensity in each Doppler bin and averaging over all possible values of τ and w. Consider an idealised filter with centre frequency f and pass-band $\Delta f = f_r/N$:

$$\begin{aligned} \langle\tau'(f)\rangle &= \int_{f-\frac{\Delta f}{2}}^{f+\frac{\Delta f}{2}} \int_0^\infty \int_0^\infty P_\tau(\tau)P_w(w)G(f', \tau, w)d\tau\, dw\, df', \\ \langle\tau'(f)^2\rangle &= \int_{f-\frac{\Delta f}{2}}^{f+\frac{\Delta f}{2}} \int_0^\infty \int_0^\infty P_\tau(\tau)P_w(w)G^2(f', \tau, w)d\tau\, dw\, df'. \end{aligned} \tag{6.48}$$

If the noise power, p'_n, in each Doppler bin can be related to the overall CNR, $\mathcal{C}$, by

$$p'_n = \frac{\langle\tau\rangle}{\mathcal{C}N}, \tag{6.49}$$

then the CNR in each Doppler bin, $\mathcal{C}'(f)$, for the idealised Doppler filter will be

$$\mathcal{C}'(f) = \frac{\langle\tau'(f)\rangle}{p'_n}. \tag{6.50}$$

It has been found that the amplitude statistics in individual Doppler bins can usually be fitted quite well to a K+noise distribution. This suggests that the intensity moments given by (6.48) can be fitted to a gamma distribution to determine the shape parameter $\nu'(f)$ in each Doppler bin,

$$\nu'(f) = \left(\frac{\langle \tau'(f)^2 \rangle}{\langle \tau'(f) \rangle^2} - 1 \right)^{-1}. \tag{6.51}$$

With the single-pulse target, clutter and noise levels estimated from the radar range equation, it is now possible to calculate the detection performance as a function of Doppler frequency at a given range. Also, note that this approach can readily be extended to include variations of the basic spectrum model, such as the bimodal model described in Chapter 2.

Some examples of the coherent detection performance are now presented using the parameters in Table 6.2, which were derived from data reported in [46,47]. Figure 6.6(a) shows the variation of shape parameter as a function of Doppler frequency using (6.51), while Figure 6.6(b) shows the CNR as a function of Doppler frequency,

Table 6.2 Parameter values used for the example in Section 6.4.1

Parameter	Value
Wavelength, λ	0.032 m
Pulse repetition frequency, f_r	1 kHz
Number of pulses, N	32
K-distribution shape parameter, ν	4.3
Doppler centroid offset, A_c	54.4 Hz
Doppler centroid slope, B_c	28.5 Hz
Doppler width mean m_w	54.4 Hz
Doppler width standard deviation, σ_w	15 Hz
Clutter-to-noise ratio, C_{dB}	18 dB

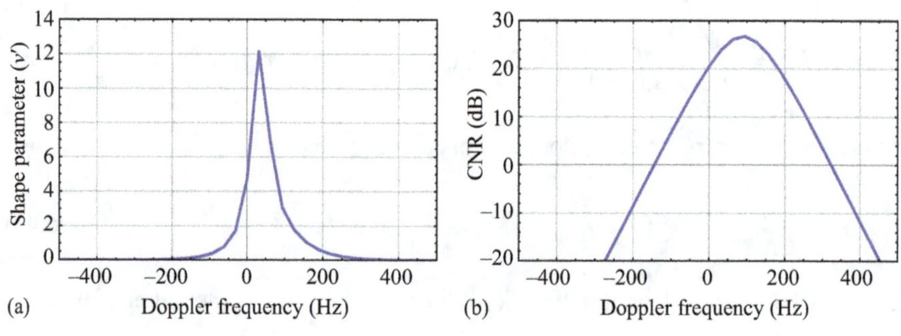

Figure 6.6 (a) K+noise distribution shape parameter against Doppler frequency, (b) CNR against Doppler frequency

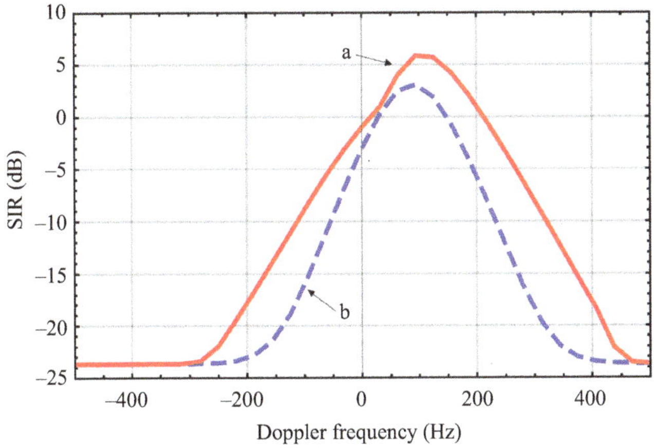

Figure 6.7 SIR required to achieve $P_d = 0.5$, $P_{fa} = 10^{-4}$ for a non-fading target with CNR = 18 dB: (a) range varying Doppler model, (b) exponential clutter model (whitened spectrum)

with the peak CNR occurring at 83 Hz, which is expected from the value of mean Doppler shift, $A_c + B_c$.

Figure 6.7 then shows the SIR required to achieve $P_d = 0.5$ and $P_{fa} = 10^{-4}$ for a non-fading target as a function of Doppler frequency, where the SIR is defined as the ratio of mean target power to mean interference power in a single-pulse return. Also shown in Figure 6.7 is the SIR that would be required at the output of an ideal whitening filter (see Section 6.4.2), where the statistics in each Doppler bin will have an exponential intensity distribution. The challenge when using a simple Doppler analysis is to set a threshold in each Doppler bin in order to maintain a CFAR. One approach is to run a separate CA-CFAR along range in each Doppler bin. The required threshold multipliers for each Doppler bin would then have to be estimated using methods such as those described in Chapter 7. As with performance prediction of a CA-CFAR used in a non-coherent radar, a CFAR loss could be estimated for each Doppler bin to account for the non-ideal behaviour of the detection thresholds.

6.4.2 Adaptive processing

The performance of adaptive pre- or post-Doppler whitening filters can also be studied using these models [41,48]. This can be achieved analytically using models for the spectrum or covariance. Consider the signal vector (a burst of pulse returns), $\mathbf{s} = A\mathbf{s}_t$, where A is an unknown complex amplitude target signal and $\mathbf{s}_t$ is the normalised steering vector:

$$\mathbf{s}_t = \frac{1}{\sqrt{N}}\left[1, \exp\left(-j2\pi f T_r\right), \ldots, \exp\left(-j2\pi f (N-1) T_r\right)\right]^T \tag{6.52}$$

where $-1/(2T_r) < f \leq 1/(2T_r)$ is the Doppler frequency of the target signal, T_r is the pulse repetition interval and $(\cdot)^T$ denotes the transpose. The clutter returns are represented by a compound Gaussian process with $\mathbf{c} = \sqrt{\tau}\mathbf{x}$, where τ is the slowly varying underlying intensity (assumed to be constant during the coherent processing interval) and $\mathbf{x}$ is the fast-varying speckle component (multivariate zero mean complex Gaussian random process, $\mathcal{N}(0, 1/2) + j\mathcal{N}(0, 1/2)$).

To demonstrate the modelling approaches, consider the normalised adaptive matched filter (NAMF) detector [42]:

$$\frac{N\left|\mathbf{s}_t^H\mathbf{R}^{-1}\mathbf{h}\right|^2}{(\mathbf{s}_t^H\mathbf{R}^{-1}\mathbf{s}_t)(\mathbf{h}^H\mathbf{R}^{-1}\mathbf{h})} \underset{H_1}{\overset{H_0}{\gtrless}} \gamma \tag{6.53}$$

where $(\cdot)^H$ represents the Hermitian transpose, γ is a threshold that determines the probability of declaring a detection and $\mathbf{R}$ is the covariance matrix of the clutter. The term $\left|\mathbf{s}_t^H\mathbf{R}^{-1}\mathbf{h}\right|^2$ represents a whitening filter for the data vector, $\mathbf{h}$, in a single range cell with the target steering vector, $\mathbf{s}_t$, acting as a filter centred on the frequency f. The other two terms act to normalise the result with $\left(\mathbf{h}^H\mathbf{R}^{-1}\mathbf{h}\right)$ being a measure of the multilook intensity after whitening. If the received signal in the absence of a target can be represented as a compound Gaussian SIRP, then the two components of the clutter are mutually independent [49] and

$$\left\langle \mathbf{h}\mathbf{h}^H \right\rangle = \langle\tau\rangle \left\langle \mathbf{x}\mathbf{x}^H \right\rangle + p_n\mathbf{I}. \tag{6.54}$$

If the clutter is not accurately described as an SIRP, then the texture and speckle components of the clutter are not mutually independent and the covariance matrix will not totally represent the clutter correlation in the CUT. In [50], Watts has analysed the sensitivity of the NAMF to such non-stationary behaviour by postulating that an ideal detector can be represented using an average covariance, $\mathbf{R}_{av}$, with the texture, τ, being constant over the processing period. The performance of the NAMF can then be assessed in terms of maintaining a constant P_{fa}, when the data in the CUT has a covariance matrix, $\mathbf{R}_k$, that is different from $\mathbf{R}_{av}$.

The baseline against which performance will be assessed is the fully matched case, $\mathbf{R}_k = \mathbf{R}_{av}$, where the threshold to achieve a desired value of P_{fa} is independent of $\mathbf{R}$ and $\mathbf{s}_t$. If the local mean level of the compound-Gaussian clutter, τ, is known, then the optimum detector is

$$\frac{\left|\mathbf{s}_t^H\mathbf{R}^{-1}\mathbf{h}\right|^2}{(\mathbf{s}_t^H\mathbf{R}^{-1}\mathbf{s}_t)(\tau + p_n)} \underset{H_1}{\overset{H_0}{\gtrless}} \gamma \tag{6.55}$$

and the test statistic will have an exponential distribution independent of f and τ. If the normalising factor $(\tau + p_n)$ is replaced by $(\mathbf{h}^H\mathbf{R}^{-1}\mathbf{h})/N$, the threshold is still independent of f and τ, but the distribution will no longer be exponential.

Now consider the case when the local CUT has a covariance matrix, $\mathbf{R}_k$, that is different from its long-term average, $\mathbf{R}_{\mathrm{av}}$, while still assumed to be constant over the observation period. The NAMF is now written as

$$\frac{N\left|\mathbf{s}_{\mathrm{t}}^{H}\mathbf{R}_{\mathrm{av}}^{-1}\mathbf{h}\right|^{2}}{(\mathbf{s}_{\mathrm{t}}^{H}\mathbf{R}_{\mathrm{av}}^{-1}\mathbf{s}_{\mathrm{t}})(\mathbf{h}^{H}\mathbf{R}_{\mathrm{av}}^{-1}\mathbf{h})}\underset{H_1}{\overset{H_0}{\gtrless}}\gamma. \tag{6.56}$$

The test still attempts to normalise using the local clutter plus noise power, $\langle|g|^2\rangle = \tau + p_{\mathrm{n}}$, except that the normalising factor $(\mathbf{h}^H\mathbf{R}_{\mathrm{av}}^{-1}\mathbf{h})$ no longer has a mean value of $N\langle|g|^2\rangle$. Instead, it can be written as

$$\left\langle\mathbf{h}^{H}\mathbf{R}_{\mathrm{av}}^{-1}\mathbf{h}\right\rangle = \mathrm{Tr}\left[\mathbf{R}_{\mathrm{av}}^{-1}\left\langle\mathbf{h}\mathbf{h}^{H}\right\rangle\right] = \mathrm{Tr}\left[\mathbf{R}_k\mathbf{R}_{\mathrm{av}}^{-1}\right]\left\langle|\mathbf{h}|^2\right\rangle \tag{6.57}$$

where Tr[·] represents the trace of a matrix. In (6.56), the term $\left|\mathbf{s}_{\mathrm{t}}^{H}\mathbf{R}_{\mathrm{av}}^{-1}\mathbf{h}\right|^{2}$ is a whitening filter with a constant value and is independent of frequency if $\mathbf{R}_k = \mathbf{R}_{\mathrm{av}}$. However, if the covariance matrices are mismatched, then this term will have a value that is dependent on frequency. To assess this dependency, assume that the spectrum of the local clutter in the CUT is $G_k(f)$ and the average spectrum is $G_{\mathrm{av}}(f)$. It was proposed in [50] that the action of the whitening filter can be equated to filtering in the frequency domain, with a filter of length N. If the CNR in the CUT is $\mathcal{C}_k$, and the average CNR is $\mathcal{C}_{\mathrm{av}}$, then the whitening filter would produce a variation with f of the form

$$Q_k\left(f\right) = \int_{f_1}^{f_2}\frac{G_k\left(f'\right)+\left(f_{\mathrm{r}}\mathcal{C}_k\right)^{-1}}{G_{\mathrm{av}}\left(f'\right)+\left(f_{\mathrm{r}}\mathcal{C}_{\mathrm{av}}\right)^{-1}}\left(\frac{1+\mathcal{C}_{\mathrm{av}}^{-1}}{1+\mathcal{C}_k^{-1}}\right)\frac{1}{f_2-f_1}df'. \tag{6.58}$$

The first term of the integral is essentially the ratio of the actual and assumed whitening filters with the gains corrected to take account of the presence of the (white) noise power and the two CNR terms are included to normalise the ratios. The frequency interval $(f_2 - f_1)$ is defined for an idealised filter of width f_{r}/N, centred on the Doppler frequency of interest, f, so that $f_1 = f - f_{\mathrm{r}}/(2N)$ and $f_2 = f + f_{\mathrm{r}}/(2N)$.

If the PDF of the test statistic for the mismatched case is approximately exponential but contains scaling errors [50], then for a CUT containing Gaussian clutter and thermal noise, an approximate expression for P_{fa} can be written as a function of f with the threshold, γ. This can be set to give a desired value of probability of false alarm, $P_{\mathrm{fa,ideal}} = \exp(-\gamma)$ on the assumption that $\mathbf{R}_k = \mathbf{R}_{\mathrm{av}}$,

$$P'_{\mathrm{fa}}(f) \approx \exp\left(\ln(P_{\mathrm{fa,ideal}})\frac{F_{\mathrm{a}}(k)}{Q_k\left(f\right)}\right) \tag{6.59}$$

where $F_{\mathrm{a}}(k) = \mathrm{Tr}\left[\mathbf{R}_k\mathbf{R}_{\mathrm{av}}^{-1}\right]/N$. This expression normalises for the interference power and assumes that the values of $\mathbf{R}$ are normalised. It may be noted that even in the noise region of the spectra, the test statistic will be biased above or below the value expected for a matched filter. This bias level is given by (6.57) and will result in an incorrect value of P_{fa}. It is also interesting to note that if the normalised covariance matrices are matched to the clutter spectra, but $C_k \neq C_{av}$, there will also be an error in the threshold. This would be true even with an idealised SIRP in compound Gaussian clutter, where the local CNR will depend on the local value of τ. In any event, the CNR will vary with τ and needs to be correctly estimated to set the correct threshold.

The probability of detection can be analysed in a similar manner to the probability of false alarm. In this case, the CUT is assumed to contain a target with a given SIR and the whitening filter is applied to the target-plus-interference. If the whitening filter is analysed in the frequency domain, then the filter output may be approximated by

$$Q_k(f, S) = Q_k(f) + \int_{f_1}^{f_2} \frac{W(2\pi N(f' - f)/f_r)\, S(1 + (f_r C_{av})^{-1})}{G_{av}(f') + (f_r C_{av})^{-1}} df' \tag{6.60}$$

where $W(f)$ represents the spectrum of the target which is modelled as a normalised sinc function:

$$W(f) = \left(\frac{\sin f}{f}\right)^2 \Big/ \int_{-f_r/2}^{f_r/2} \left(\frac{\sin f'}{f'}\right)^2 df'. \tag{6.61}$$

The normalising factor $F_a(k)$ is unchanged, but now $\mathbf{R}_k$ will include the target. Under matched conditions, the whitening filter and normalising factor are assumed to have values of Q_m and F_m, respectively, which are both equal to 1 in the absence of a target. If the covariance matrix of the interference in the cell containing the target does not match $\mathbf{R}_{av}$, then they are assumed to have values of Q_e and F_e and the test statistic will be modified by a factor, $F_m Q_e/(F_e Q_m)$. The PDF of the NAMF statistic for target-plus-interference with a non-fading target will be Rician and the probability of detection could be calculated if required. However for a quick assessment, the change in threshold should approximately be the same as the detection loss for $P_d \approx 0.5$.

The models given previously can be investigated using the evolving Doppler spectra model described in Section 6.4.1. Given a Gaussian-shaped power spectrum for the clutter with $\tau = 1$, the ACF of the clutter-plus-noise can be written as

$$\rho(n) = \begin{cases} \exp\left(-j2\pi m_f T_r n - \frac{1}{2}(2\pi T_r s n)^2\right), & n = -(N-1), \ldots, (N-1), n \neq 0 \\ (1 + 1/C)^{-1}, & n = 0 \end{cases} \tag{6.62}$$

with covariance matrix,

$$\mathbf{R} = \begin{bmatrix} \rho(0) & \cdots & \rho(N-1) \\ \vdots & \ddots & \vdots \\ \rho(-N+1) & \cdots & \rho(0) \end{bmatrix}. \tag{6.63}$$

Performance can be assessed using the parameter values in Table 6.2, except the PRF is set to 600 Hz and the CNR is 20 dB. The choice of PRF results in some aliasing, which has been included in the modelling of the spectra. The texture, τ, has a gamma PDF and the value of P_{fa}, averaged over all possible values of τ and w, can be derived from

$$\langle P'_{\mathrm{fa}}(f)\rangle = \int_0^\infty \int_0^\infty P'_{\mathrm{fa}}(f)P_w(w)P_\tau(\tau)d\tau dw \tag{6.64}$$

where $P'_{\mathrm{fa}}(f)$ is given by (6.59) with

$$\begin{aligned} G_{\mathrm{av}}(f) &= \frac{1}{\sqrt{2\pi}w_{\mathrm{av}}}\exp\left(-\frac{\left(f-m_{\mathrm{f,av}}\right)^2}{2w_{\mathrm{av}}^2}\right), \\ G_k(f) &= \frac{1}{\sqrt{2\pi}w}\exp\left(-\frac{\left(f-m_{\mathrm{f}}(\tau)\right)^2}{2w^2}\right). \end{aligned} \tag{6.65}$$

This equation has been evaluated using the values in Table 6.2 with

- $m_{\mathrm{f,av}} = A_{\mathrm{c}} + B_{\mathrm{c}}$, the average value of mean Doppler shift,
- $w_{\mathrm{av}} = m_{\mathrm{w}}$, the average standard deviation of the spectrum,
- $m_{\mathrm{f}}(\tau) = A_{\mathrm{c}} + B_{\mathrm{c}}\bar{\tau}$, the local value of mean Doppler shift,
- $C_{\mathrm{av}} = 20$ dB; $C_k = \tau C_{\mathrm{av}}$,
- $\mathbf{R}_k$ and $\mathbf{R}_{\mathrm{av}}$ are derived using these local and average values, respectively.

Figure 6.8(a) shows an example of the variation of $\langle P'_{\mathrm{fa}}(f)\rangle$ after averaging over all values of τ and w. The results show the average P_{fa} as being higher than the desired value for almost all Doppler frequencies. This would not be acceptable in a practical radar application and the threshold would need to be adjusted to maintain the desired

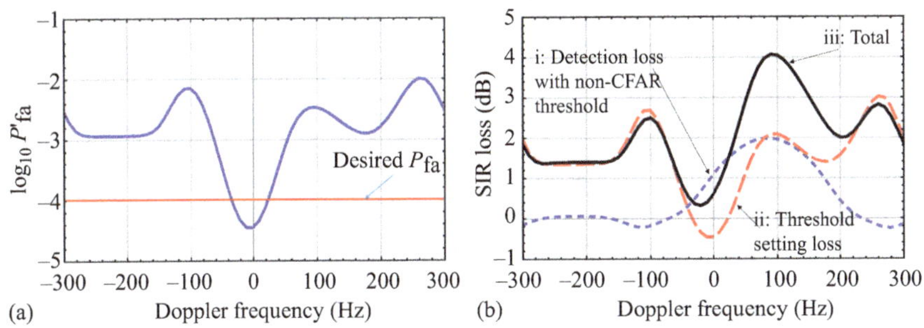

Figure 6.8 (a) Average P_{fa} against Doppler frequency with a fixed threshold for $P_{\mathrm{fa,ideal}} = 10^{-4}$; (b) detection loss against Doppler frequency. i: Loss due to incorrect threshold setting, ii: detection processing loss, iii: combined losses

average value of P_{fa}. The amount by which it would need to be raised relative to the ideal value will be approximately the detection loss that would be incurred for $P_d \approx 0.5$. Assuming the same fixed threshold, this loss can be determined by the mismatched covariance factor, $F_m Q_e/(F_e Q_m)$. This loss is plotted in Figure 6.8(b) (line 'ii'), assuming a non-fading target along with the loss associated with raising the threshold to control the P_{fa} (line 'i'). The peak detection loss in this example is almost 3 dB and even in the noise region of the spectrum, the loss is about 2 dB. The total loss associated with both detecting and controlling P_{fa} can be approximated by the sum of the two losses and is plotted in Figure 6.8(b) (line 'iii').

6.5 Assessing end-to-end radar performance

The end-to-end modelling of radar performance is important as it shows how the radar is expected to perform for different targets as a function of range, viewing geometries and environmental conditions. The radar range equation can provide estimates of the target, clutter and noise power levels with the clutter characteristics predicted using the empirical models introduced in Chapter 4. To accurately assess the performance of a real system, detailed knowledge of the processing losses are needed. For example, when a CA-CFAR is used, the losses incurred will be dependent on the clutter and target statistics, the length of the clutter range profile, any spatial correlation in the clutter and the ability to determine the appropriate threshold multiplier (see [1, Ch. 13]). Once an understanding of the signal processing at the point of detection and the desired false alarm level is known, then the techniques described earlier can be used to predict performance. In the examples given in Section 6.5.1, performance is expressed as the variation with a range of the minimum target RCS required to achieve a minimum desired probability of detection at an acceptable false alarm rate. Using this form of presentation, the performance of coherent and non-coherent detection processing can be compared [51]. The methods described are quite general and can be applied to other choices of signal processing and waveforms as appropriate. In addition, the results can be presented for different target fluctuation models, viewing geometries and environmental conditions. In Section 6.5.2, further examples are shown with Monte Carlo methods used to assess simulated radar performance. This has the advantage of incorporating the effects of range- and time-varying spectra on estimates of the covariance matrices, but requires large simulated data samples to accurately determine the performance for each set of conditions. Finally, a discussion of the main results is given in Section 6.5.3.

6.5.1 Radar performance examples

The analysis methods in this chapter for both coherent and non-coherent processing are now used to demonstrate end-to-end radar performance. These examples were presented in [51] with the radar parameters in Table 6.3 broadly based on the example in [1, Ch. 12]. In general, the processing losses for the clutter and target are different, with the latter including effects such as range and Doppler cusping losses, mismatched filtering and so on. A nominal value has been used here for the total loss, L_t, combining

Table 6.3 Radar parameters for the performance prediction example in Section 6.5

Parameter	**Value**
Transmitter frequency, f_{RF}	9.5 GHz
Peak power, P_t	20 kW
Pulse length, T_p	25 μs
Range resolution, ΔR	2 m
Antenna gain (one way), $G_{t,dB}$	33 dB
Azimuth 3 dB beamwidth, θ_{3dB}	2.0°
Elevation 3 dB beamwidth, ϕ_{3dB}	8.0°
Pulse repetition frequency, f_r	1.8 kHz
Number of pulses, N	10
System noise figure, $F_{n,dB}$	4.0 dB
Target processing losses, $L_{t,target,dB}$	5.0 dB
Clutter processing losses, $L_{t,clutter,dB}$	2.0 dB
Range scale (max. range), $R_{max,NM}$	40 NM

losses due to atmospheric propagation, signal processing, radome and RF hardware. At different aircraft heights, the antenna tilt is assumed to be set so that maximum gain is directed at the maximum instrumented range, R_{max}. This means that at high altitudes, the range coverage is restricted by the elevation 3 dB beamwidth, ϕ_{3dB} and will appear as a 'hole' at short ranges below the aircraft. In practice, a radar might undertake general surveillance with the antenna tilt set in this way, increasing the tilt if tracking a target into short ranges.

The performance of a non-coherent radar flying at either 1,500 ft (457 m) or 20,000 ft (6,096 m) is first examined to illustrate the loss in performance when flying at higher altitudes. The non-coherent processing is assumed to comprise pulse-to-pulse integration over $N = 10$ pulses, with pulse-to-pulse frequency agility of sufficiently large steps to decorrelate the clutter speckle. For a range resolution of $\Delta R = 2$ m, the frequency steps would need to be at least 75 MHz, implying a total agile bandwidth of 750 MHz. The integrated returns are thresholded and then integrated over three scans using a binary integrator with a two-out-of-three detection confirmation to better discriminate against clutter spikes. The probabilities of false alarm and detection required at the first threshold are determined following the binary integrator, assuming independent detections from scan-to-scan. A Swerling case 2 target model is assumed given the pulse-to-pulse frequency agility with the target being an isotropic scatterer having the same RCS for both vertical and horizontal polarisations.

Figure 6.9 compares the non-coherent detection results for vertical and horizontal polarisations and sea state (SS) 4. The minimum detectable RCS is determined to achieve a final probability of detection, $P_d = 0.5$, with a final probability of false alarm, $P_{fa} = 10^{-5}$. Also shown are the noise-limited conditions when no clutter is present. It can be seen that performance is significantly degraded when operating at the higher altitude, especially at longer ranges. At 1,500 ft altitude, the vertical polarisation shows a significant improvement over the horizontal and in this example,

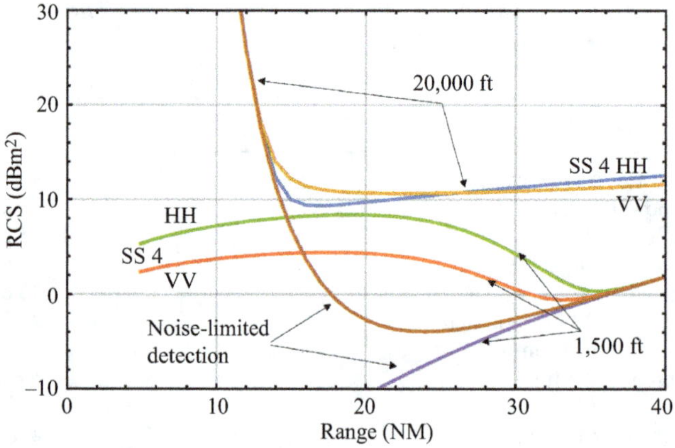

Figure 6.9 Minimum detectable RCS against range for a Swerling 2 target, $P_d = 0.5$, $P_{fa} = 10^{-5}$ with non-coherent processing in sea state 4 [51]. © Commonwealth of Australia 2017

the performance at short ranges is also non-existent at 20,000 ft altitude, due to the antenna tilt setting.

In the next results, coherent processing is compared with non-coherent processing. For the former, a simple Doppler filtering method is used with a coherent dwell of $N = 32$ pulses and a PRF of 1.8 kHz. It is assumed that any Doppler shifts due to platform motion are corrected from burst to burst in the signal processing so that stationary point targets on the sea surface always have a Doppler of approximately 0 Hz at the centre of the beam. Also, the spreading of the clutter Doppler due to the finite antenna azimuth beamwidth and platform motion must be modelled according to the radar look direction. The subsequent detection performance assumes that an independent CFAR detection threshold is set in each Doppler bin, based on the CNR and amplitude statistics (as observed over range and/or time at that Doppler frequency). For fixed frequency operation, a Swerling case 1 target model is used with the final probability of false alarm at the radar display set to $P_{fa} = 10^{-5}$. The Doppler analysis is undertaken over a 32 point fast Fourier transform corresponding to the assumed dwell time, and hence the probability of false alarm at each Doppler bin was set to $P_{fa}/32$, assuming that the Doppler bins are collapsed following detection.

Figure 6.10 shows the coherent detection performance at 20,000 ft for a selection of Doppler frequencies. A Doppler shift of 100 Hz is equivalent to a radial velocity of about 1.6 m/s in X-band. The clutter is modelled for a horizontal polarisation looking up-wind and up-swell in sea state 4 and assumes no additional effects from spectrum spreading due to platform motion (i.e. the radar is forward looking). Also shown in Figure 6.10 are the non-coherent detection results, assuming pulse-to-pulse integration with frequency agility, followed by scan-to-scan binary integration. It can be seen that the coherent detection performances for a stationary target and for a range of positive Doppler frequencies are up to 7 dB worse than the non-coherent results.

This is further illustrated in Figure 6.11, which shows detection performance as a function of Doppler frequency at ranges of 15, 25 and 40 NM, together with the non-coherent detection performance. Clearly the coherent detector works better for targets with Doppler shifts that take them outside the clutter spectrum. The best performance with the modelled antenna geometry occurs at 25 NM, while at 15 NM, the performance is worse because the target is close to the edge of the radar elevation beam. It may be noted that the variation of noise-limited performance with range is also affected by the changing antenna gain. Within the clutter spectrum, the better

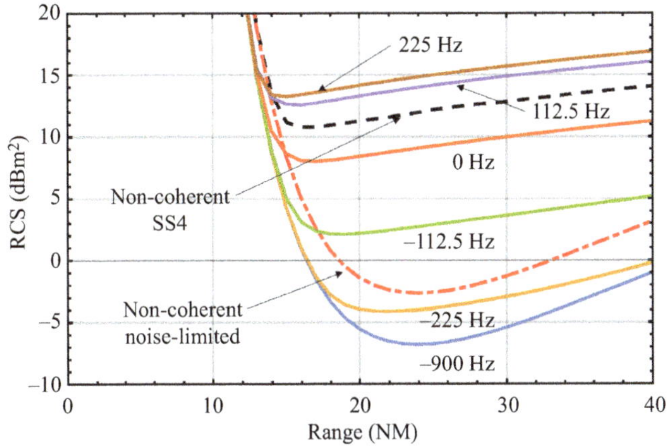

Figure 6.10 *Minimum detectable RCS against range for $P_d = 0.5$, $P_{fa} = 10^{-5}$ with coherent and non-coherent processing in sea state 4, up-wind, up-swell, height of 20,000 ft and $\Delta R = 2$ m [51].*

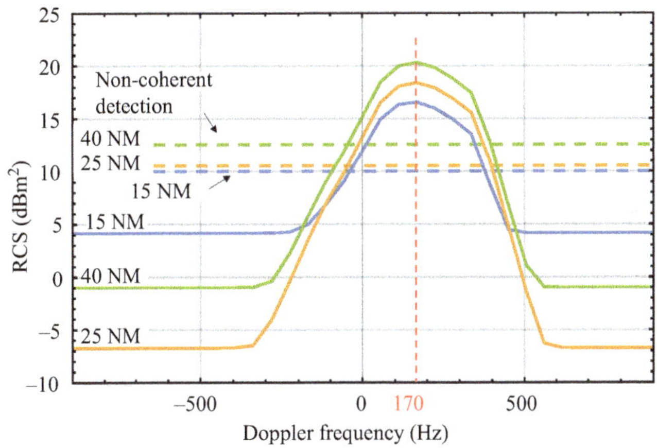

Figure 6.11 *Minimum detectable RCS against Doppler frequency for $P_d = 0.5$, $P_{fa} = 10^{-5}$ with coherent and non-coherent processing in sea state 4, up-wind, up-swell, height of 20,000 ft and $\Delta R = 2$ m [51].*

performance of the non-coherent detector is due to the use of a very wide bandwidth compared with the coherent detector and the use of scan-to-scan integration. Scan-to-scan processing is assumed to decorrelate the texture component of the clutter, which is assumed to be constant during a beam dwell. Of course the coherent detector could also employ scan-to-scan integration in each Doppler bin or following detection and collapsing in Doppler for each scan, but that would require a slower update rate.

It is instructive to consider why the coherent detection performance is worst around 170 Hz. Figure 6.12 shows the average CNR, $C'(f)$, as a function of Doppler frequency at a range of 30 NM where the single-pulse CNR, C, is 8.5 dB. It can be seen that the mean Doppler shift is close to 50 Hz. Figure 6.13 then shows the clutter shape parameter $\nu'(f)$ (in the absence of added noise) as a function of Doppler frequency, with the texture having a shape parameter of $\nu = 2.4$. It can be seen that $\nu'(f)$ has a

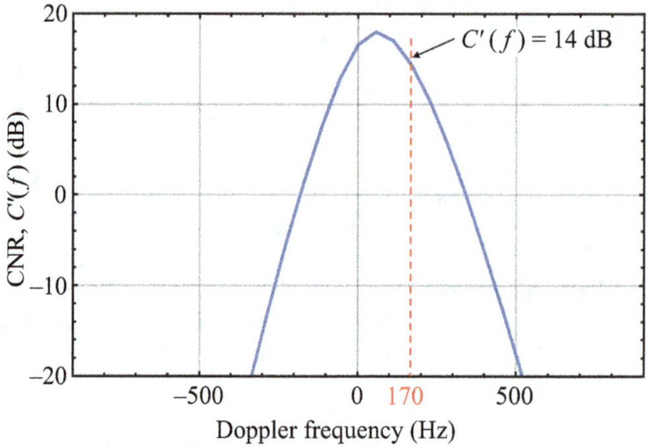

Figure 6.12 Clutter-to-noise ratio against Doppler frequency at 30 NM range

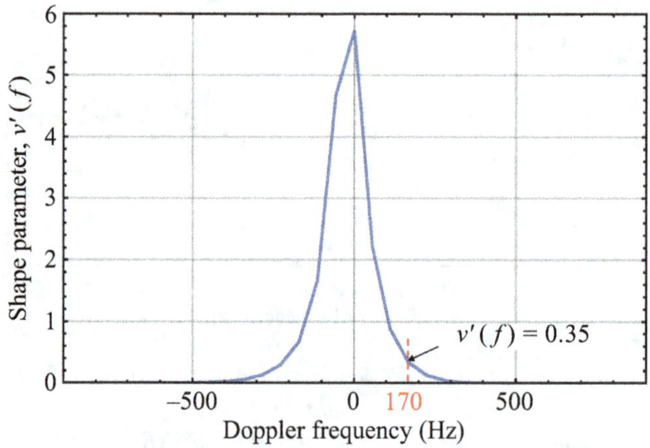

Figure 6.13 K+noise distribution shape parameter against Doppler frequency at 30 NM range

peak value of about 5.5 and is close to 0 Hz. However, at a Doppler shift of 170 Hz, its value has reduced to about 0.35, indicating spiky behaviour. The CNR, $C'(f)$, at this frequency is still quite high at about 14 dB, compared with its peak value of 18 dB at 50 Hz, and due to the much smaller value of $\nu'(f)$, the detection performance is worse.

As shown in Figure 6.11, the performance of Doppler filtering may be worse than non-coherent processing for slow moving targets. This is because the clutter energy is confined to a narrow band of Doppler frequencies around 0 Hz. If the main role of the detector is to detect slow moving targets, some improvement can be made by narrowing the Doppler filter width with respect to the width of the Doppler spectrum of the clutter. For example, if the number of pulses in a beam dwell is fixed, this could be achieved by lowering the radar PRF (see Chapter 7). Figures 6.14 and 6.15 show the detection performance for the same conditions as Figure 6.11, but with PRF values of 1,000 and 100 Hz, respectively. Some improvement is observed for a PRF of 1,000 Hz, while the lower PRF of 100 Hz shows detection performance that is almost independent of Doppler frequency. This is because the clutter energy is now uniformly spread over the unambiguous Doppler space. In that case, if the original time-domain data has a CNR of C and the clutter component observed over range is K-distributed with a shape parameter ν, then each of the Doppler filter outputs will have the same values of C and ν. However, if the target has a much narrower spectrum width, it will not be spread by aliasing and its energy in a single Doppler bin will be coherently increased by a factor equal to the number of returns integrated. It can be seen from Figure 6.15 that the detection performance is about 3.5 dB better than the non-coherent result. Of course, for a fast moving target, the coherent detection performance is still much worse than when filtering with a higher PRF, as shown in Figure 6.11.

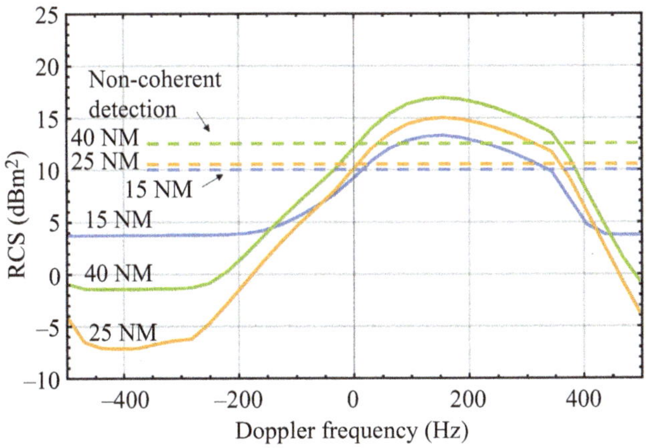

Figure 6.14 Minimum detectable RCS against range for a Swerling 1 target, $P_d = 0.5$, $P_{fa} = 10^{-5}$ with non-coherent processing and Doppler filtering with a PRF of 1,000 Hz in sea state 4, up-wind, up-swell, height of 20,000 ft and $\Delta R = 2$ m. Green: 40 NM, brown: 25 NM, blue: 15 NM

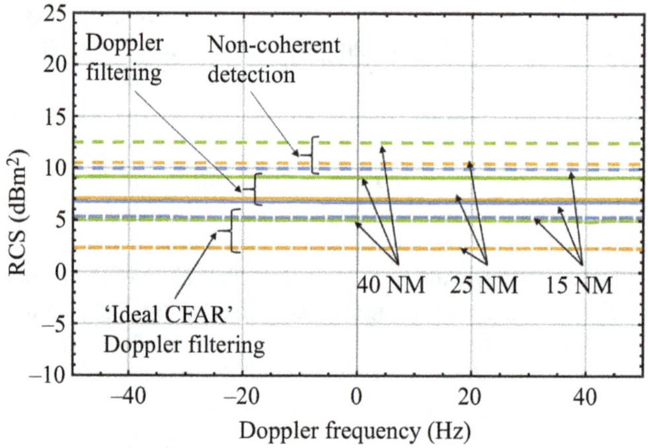

Figure 6.15 Minimum detectable RCS against range for a Swerling 1 target, $P_d = 0.5$, $P_{fa} = 10^{-5}$ with non-coherent processing, standard Doppler filtering and 'ideal CFAR' Doppler processing with a PRF of 100 Hz in sea state 4, up-wind, up-swell, height of 20,000 ft and $\Delta R = 2$ m. Green: 40 NM, brown: 25 NM, blue: 15 NM

The detection results given here assume a fixed threshold in each Doppler bin, derived from a knowledge of the global CNR and K-distribution shape parameter. Performance with a fixed threshold based on global statistics could be improved if a locally adaptive threshold could be used with an estimate of the local interference level in each range cell. If the target environment is relatively sparse, it might be assumed that no more than one Doppler bin contains a target. In this case, if the largest bin is assumed to contain the target, then an estimate of the local interference level could be made by considering all but the largest bin. The limiting case of this would be exact knowledge of the local interference level, which would be equivalent to the 'ideal CFAR' performance presented in [1,52]. The accuracy of the local interference level estimate will depend on the number of independent samples available. If the returns in surrounding range cells are spatially correlated, they might also be used to improve the estimate. In that event, the performance of a CA-CFAR in the time domain would also be improved [53]. Figure 6.15 shows this 'ideal CFAR' performance for a detector using Doppler filtering with a PRF of 100 Hz. The adaptive 'ideal CFAR' threshold provides an additional improvement that is dependent on the clutter statistics. It will be greater for spikier clutter (lower values of shape parameter) and for higher values of CNR [52]. For example, in Figure 6.15, the performance at 25 NM range is about 5 dB better than when using a fixed threshold, while at 15 NM where the CNR is low, the additional performance is only about 2 dB. It must also be remembered that the results are given for 'ideal CFAR' performance and in practice this is unlikely to be achieved. Also, operating a radar with a PRF of 100 Hz would often be quite impractical. However, it is noted [54] that by using an electronically scanned antenna capable of beam steering from pulse-to-pulse, it may be possible to collect coherent returns from a given direction at a desired low sample rate, while maintaining a suitable overall scan update rate.

6.5.2 Simulated performance prediction

The transient behaviour of detectors, especially when the clutter characteristics show spatial correlation over range, may be best assessed using simulation and Monte Carlo modelling. For non-coherent processing, a good example is the CA-CFAR detector, which predicts the mean level of the CUT from surrounding range cells. The statistical nature of such an estimate leads to a CFAR loss when compared to an exact knowledge of the mean level in the CUT. If there is a long sea swell pattern that is highly spatially correlated, a better estimate of the local value of intensity can be obtained using a shorter clutter range profile. In this case, there is a potential 'CFAR gain' as opposed to the usual loss [1, Chapter 13] and determining the performance can only be achieved with Monte Carlo methods.

Regarding coherent detectors, the analytic methods provide results that are averaged over all parameter values. Also the performance of both pre-Doppler and post-Doppler whitening filters cannot be easily modelled analytically as they are dependent on the spatial correlation of the data. As discussed in Chapter 5, sea clutter with the appropriate characteristics can be simulated either as discrete spectra varying over range or as continuous time-series data. These methods allow simulation of the range- or time-varying local power spectrum that is required for predicting performance using the adaptive detection methods. This means that detection performance, including some of the specific signal processing losses, can now be directly assessed using Monte Carlo simulations [41,55]. The main difficulty of this approach is the need to generate very large data samples to accurately assess low false alarm rates.

Once the means of simulating realistic data is available, it is straightforward to investigate the performance of different detection schemes. Consider the covariance matrix $\mathbf{R}$ that is required for the NAMF algorithm in (6.53). For a the kth range bin, the sample covariance matrix (SCM) estimate can be estimated from K range bins around the CUT,

$$\widehat{\mathbf{R}}_{\mathrm{SCM}}(k) = \frac{1}{K} \sum_{\substack{n=k-K/2,\\ n\neq k}}^{k+K/2} \mathbf{h}_n \mathbf{h}_n^H. \tag{6.66}$$

Alternatively, if the texture varies over range, the normalised sample covariance matrix (NSCM) can be estimated by

$$\widehat{\mathbf{R}}_{\mathrm{NSCM}}(k) = \frac{1}{K\widehat{\tau}} \sum_{\substack{n=k-K/2,\\ n\neq k}}^{k+K/2} \mathbf{h}_n \mathbf{h}_n^H; \qquad \widehat{\tau} = \frac{1}{KN} \sum_{\substack{n=k-K/2,\\ n\neq k}}^{k+K/2} \mathbf{h}_n^H \mathbf{h}_n \tag{6.67}$$

where $\widehat{\tau}$ is the estimated value of $\langle \tau \rangle$ and $\mathbf{h}_k$ is a vector of N coherent samples from the kth range cell. To demonstrate the Monte Carlo performance, consider simulated clutter with complex Gaussian samples, a fixed covariance and a CNR of 30 dB. Figure 6.16 illustrates the variation of P_{fa} for both an ideal exponential distribution and the NAMF using the NSCM estimate with $N = 32$ samples.

It is important to have confidence that the models are sufficiently representative of the real clutter to give reliable predictions of performance. For example, Figure 6.17 shows the performance of the NAMF using an SCM covariance estimate with two datasets. The first is the Ingara X-band data set and the second is a simulation using model parameters derived from the original data. The first two results show the

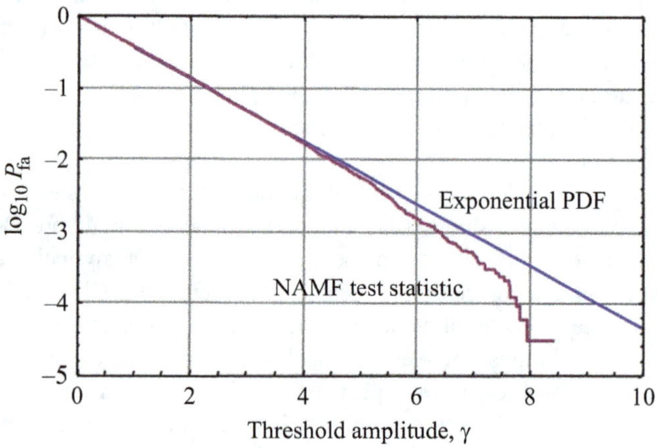

Figure 6.16 P_{fa} *as a function of threshold, γ, for the NAMF detector and the ideal test statistic for an exponential PDF [50].* © *IEEE 2013*

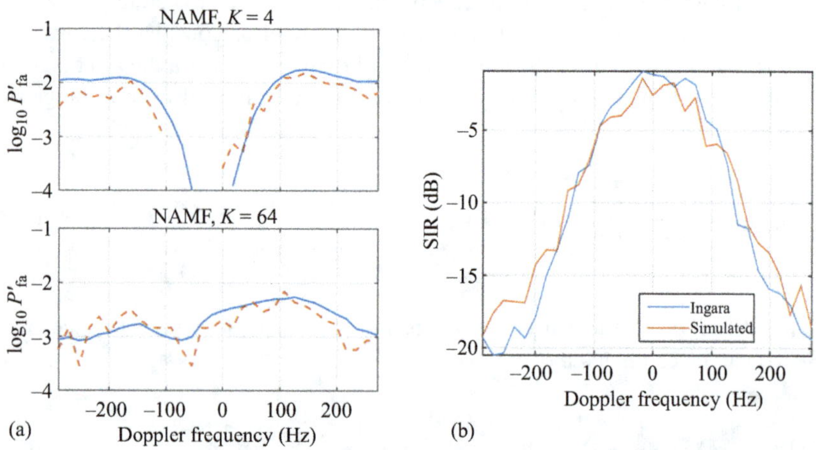

Figure 6.17 *Comparison of real and simulated detection performance for a desired* $P_{fa} = 10^{-3}$. *(a) Actual* P'_{fa} *against Doppler frequency for* $K = 4$ *and 64 [48]. (b) Detection performance measured by the SIR required to achieve* $P_d = 0.5$. © *Commonwealth of Australia 2017*

variation of $P'_{\rm fa}$ across Doppler for different amounts of averaging, $K = 4$ and 64, while the second shows the minimum required SIR for a $P_{\rm d} = 0.5$ and a $P_{\rm fa} = 10^{-3}$. While the performance is not identical, the trends in the variation of $P'_{\rm fa}(f)$ and the detection performance are well reproduced by the model.

Another example which cannot be easily achieved using real data is the relationship between spatial correlation and the accuracy of the local covariance matrix estimate [56]. To achieve this, the range correlation is modelled as a negative exponential function with a correlation length of $\rho_{\rm L}$ range samples:

$$\rho(k) = \exp\left(-\frac{k}{\rho_{\rm L}}\right). \tag{6.68}$$

The other model parameters are taken from an example sea clutter Doppler spectra reported in [57], with $\nu = 0.5$, $A_{\rm c} = 50$ Hz, $B_{\rm c} = 6$ Hz, CNR = 20 dB, $m_{\rm w} = 60$ Hz and $\sigma_{\rm w} = 4.7$ Hz. The coherent processing interval was $N = 16$ pulses, the PRF was 1 kHz and the number of range cells used to estimate the covariance was $K = 64$. For each value of range correlation coefficient, $\rho_{\rm L}$, data was simulated as coherent bursts of N pulses over groups of 30,000 range cells. Figure 6.18(a) shows $P'_{\rm fa}$ as a function of Doppler frequency for a fixed threshold, γ, using the SCM estimator in (6.66) and $\rho_{\rm L} = 0$, 5, 25, 50 and 100 range cells. The threshold was set assuming an ideally matched spectrum to achieve a $P_{\rm fa} = 10^{-3}$. For this example, the NAMF with the SCM estimator is not CFAR, showing a factor of 3 variation in $P'_{\rm fa}$ over Doppler for $\rho_{\rm L} = 0$. As $\rho_{\rm L}$ increases, the variation in $P'_{\rm fa}$ reduces and the average value approaches the desired value. Figure 6.18(b) shows the SIR required to achieve $P_{\rm d} = 0.5$, as a function of Doppler frequency for both the SCM and NSCM estimators and with $\rho_{\rm L} = 0$ or 100.

A similar example of predicted end-to-end radar performance using simulated data is now given using the example from Section 6.5.1, but with a finer range resolution of $\Delta R = 0.75$ m [51]. For this example, a Swerling 1 target model has

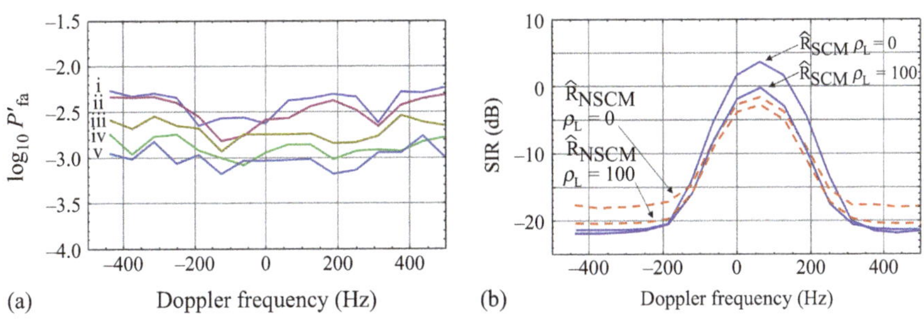

Figure 6.18 (a) Measured false alarm rate against Doppler frequency for the NAMF detector using $\widehat{\boldsymbol{R}}_{\rm SCM}$: i: $\rho_{\rm L} = 0$; ii: $\rho_{\rm L} = 5$; iii: $\rho_{\rm L} = 25$; iv: $\rho_{\rm L} = 50$; v: $\rho_{\rm L} = 100$; (b) minimum required SIR against Doppler frequency for $P_{\rm d} = 0.5$, using the NAMF detector with $\widehat{\boldsymbol{R}}_{\rm NSCM}$ and $\widehat{\boldsymbol{R}}_{\rm SCM}$, $\rho_{\rm L} = 0$ and 100 [56]. © IEEE 2015

been used with an RCS of 0 dBm^2 and no spatial correlation of the clutter texture was modelled. Three ranges are compared, 15, 25 and 40 NM, with the NAMF utilising an SCM covariance estimate and 128 range cells either side of the CUT ($K = 256$). Given the assumption here of no spatial correlation, the SCM should give a good estimate of the noise levels in the data but will generally be mismatched to the clutter spectrum because of the range-varying nature of the spectra. In general, this mismatch will result in the NAMF not producing CFAR behaviour. To achieve the desired P_{fa} of 10^{-3} at the display, the threshold in each Doppler bin can be adjusted to give $P_{\text{fa}} = 10^{-3}/32$ assuming collapsing in Doppler after detection. Such adjustment of the threshold in a practical radar may require the application of a CA-CFAR threshold, combined with an assessment of the statistics of the interference following incomplete whitening.

The results were compared with the performance of the Doppler filtering detector as described in Section 6.4.1 using the same scenario and radar parameters, except that the range resolution is 0.75 m and non-coherent processing assumes fixed frequency operation. An allowance was made for differences in the signal processing losses by adjusting the SIR as a function of range when determining the minimum detectable RCS. This ensures that the noise-limited performances of the two methods were approximately the same, which is a reasonable assumption for illustrative purposes. A more accurate analysis would require a detailed assessment of comparative processing losses in the two detectors. The final performance comparison is shown in Figure 6.19, where the simulated results are 'noisy', reflecting the limited size of the data samples used in the Monte Carlo simulation. However, it is interesting to

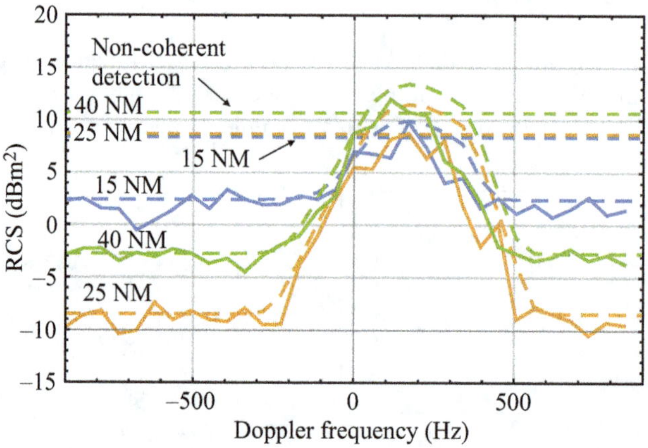

Figure 6.19 Minimum detectable RCS against range for $P_{\text{d}} = 0.5$, $P_{\text{fa}} = 10^{-5}$ with coherent and non-coherent processing in sea state 4, up-wind, up-swell, a height of 20,000 ft and $\Delta R = 0.75$ m. Solid line: simulation of NAMF with SCM estimator, dashed line: Doppler filter detector, fine dashed line: non-coherent detection [51].

note the improved performance predicted for the NAMF detector compared with the Doppler filter detector in the main body of the Doppler spectrum. The worst case Doppler is at 225 Hz, where the NAMF shows about 3 dB better detection sensitivity. This improvement is maintained at the edge of the spectrum, around 300–400 Hz. As expected, performances in the noise-limited regions are the same for both detectors.

6.5.3 Discussion

The comparison of coherent and non-coherent detectors must be treated with care, as there are many parameters that influence the results. For the relatively simple examples given here, it was assumed that the radar was looking up-wind and up-swell and platform motion effects were small. Coherent detection will vary with look direction and platform motion effects must also be considered when looking to either side of the aircraft track. The choice of PRF will also affect how well the clutter spectrum is resolved when detecting targets at similar Doppler frequencies. For fast moving targets, coherent processing gives the potential for very large processing gains compared with non-coherent processing. For targets with Doppler shifts that are comparable to the clutter, coherent and non-coherent processing gave similar results in the examples examined here and in some cases non-coherent processing may actually be better. For the Doppler analysis scheme, the worst detection is likely to occur at the edges of the clutter Doppler spectrum, where the range-varying nature can make the observed statistics over range very spiky. If only slow moving targets are of interest, the coherent detection schemes that exploit aliased clutter spectra may give significant advantage. In general, Doppler whitening filters may perform better than the simple Doppler analysis discussed here, but that also depends on the ability to estimate the local clutter spectrum, which can vary widely with environmental conditions. In addition, the added complexity of implementing a whitening filter would need to be weighed against simpler Doppler analysis methods, particularly if extra correction of the detection thresholds is required in each Doppler bin to ensure CFAR performance.

References

[1] K. D. Ward, R. J. A. Tough, and S. Watts, *Sea clutter: Scattering, the K-distribution and radar performance*, 2nd ed. The Institute of Engineering Technology, London, 2013.

[2] F. E. Nathanson, J. P. Reilly, and M. N. Cohen, *Radar design principles*, 2nd ed. McGraw-Hill, New York, 1991.

[3] M. I. Skolnik, *Radar handbook*, 3rd ed. McGraw-Hill, New York, 2008.

[4] D. K. Barton, *Radar system analysis and modeling*. Artech House, Norwood, 2005.

[5] S. Watts and G. Knight, "Performance prediction for scanning radars with correlated returns from K-distributed sea clutter," *Radar 2014 – International Conference on Radar Systems*, 2004, pp. 1–6.

[6] K. D. Ward, R. J. A. Tough, and P. W. Shepherd, "Modelling sea clutter: Correlation, resolution and non-Gaussian statistics," *IEEE International Radar Conference*, 1997, pp. 95–99.
[7] M. Schwartz, "A coincidence procedure for signal detection," *IRE Transactions on Information Theory*, vol. 2, no. 4, pp. 135–139, 1956.
[8] J. L. Marcum, "A statistical theory of target detection by pulsed radar," *IRE Transactions*, vol. IT-6, pp. 59–144, 1960.
[9] W. Weinstock, "Target cross section models for radar system analysis," Ph.D. dissertation, 1964.
[10] P. Swerling, "Probability of detection for fluctuating targets," *IRE Transactions*, vol. IT-6, pp. 269–308, 1960.
[11] D. A. Shnidman, "Efficient evaluation of the probabilities of detection and the generalized Q-function," *IEEE Transactions on Information Theory*, vol. 22, pp. 746–751, 1976.
[12] D. A. Shnidman, "The calculation of the probability of detection and the generalized Marcum Q-function," *IEEE Transactions on Information Theory*, vol. 35, no. 2, pp. 389–400, 1989.
[13] D. A. Shnidman, "Radar detection probabilities and their calculation," *IEEE Transactions on Aerospace and Electronic Systems*, vol. 31, no. 3, pp. 928–950, 1995.
[14] D. P. Meyer and H. A. Mayer, *Radar target detection*. Academic Press, New York, 1973.
[15] J. N. Briggs, *Target detection by marine radar*. The Institution of Electrical Engineers, London, 2004.
[16] M. R. Allen and H. Urkowitz, "Radar detection performance limitations in sea-clutter for conventional noncoherent integration," *IEEE National Radar Conference*, 1993, pp. 260–263.
[17] L. Rosenberg and S. Bocquet, "Non-coherent radar detection performance in medium grazing angle X-band sea-clutter," *IEEE Transactions on Aerospace and Electronic Systems*, vol. 53, no. 2, pp. 669–682, 2017.
[18] K. D. Ward, "A radar sea clutter model and its application to performance assessment," *International Radar Conference*, 1982, pp. 203–207.
[19] S. Bocquet, "Calculation of radar probability of detection in K distributed sea clutter and noise," Defence Science Technology Organisation, Technical Note DSTO-TN-1000, 2012, revised Edition.
[20] S. Watts, "Radar detection prediction in K-distributed sea clutter and thermal noise," *IEEE Transactions on Aerospace and Electronic Systems*, vol. AES-23, no. 1, pp. 40–45, 1987.
[21] T. S. Edrington, "The amplitude statistics of aircraft radar echoes," *IEEE Transactions on Military Electronics*, vol. 9, no. 1, pp. 10–16, 1965.
[22] A. Farina, F. Gini, M. S. Greco, and P. H. Y. Lee, "Improvement factor for real sea-clutter Doppler frequency spectra," *IEE Proceedings of Radar, Sonar and Navigation*, vol. 143, no. 5, pp. 341–344, 1996.

[23] D. K. Barton, "Simple procedures for radar detection calculations," *IEEE Transactions on Aerospace and Electronic Systems*, vol. AES-5, no. 5, pp. 837–846, 1969.

[24] I. Kanter, "Exact detection probability for partially correlated Rayleigh targets," *IEEE Transactions on Aerospace and Electronic Systems*, vol. AES-22, no. 2, pp. 184–196, 1986.

[25] S. Kotz and J. Neumann, "On the distribution of precipitation amounts for periods of increasing length," *Journal of Geophysical Research*, vol. 68, no. 12, pp. 3635–3640, 1963.

[26] L. Rosenberg, "The effect of temporal correlation with K and KK-distributed sea-clutter," *IEEE Radar Conference*, 2012, pp. 303–308.

[27] L. Rosenberg and S. Bocquet, "Radar detection performance in medium grazing angle X-band sea-clutter," Defence Science and Technology Group, Technical Report DST-Group-TR-3193, Dec. 2015.

[28] L. Rosenberg and S. Bocquet, "Robust performance prediction modelling for compound distributions with temporal correlation," *International Radar Conference*, 2013, pp. 388–393.

[29] M. A. Weiner, "Detection probability for partially correlated chi-square targets," *IEEE Transactions on Aerospace and Electronic Systems*, vol. 24, no. 4, pp. 411–416, 1988.

[30] X.-Y. Hou and N. Morinaga, "Detection performance of Rayleigh fluctuating targets in correlated Gaussian clutter plus noise," *The Transactions of the Institute of Electronics, Information and Communication Engineers*, vol. E. 71, no. 3, pp. 208–217, 1988.

[31] S. Bocquet, J. Zuk, and L. Rosenberg, "Non-coherent radar detection probability in compound sea clutter with correlated speckle," *IEEE Radar Conference*, 2018, pp. 0140–0145.

[32] J. Zuk, S. Bocquet, and L. Rosenberg, "New saddle-point technique for non-coherent radar detection with application to correlated targets in uncorrelated clutter speckle," *IEEE Transactions on Signal Processing*, vol. 67, no. 8, pp. 2221–2233, 2019.

[33] J. Zuk, "Correlated non-coherent radar detection for gamma-fluctuating targets in compound clutter," *IEEE Transactions on Aerospace and Electronic Systems*, 2021. [Online]. Available: https://doi.org/10.36227/techrxiv.14789673.v1

[34] S. Kotz and J. W. Adams, "Distribution of sum of identically distributed exponentially correlated gamma variates," *The Annals of Mathematical Statistics*, vol. 35, pp. 277–283, 1964.

[35] P. G. Moschopoulos, "The distribution of the sum of independent gamma random variables," *Annals of the Institute of Mathematical Statistics, Part A*, vol. 37, pp. 541–544, 1985.

[36] M. Alouini, A. Abdi, and M. Kaveh, "Sum of gamma variates and performance of wireless communication systems over Nakagami-fading channels," *IEEE Transactions on Vehicle Technology*, vol. 50, no. 6, pp. 1471–1480, 2001.

[37] C. W. Helstrom, "Detection probabilities for correlated Rayleigh fading signals," *IEEE Transactions on Aerospace and Electronic Systems*, vol. 28, no. 1, pp. 259–267, 1992.
[38] H. Robbins and E. J. G. Pitman, "Application of the method of mixtures to quadratic forms in normal variates," *Annals of Mathematical Statistics*, vol. 20, pp. 552–560, 1949.
[39] C. W. Helstrom, "Approximate evaluation of detection probabilities in radar and optical communications," *IEEE Transactions on Aerospace and Electronic Systems*, vol. 14, no. 4, pp. 630–640, 1978.
[40] A. De Maio and M. S. Greco, Eds., *Modern radar detection theory*. SciTech, London, 2016.
[41] S. Watts and L. Rosenberg, "Coherent radar performance in sea clutter," *IEEE Radar Conference*, Johannesburg, 2015, pp. 103–108.
[42] E. Conte, M. Lops, and G. Ricci, "Asymptotically optimum radar detection in compound-Gaussian clutter," *IEEE Transactions on Aerospace and Electronic Systems*, vol. 31, no. 2, pp. 617–625, 1995.
[43] E. Conte, A. De Maio, and G. Ricci, "Covariance matrix estimation for adaptive CFAR detection in compound-Gaussian clutter," *IEEE Transaction on Aerospace and Electronic Systems*, vol. 38, no. 2, pp. 415–426, 2002.
[44] F. Gini and M. Greco, "Covariance matrix estimation for CFAR detection in correlated heavy-tailed clutter," *Signal Processing, Special Issue*, vol. 82, no. 12, pp. 1847–1859, 2002.
[45] M. Greco, P. Stinco, F. Gini, and M. Rangaswamy, "Impact of sea clutter nonstationarity on disturbance covariance matrix estimation and CFAR detector performance," *IEEE Transaction on Aerospace and Electronic Systems*, vol. 46, no. 3, pp. 1502–1513, 2010.
[46] S. Watts, L. Rosenberg, S. Bocquet, and M. Ritchie, "The Doppler spectra of medium grazing angle sea clutter; part 2: Exploiting the models," *IET Radar Sonar and Navigation*, vol. 10, no. 1, pp. 32–42, 2016.
[47] S. Watts, L. Rosenberg, S. Bocquet, and M. Ritchie, "The Doppler spectra of medium grazing angle sea clutter; part 1: Characterisation," *IET Radar Sonar and Navigation*, vol. 10, no. 1, pp. 24–31, 2016.
[48] L. Rosenberg and S. Watts, "Coherent detection in medium grazing angle sea-clutter," *IET Radar Sonar and Navigation*, vol. 11, no. 9, pp. 1340–1348, 2017.
[49] Y. Dong, "Optimal detection in a K-distributed clutter environment," *IET Radar Sonar and Navigation*, vol. 6, no. 5, pp. 283–292, 2012.
[50] S. Watts, "The effects of covariance matrix mismatch on adaptive CFAR performance," *International Radar Conference*, Sep. 2013, pp. 324–329.
[51] S. Watts and L. Rosenberg, "A comparison of coherent and non-coherent radar detection performance in radar sea clutter," *International Conference on Radar Systems*, 2017, pp. 1–6.
[52] S. Watts, "Radar detection prediction in sea clutter using the compound K-distribution model," *IEE Proceedings F (Communications, Radar and Signal Processing)*, vol. 132, no. 7, pp. 613–620, 1985.

[53] S. Watts, "Cell-averaging CFAR gain in spatially correlated K-distributed clutter," *IEE Proceedings on Radar, Sonar and Navigation*, vol. 143, no. 5, pp. 321–327, 1996.
[54] B. J. Wilcox, S. Stevenson, and C. P. Mountford, "Coherent sea clutter modelling," *International Radar Conference*, 2017, pp. 1–6.
[55] R. Palamà, L. Rosenberg, and H. Griffiths, "Performance evaluation of two multistatic radar detectors on real and simulated sea-clutter data," *International Radar Conference*, 2018, pp. 1–6.
[56] S. Watts, "Modelling of coherent detectors in sea clutter," *International Radar Conference*, 2015, pp. 0105–0110.
[57] S. Watts, "Modeling and simulation of coherent sea clutter," *IEEE Transactions on Aerospace and Electronic Systems*, vol. 48, no. 4, pp. 3303–3317, 2012.

Chapter 7
Maritime target detection techniques

7.1 Introduction

This chapter describes some of the standard techniques that may be used to detect returns from targets against a background of sea clutter and added thermal noise. The goal of the radar signal processing is to maximise the chance of detecting a target, while controlling false alarms from clutter and noise at an acceptable level. The ideal outcome is a constant false alarm rate (CFAR) and the challenge is to achieve CFAR performance when the clutter characteristics are unknown 'a priori' and varying considerably over the region of sea being searched by the radar.

In every radar system, the clutter power levels will vary with range according to the radar range equation. At far ranges, the received signal may be noise-limited and the goal of target detection is to maximise the signal-to-noise ratio from the target. For a single pulse, this can be achieved in the radar receiver by matched filtering with the pulse waveform [1, Ch. 8]. The detectability of the target may be further enhanced by integrating returns from successive pulses as a radar scans its antenna beam over a target and from multiple scans. In terms of target detectability against noise, the most efficient form of pulse-to-pulse integration is achieved through Doppler processing. Alternatively, non-coherent processing of the returns may also be undertaken, using a non-linear detection process to extract the envelope of the received signal. Following coherent and/or non-coherent processing, the intensity or amplitude of the processed radar returns within each range gate is usually compared with a threshold to determine the presence or absence of a target. This detection threshold can be set at a level determined by the receiver noise power, which can be continuously measured by the radar processor. Noise returns that exceed this threshold will produce false alarms.

When the radar returns include sea clutter, similar processing steps may be used, but maintaining CFAR performance will be much more difficult. Care must also be taken to accommodate or control the dynamic range of signals in the radar receiver over the entire illuminated range to prevent the signal saturating prior to digitisation. This can be achieved through an automatic gain control or by using a logarithmic or linear fast-time control to reduce the intensity of sea clutter in near ranges. A good understanding of the characteristics of sea clutter is essential to assist with the design of suitable detection schemes. For example, the level at which the final detection threshold should be set to achieve CFAR performance must be continuously adjusted as the clutter amplitude statistics and power levels vary. These characteristics will not be known 'a priori' and must be estimated as part of the radar signal processing. Incorrect estimation of these conditions will result in the detection threshold being set incorrectly. If it is too low, the radar will have excessive false alarms which could overload the subsequent signal and data processing (e.g. the tracker), while setting the threshold too high will result in a loss of target detectability.

While a detection threshold based on amplitude alone may be optimal in thermal noise, controlling false alarms in clutter can be very difficult. As discussed in Chapter 2, the clutter amplitude statistics may be non-Gaussian and correlated both spatially and temporally. In addition, the sea clutter returns may contain isolated sea spikes, often with a lifetime of seconds. These spikes could appear target-like and it may not be possible to discriminate between them and real targets on the basis of amplitude alone. Various additional discriminants may be needed, especially when looking for small targets on the sea surface. Targets and clutter will have distinct Doppler spectra, and Doppler processing can sometimes be used to aid discrimination. An obvious example is a target with a high radial speed relative to the radar and a Doppler shift that places it well clear of the clutter spectrum. However, the benefit of Doppler processing may be less clear for slow moving or stationary targets. The choice of radar polarisation can have a strong effect on performance, and polarisation characteristics may sometimes be used as a discriminant. However, this has not been widely used in airborne maritime radars, mainly due to the high implementation cost. An important discriminant between sea clutter spikes and real targets is their relative persistence. Most large spikes have a lifetime of less than 2 or 3 s [2–4]. Provided that the target can be observed over a longer exposure time, it should be possible to reject false alarms from spikes. Other discriminants have been proposed such as range extent and range signature [2], which are suitable when the radar has a fine range resolution (e.g. <1 m).

The following sections describe a number of techniques for detecting targets in sea clutter. They are broadly subdivided into non-coherent and coherent signal processing methods. Section 7.2 provides a summary of the commonly used CFAR and scan-to-scan processing techniques used for non-coherent processing, while Section 7.3 focusses on the development of CFAR methods designed for Pareto distributed clutter. Section 7.4 then covers an introduction to coherent processing in both the time and frequency domains. One of the key difficulties with coherent processing is maintaining performance in heterogeneous clutter. Section 7.5 explores this topic with a number of algorithms designed to operate in such an environment. Predicting the performance of these different detectors is covered in Chapter 6.

7.2 Non-coherent detection

Non-coherent detection involves processing the envelope of the radar returns, typically obtained by taking the modulus (linear detector) or modulus squared (square law detector) of the coherent signal. In modern radars, this is achieved as part of the digital signal processing at an intermediate frequency or at base band. A detailed review of non-coherent detection techniques can be found in [3]. A typical non-coherent processing chain was shown in Chapter 6 and involves sampling, square law detection, pulse-to-pulse integration and scan-to-scan integration with detection based on an amplitude threshold. In order to set the detection threshold with an acceptable false alarm rate, the radar must determine the characteristics of the clutter-plus-noise (interference) probability density function (PDF). If the clutter is modelled by a compound Gaussian distribution, the probability of false alarm, P_{fa}, for a fixed threshold, γ, can be written as

$$P_{\mathrm{fa}}(\gamma) = P_{\gamma}\left(\gamma|N\right) = \int_{0}^{\infty} \Gamma\left(N, \frac{\gamma}{\tau + p_{\mathrm{n}}}\right) \frac{P_{\tau}(\tau)}{\Gamma\left(N\right)} d\tau \tag{7.1}$$

where N is the number of pulses integrated (assuming independent samples of the clutter speckle from pulse-to-pulse), p_{n} is the thermal noise power and $P_{\tau}(\tau)$ is the PDF of the clutter texture. If the mean of the clutter texture is $\langle\tau\rangle$ and the clutter-to-noise ratio (CNR), $C = \langle\tau\rangle / p_{\mathrm{n}}$, then the threshold, γ, is often expressed in terms of the level relative to the mean clutter-plus-noise intensity, $\langle z\rangle$,

$$\gamma = \eta \langle z\rangle = \eta \langle\tau\rangle \left(1 + \frac{1}{C}\right) \tag{7.2}$$

where η is called the threshold multiplier. The appropriate value of γ must be determined by the radar and involves inverting (7.1) using knowledge of the clutter power $\langle\tau\rangle$ and the appropriate value of threshold multiplier, η, which will in turn depend on the statistics of the clutter and the CNR. The following section provides an overview of methods to estimate these parameters.

7.2.1 Clutter statistics

The interference amplitude statistics are described in terms of their PDF. As discussed in Chapter 2, there are various models for the clutter distribution, with the most useful being the compound-Gaussian models. These models can be used to guide the design of detection algorithms, with the goal of being robust to the wide range of clutter characteristics encountered in practice.

The appropriate interference PDF for the cell under test (CUT) can be estimated by analysing data from the surrounding range and azimuth cells. This region should be as large as possible but will be limited so that each bin has the same PDF. Successive pulses from the same area can be used to improve the accuracy of any estimate, especially if pulse-to-pulse frequency agility is used. There are a number of techniques for estimating the clutter parameters from the measured data. Methods based on fitting moments and using the $z \log z$ estimator were given in Chapter 2 for a number of distributions. Another alternative is to directly measure the clutter complementary cumulative distribution function (CCDF) and then find a distribution model that provides the best fit. This can be achieved by a least-squares minimisation between the CCDFs of the data and model, with a better match to the distribution tail achieved by performing the minimisation in the log domain. This approach is particularly useful as it allows a distribution model to extrapolate to lower false alarm levels where there may not be enough real data. However, the accuracy to which the interference PDF can be estimated will depend on the size of the data samples available.

Whatever the method used, the accuracy to which a given PDF model parameter can be estimated must be interpreted in terms of the effects of the resulting error on any threshold that is set. If the estimated shape parameter is incorrect, any threshold set assuming that value will be incorrect. A threshold error will either result in a loss of sensitivity (threshold set too high) or an increase in P_{fa} (threshold set too low). The impact of setting a threshold in K-distributed clutter-plus-noise based on an incorrect value of ν can be assessed by expanding (7.1) with the texture PDF modelled by a gamma distribution,

$$P_\eta(\eta|\nu, N) = \int_0^\infty \Gamma\left(N, \frac{\eta(1+p_\mathrm{n})}{\tau + p_\mathrm{n}}\right) \frac{\nu^\nu \tau^{\nu-1}}{\Gamma(N)\Gamma(\nu)} \exp(-\nu\tau)\, d\tau \tag{7.3}$$

where the clutter mean $\langle\tau\rangle = 1$. For an ideal value of shape parameter, ν and a specified P_{fa}, the threshold multiplier η can be determined by inverting (7.3). With this value of η fixed, the actual P_{fa} based on the estimated shape parameter $\hat{\nu}$ can be determined by again solving (7.3). Figure 7.1 shows the error in P_{fa},

$$\Delta P_{\mathrm{fa}} = \frac{P_\eta(\eta|\hat{\nu}, N)}{P_\eta(\eta|\nu, N)} \tag{7.4}$$

where the specified $P_{\mathrm{fa}} = 10^{-6}$, $N = 1$, $C = 30$ dB, $\nu = 0.1, 1, 3, 10$ and $0.5\nu \le \hat{\nu} \le 1.5\nu$. The equivalent results for $N = 10$ are shown in Figure 7.2. In both cases, $\Delta P_{\mathrm{fa}} = 1$ signifies that the achieved P_{fa} is an order of magnitude higher than the required P_{fa}.

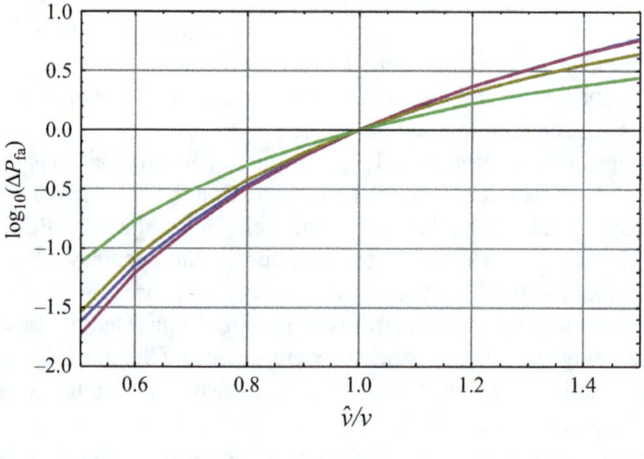

Figure 7.1 $\log_{10}(\Delta P_{\mathrm{fa}})$ *for $N = 1$: (—) $\nu = 10$, (—) $\nu = 3$, (—) $\nu = 1$, (—) $\nu = 0.1$*

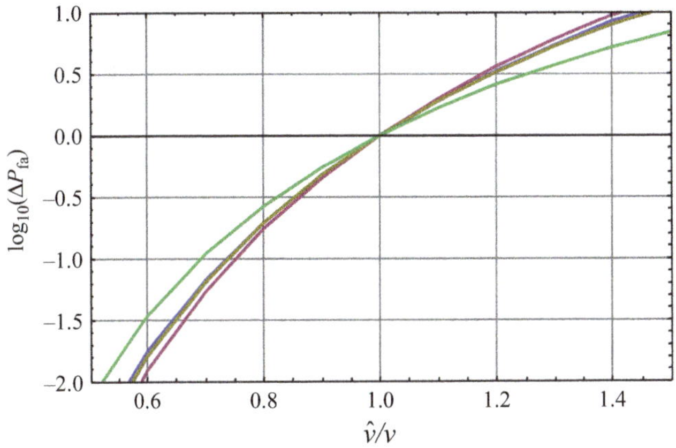

Figure 7.2 $log_{10}(\Delta P_{fa})$ *for N = 10: (—) ν = 10, (—) ν = 3, (—) ν = 1, (—) ν = 0.1*

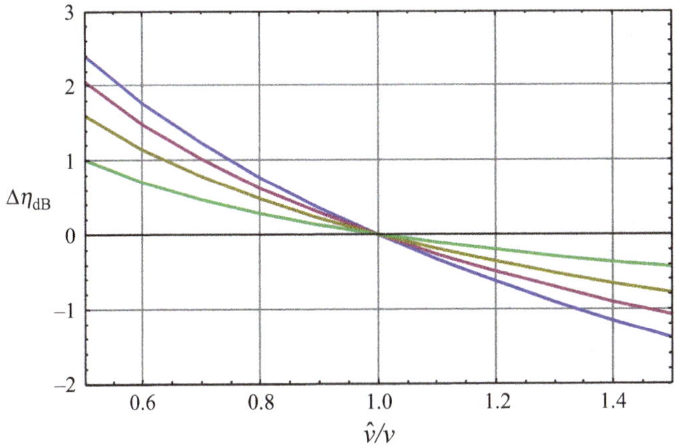

Figure 7.3 $\Delta\eta_{dB}$ *for N = 1: (—) ν = 10, (—) ν = 3, (—) ν = 1, (—) ν = 0.1*

The variation in sensitivity in terms of the threshold multiplier can similarly be expressed by

$$\Delta\eta_{dB} = 10\log_{10}\left(\frac{\eta(\hat{\nu})}{\eta(\nu)}\right) \tag{7.5}$$

where the value of η achieved using the estimated shape, $\eta(\hat{\nu})$, is compared with the correct value, $\eta(\nu)$. Here a positive value of $\Delta\eta$ in dB signifies that the threshold multiplier is higher than required and hence there is an equivalent loss in detection sensitivity. Figures 7.3 and 7.4 show the values of $\Delta\eta$ associated with errors in the estimation of the shape parameter for $N = 1$ and 10, respectively.

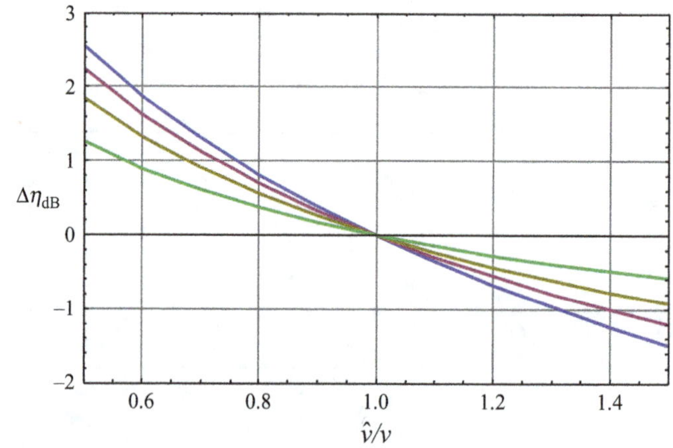

Figure 7.4 $\Delta\eta_{dB}$ *for N = 10: (—) ν = 10, (—) ν = 3, (—) ν = 1, (—) ν = 0.1*

For the examples given here, it can be seen that if $\log_{10}(\Delta P_{fa}) \approx -1.5$ (i.e. the threshold is too high), there is a loss of sensitivity, given approximately by $\Delta\eta_{dB} \lesssim 1.5$ dB. The more serious problem for the radar in practice will be setting a threshold too low. For example, $\log_{10}(\Delta P_{fa}) \approx 1.5$ equates with the false alarm rate being increased by a factor of 30. However, as ν increases, C reduces and the effects of errors become less severe. For small values of ν, the value of C has less influence.

The thermal noise level may vary as the temperature of the equipment varies and its performance degrades over time. Many modern radars will continuously monitor the receiver thermal noise level to aid in setting detection thresholds and receiver gain. It can be measured from Doppler filters centred on frequencies well away from the clutter spectra or when the receiver input is muted. It is also possible to determine the noise level from analysis of the interference returns. For K-distributed clutter-plus-noise, the analysis of the first three intensity moments can provide estimates of the clutter and noise parameters [3]:

$$\hat{\nu} = \frac{18\left(\langle z^2\rangle - 2\langle z\rangle^2\right)^3}{\left(12\langle z\rangle^3 - 9\langle z\rangle\langle z^2\rangle + \langle z^3\rangle\right)^2}, \tag{7.6}$$

$$\hat{p}_n = \langle z\rangle - \left(\frac{\hat{\nu}}{2}\left(\langle z^2\rangle - 2\langle z\rangle^2\right)\right)^{1/2}. \tag{7.7}$$

Similar expressions can be derived for other compound Gaussian models, such as the Pareto distribution [5–9]. The problem with this method for estimating the noise power is the large number of independent and identically distributed data samples that are required to accurately estimate the third-order moments.

7.2.2 *Constant false alarm rate detectors*

A standard technique for estimating the local clutter intensity or amplitude is the cell-averaging (CA)-CFAR [10], which is illustrated in Figure 7.5. Here the mean

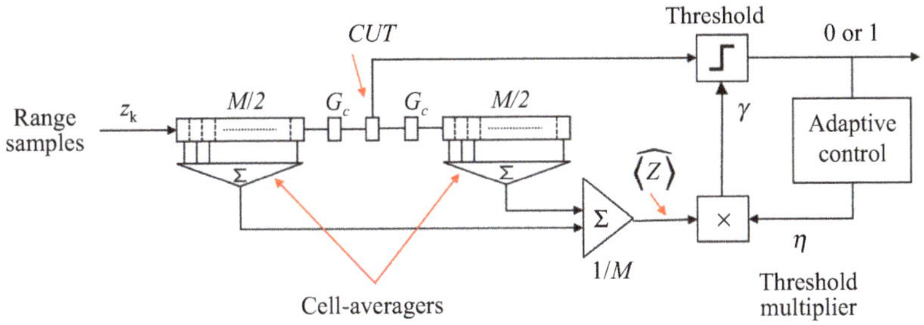

Figure 7.5 Cell-averaging CFAR with M samples in the clutter range profile and $2G_c$ guard cells

level of the CUT is estimated from the average of M surrounding range cells, also known as the clutter range profile (CRP). Guard cells, G_c, can also be used to avoid contamination from any range-extensive targets. The threshold, γ, to be applied to the CUT is determined from an estimate of interference power and a threshold multiplier, η, as in (7.2). However, due to statistical variations of the estimated mean level, the value of η cannot be calculated directly by inverting (7.1).

If the CRP is defined by the region $\boldsymbol{\Omega}_k = [k - M/2 - G_c, \ldots, k - G_c + 1, k + G_c + 1, \ldots, k + M/2 + G_c]$, then local mean for the kth range bin is estimated from

$$\widehat{\langle z_k \rangle} = \frac{1}{M} \sum_{m \in \boldsymbol{\Omega}_k} z_m \tag{7.8}$$

with the threshold, $\gamma_k = \eta\widehat{\langle z_k \rangle}$. Because the sample size, M, used to estimate the mean level is finite, the estimated mean level, $\widehat{\langle z_k \rangle}$, will fluctuate from one estimate to the next and have its own statistics. This means that the threshold, γ_k, will also fluctuate and will require a larger value of η for a given P_{fa} than if the true value of z_k were known. This fluctuation in threshold is known as a CFAR loss and is formally defined as the 'increase in signal power required to achieve a given probability of detection, P_d, for a given P_{fa}, compared with the ideal fixed threshold having complete knowledge of the clutter and noise statistics'. The CFAR loss in sea clutter can be significant, especially in spiky conditions [3]. Figure 7.6 illustrates the CFAR loss for K-distributed clutter with $\nu = 0.2$, and for $C = 10$ dB and $C = \infty$.

From Figure 7.6, it can be seen that if $M > 32$, the CFAR loss may be tolerable. It will also be less for larger values of ν [3]. However, the loss can become very large if shorter values of M are used, especially in the absence of added thermal noise. For example, with $\nu = 0.2$, $M = 8$ and $C = \infty$, the CFAR loss for $P_{fa} = 10^{-6}$ is about 24 dB and with $C = 10$ dB, the loss is about 6 dB. By comparison, the equivalent CFAR loss in thermal noise alone would be about 4 dB.

The CA-CFAR shown in Figure 7.5 uses cell-averagers either side of the CUT and is known as a double-sided averager. It is also possible to have single-sided averagers that only use cells that are up- or down-range from the CUT. Using surrounding cells

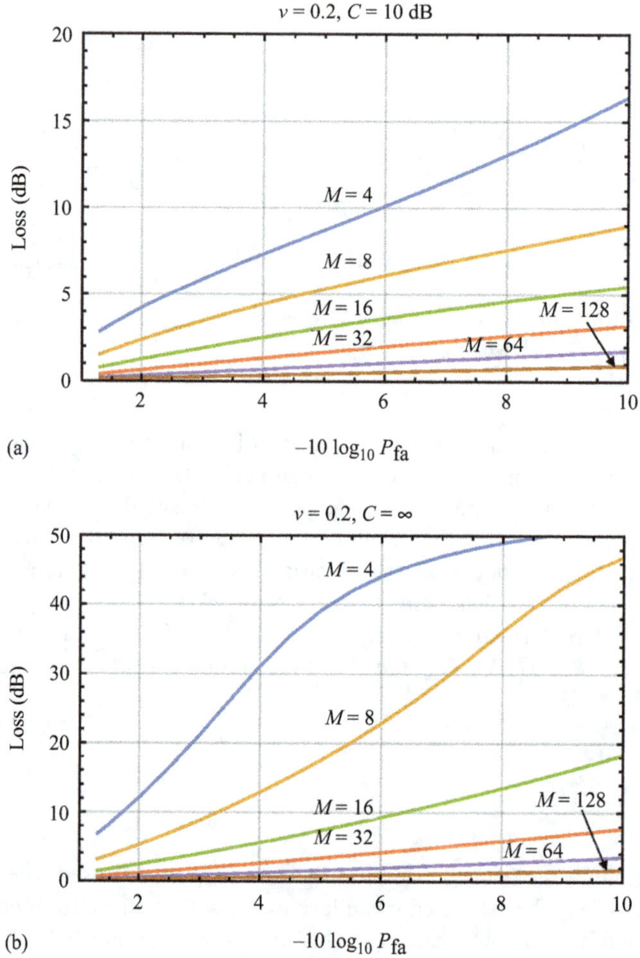

Figure 7.6 CFAR loss in K-distributed clutter with ν = 0.2 and M = 4, 8, 16, 32, 64 and 128 samples: (a) C = 10 dB and (b) C = ∞ [3]

makes the assumption that they are independent of the CUT and have the same mean level. This requirement will limit the number of cells that can be used in a practical application, as the clutter reflectivity and clutter patch size vary over range. The effect of correlation between the estimated local mean and the CUT is discussed in Section 7.2.3. There is also an implicit assumption that the surrounding cells do not include too many targets. However, it is inevitable that if a target is present over the ranges covered by the cell-averager, it will be included in the CRP and slightly bias the local estimate of the CUT. If the target density is low, it can usually be assumed that no targets are present in the cell-averager cells when a target is present in the CUT.

A conventional double-sided CA-CFAR can also cause problems when there is a step change of clutter intensity over range. As the leading (down-range) cells of the CRP approach the increase, the threshold at the CUT becomes too high for the required P_{fa}, and there is an increasing loss in detection sensitivity. Then once the CUT enters the region of high clutter, the threshold is immediately too low, resulting in an increase of false alarms. Such step changes in clutter level may be observed in an airborne maritime radar that is illuminating the interface between land and sea. This would limit the ability to detect maritime targets close to shore lines or in littoral regions where islands can create 'shadows' by reducing the impact of the wind. While the radar may not be intending to detect land targets, the excessive false alarms may swamp the radar signal processor and overwhelm the target tracker. One solution to this problem is apply a shoreline mask, but that will require an accurate map and potentially limit the effectiveness of operation in the littoral environment. However, there are many CFAR variants aimed at mitigating these problems. Some examples include the following:

Greater of (GO)-CFAR: It uses the greater of the two mean level estimates of a double-sided CA-CFAR.
Smaller of (SO)-CFAR: It uses the smaller of the two mean level estimates of a double-sided CA-CFAR.
Ordered statistic (OS)-CFAR: The samples in the CRP are placed in ranked order of magnitude and the local clutter power is estimated from the intensity of the Jth largest sample.
Trimmed mean (TM)-CFAR: The samples in the CRP are rank ordered with T_1 trimmed from the lowest values, T_2 trimmed from the highest values and the mean of the remaining samples used to estimate the local clutter level.
Censored mean (CM)-CFAR: This is a TM-CFAR with $T_1 = 0$.

The GO-CFAR and SO-CFAR are aimed at controlling problems associated with clutter edges. The GO-CFAR suppresses the increase in false alarms and the SO-CFAR prevents a loss in sensitivity close to the edge. The TM-CFAR with a large value of T_1 and a small value of T_2 can also reduce the increase in false alarms at a clutter edge. The other CFAR variants are mainly aimed at reducing the effects of having multiple targets in the CRP. For the OS-CFAR, the $M - J$ strongest returns from the CRP are removed which could be from clutter discretes or targets. In the TM-CFAR and CM-CFAR, the value of T_2 would ideally equal the number of interfering targets.

All these variants will incur an additional CFAR loss compared with the CA-CFAR [3], although in many cases this may be quite small. For example, the GO-CFAR has an additional loss in thermal noise of <0.3 dB even for short CRPs. There are also many other variations of these configurations, aimed at improving performance in specific scenarios [11].

7.2.3 CA-CFAR performance in spatially correlated clutter

The CFAR loss results in Figure 7.6 assume that independent samples of the K-distributed clutter are present in the CRP. Any spatial correlation in the clutter texture, as described in Chapter 2, can also have a significant effect on the CA-CFAR performance. In the extreme of very strong spatial correlation, such as from a long wavelength sea swell, a CRP of appropriate length might be able to follow the local

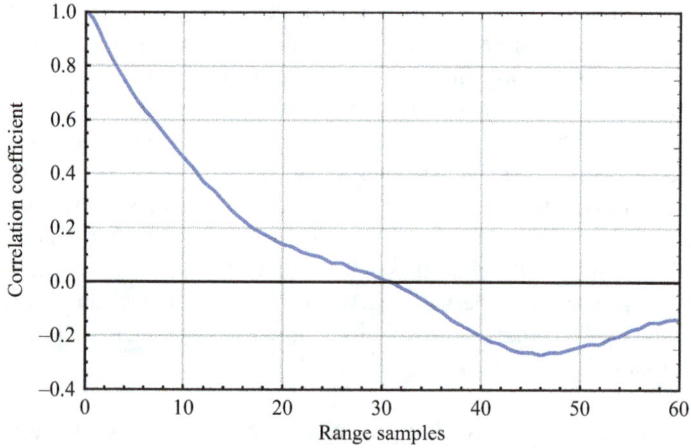

Figure 7.7 Correlation coefficient for sea clutter texture with strong spatial correlation

mean level, rather than trying to adapt to the average mean level. The limiting performance is achieved by an 'ideal CFAR' [12] that has exact knowledge of the local texture intensity, τ. In practice, ideal CFAR performance will not be achieved but it may still be possible to see some CFAR gain (i.e. a negative CFAR loss) in some conditions.

As an example, Figure 7.7 shows the texture correlation coefficient for an example data set having a range sample spacing of 3 m. Using this spatial correlation, K-distributed clutter has been simulated with a shape parameter of $\nu = 1$ and a CNR of 30 dB. Figure 7.8 shows a range slice from this simulated clutter following coherent integration over $N = 10$ pulses with independent speckle from pulse-to-pulse. It also shows the thresholds set by a double-sided CA-CFAR with a CRP of length $M = 10$ and a fixed threshold designed to achieve a $P_{fa} = 10^{-3}$. In this case, it can be seen that the CA-CFAR threshold has a lower level on average than the fixed threshold. In fact, the adaptive threshold is exhibiting a 'CFAR gain' of about 2 dB, while the threshold multiplier is set to 6.5 dB above the local mean and the fixed threshold is 8.5 dB above the global mean. The amount of CFAR gain (or loss) obtained in practice will depend on the degree of spatial correlation, the length of the CRP and its configuration (double-sided or single-sided, etc.). Figure 7.9 shows the CFAR loss for the same simulated clutter and also for data with the same amplitude statistics, but independent texture samples in adjacent range cells. For the former, it can be seen that there is a broad range of values giving a CFAR gain and a small value of M would be preferred. Conversely, in the absence of any spatial correlation, low values of M would give a significant CFAR loss. Further results for different correlation coefficients and clutter shape parameters can be found in [3].

7.2.4 Scan-to-scan processing

The presence of persistent sea spikes can make it difficult to control false alarms while detecting real targets. Discriminating between small real targets and large sea clutter

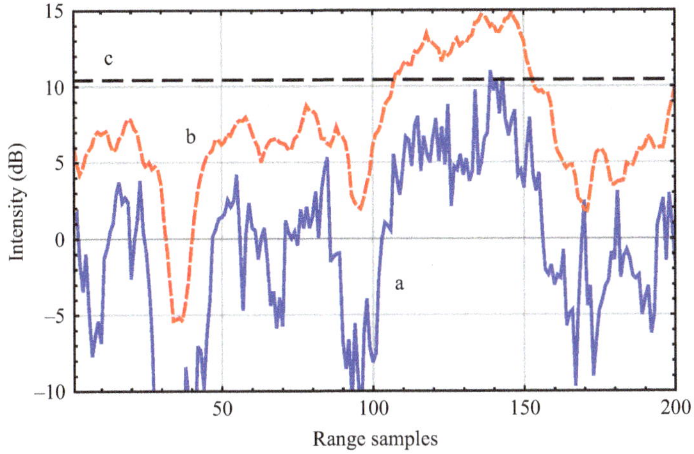

Figure 7.8 *CA-CFAR example in spatially correlated clutter: (a) K-distributed clutter, $\nu = 1$, $C = 30$ dB, $N = 10$; (b) CA-CFAR threshold, $P_{fa} = 10^{-3}$, $M = 10$, $G_c = 2$; (c) fixed threshold, $P_{fa} = 10^{-3}$*

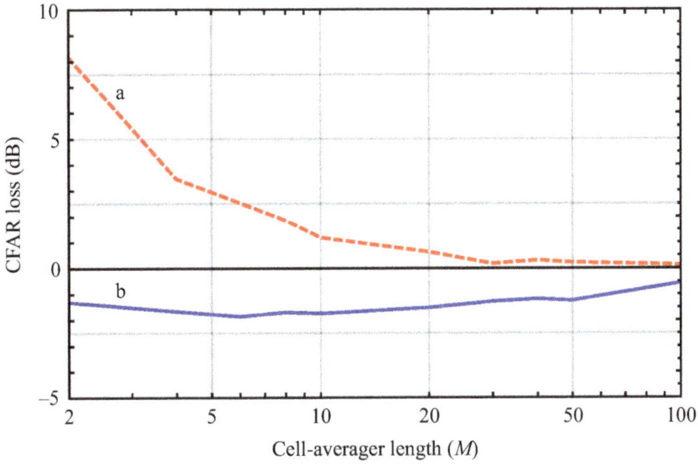

Figure 7.9 *CFAR loss as a function of clutter range profile length in K-distributed clutter, $\nu = 1$, $N = 10$, $C = 30$ dB: (a) clutter with no spatial correlation; (b) clutter with correlation as in Figure 7.7, showing CFAR gain*

spikes cannot be done reliably on the basis of amplitude alone. While the Doppler shift can be a good discriminant for targets with sufficiently high radial velocity relative to the radar, it may not be reliable for targets where the Doppler shift lies within the sea-clutter Doppler spectrum. Submarine periscope masts are a good example of small targets with a low velocity, compounded by an additional complication that they may only be exposed above the water for several seconds. Small slow moving boats or buoys will have a more persistent presence but are also likely to have Doppler shifts that place them within the clutter Doppler spectrum. In addition to the techniques already described, it may be possible to exploit other target characteristics to discriminate them from clutter spikes. One possibility is the range extent of large targets and the wake that follows a moving target. An important characteristic of sea clutter spikes is their relatively short lifetime and observing returns over a sufficiently long time to allow discrimination against more persistent targets is an important part of detection. This can be achieved with scan-to-scan processing.

Airborne radars will typically employ an antenna continuously scanning in azimuth. The speed of scanning will be a compromise between the number of pulses required in a beam dwell and the scan-to-scan update rate. When detecting targets in sea clutter, there may be benefit from processing returns over several scans, using scan-to-scan integration or track-before-detect methods. Analogue scan-to-scan processing will average the clutter fluctuation and may assist in discriminating against sea spikes by providing some decorrelation of the clutter texture. However, if the scan update rate is too fast, some of these potential benefits may be lost. Clutter spikes may typically have lifetimes of 1–2 s and if the update rate is appreciably shorter than this, the spike returns may remain correlated from scan-to-scan. Further, scan-to-scan processing is often less efficient than pulse-to-pulse processing. This loss of efficiency may occur due to the need to accommodate unknown target motion. If Doppler processing is used, this may require a longer dwell time, which may limit the scan rate that can be used. Discussion of the competing merits of different scan rates can be found in [13,14].

Binary scan-to-scan processing occurs by thresholding the signals from each scan. This is typically referred to as an M/N binary detector with a detection declared when there are at least M_b detections out of N_b previous scans. The technique is used to control false alarms while maintaining or improving the probability of detection. For an airborne radar with a scan period of 1 s, the scan-to-scan binary integration might take place over 3–10 scans, depending on the exposure time of the target and the update rate required. If it is assumed that the binary detections (false alarms and target detections) are statistically independent from scan-to-scan, then the binary integrator can be represented by

$$P_{\text{out}} = \sum_{n=M_b}^{N_b} \frac{N_b!}{(N_b - n)!n!} P_{\text{in}}^n (1 - P_{\text{in}})^{N_b - n} \tag{7.9}$$

where P_{in} is the probability of detection or false alarm at the input to the binary integrator and P_{out} is the probability at the output. Usually, when predicting performance, the probabilities of detection and false alarm at the output of the integrator represent the desired radar performance. It will then be necessary to invert (7.9) to determine the performance at the input of the scan-to-scan integrator. There will be losses due to the presence of persistent sea spikes between scans and the signal processing needs to take into account any relative target and platform motion between scans.

With the radar on a moving platform, it is important that the returns on successive scans are ground stabilised to remove changes in relative range and bearing due to platform motion. It is also important to consider the possible motion of the target when integrating over scans. The automatic radar periscope detection and discrimination (ARPDD) program [2] was designed to automate the detection of transient periscope exposures while maintaining an extremely low false alarm rate. It uses a two-stage periscope declaration process that includes conventional target detection with a moderate false alarm rate followed by signature discrimination to reduce the false alarms. This latter approach is termed 'retrospective processing' and involves looking for targets that are persistent over time. Another approach described in [15] proposes a series of range-azimuth 'tubes' that are tested for the presence of a target over multiple scans. It is also noted that if a Doppler shift is measured as part of the pulse-to-pulse processing, this can be used to aid with a plot association of successive returns that have inconsistent range-rate and Doppler shift. This is effectively a form of track-before-detect processing and is expanded further in Chapter 8.

7.2.5 Alternative methods

If the clutter is well modelled by a family of distributions, then this information can be used to describe a CFAR detector. For example, in [16], the clutter following envelope detection in the radar receiver was assumed to be modelled using either the Weibull or log–normal family of distributions. A test statistic is defined as

$$\frac{\varsigma_k - \frac{1}{M}\sum\limits_{m\in\boldsymbol{\Omega}_k}\varsigma_m}{\sqrt{\frac{1}{M}\sum\limits_{m\in\boldsymbol{\Omega}_k}\left(\varsigma_m - \frac{1}{M}\sum\limits_{m'\in\boldsymbol{\Omega}_k}\varsigma_{m'}\right)^2}} \underset{H_0}{\overset{H_1}{\gtrless}} \gamma \tag{7.10}$$

where $\varsigma_m = \ln\sqrt{z_m}$ and ς_k is the CUT, with M independent and identically distributed samples in the CRP. This target detection problem is formulated as a hypothesis test with the null hypothesis, H_0 being that the target is not present and the alternative hypothesis H_1 being that the target is present. If the clutter has either a Weibull or log–normal distribution, then the test statistic is CFAR with respect to changes in shape and scale, but requires different threshold settings to achieve the desired P_{fa}. The trade-off with this approach is the high CFAR loss. For example, with Rayleigh distributed clutter, $M = 50$ and $P_{\text{fa}} = 10^{-4}$, the loss is 2.5 dB, compared with a loss of closer to 0.5 dB for a CA-CFAR. As the clutter becomes spikier, the loss increases with a value of 10 dB for a Weibull shape parameter of $\nu_{\text{W}} = 0.5$.

Another approach to CFAR detection is to exploit the transient coherence of sea clutter [17]. This attempts to predict the sea-wave dispersion relation from a series of observations from an area of sea that can be used to build up a range/time intensity plot. The power spectral density of the range/time intensity plot is known as an ω–k plot and is used to predict the movement of wave features in order to predict the threshold required at a later time.

Rather than estimating the parameters of a PDF model directly, another method is to count false alarms following the detection threshold and then control the threshold multiplier, η, in a closed-loop feedback system to maintain a desired false alarm rate.

An early version of this was described in [18], where a leaky integrator was used to continually adjust the values of the threshold multiplier as the radar scans in azimuth. In this system, the initial threshold was adjusted to meet the desired $P_{\text{fa}} \sim 10^{-2}$. This value was intentionally set high, as counting false alarms directly can be a problem if there are not enough samples in the clutter dominated region, (i.e. there may only be 1,000 high-resolution range cells covering a range of 1–2 km). To achieve lower false alarm rates (say 10^{-4}–10^{-6}), extrapolation using an assumed model may be employed [19]. This approach can work well unless there is a model mismatch, such as with discrete sea spikes that appear target-like over short durations.

7.3 Pareto CFAR detectors

The most widely used compound model is the K-distribution which is described in terms of Rayleigh speckle modulated by a gamma distributed texture (see Chapter 2). However, it has been demonstrated that the K-distribution cannot capture the effect of discrete sea spikes, which can appear target-like and extend the tails of the standard modelled distributions [20,21]. This led to the development of other models such as the KA- and KK-distributions [20,22,23] which are described in Chapter 2. Subsequent work has found that if the underlying RCS is modelled with an inverse gamma distribution, then the compound representation becomes Pareto distributed. This model is a useful alternative as it is described with only two parameters yet can often better model the long tails due to discrete spikes present in the sea clutter distribution [6–8,21]. As a result, there has been much interest in the production and analysis of CFAR detectors under such a clutter model assumption [9,24–27].

An approach to constructing CFAR detectors based on the memoryless nonlinear transformation was introduced in [24] and then generalised in [28]. This technique shows how the original CFAR detectors, designed to operate with exponentially distributed clutter intensity, could be transformed to any other clutter distribution. The approach has the added advantage of preserving the original relationship between the probability of false alarm, P_{fa}, and the threshold multiplier. One disadvantage of this approach is that when applied to Pareto clutter, the resultant transformed detector required a priori knowledge of the Pareto scale parameter [27,29]. It turns out that this can be rectified for Pareto Type I distributions through application of sufficient statistics and for Pareto Type 2 distributions, using a Bayesian approach. Based on this, a general form for CFAR detectors can be derived for Pareto distributed clutter.

The Pareto Type I distribution has the following cumulative distribution function for the intensity, $\bar{z}$ [26],

$$F(\bar{z}) = 1 - \left(\frac{b_{\text{p}}}{\bar{z}}\right)^{\nu_{\text{p}}}, \quad \bar{z} \geq b_{\text{p}} \tag{7.11}$$

where ν_{p} is the shape and b_{p} is the scale. This form of the Pareto distribution model has been widely used as it simplified the development of the Pareto CFAR detectors. However, real data will have backscatter intensity values between 0 and b_{p} and any detectors based on this model require the backscatter intensity to be offset by an estimate of the scale parameter. An alternative is to use the Pareto Type II distribution for the random variable, z,

$$F(z) = 1 - \left(\frac{b_{\mathrm{p}}}{z + b_{\mathrm{p}}}\right)^{\nu_{\mathrm{p}}}, \quad z \geq 0 \tag{7.12}$$

which is the form introduced in Chapter 2. These two distributions are related via the fact that $z + b_{\mathrm{p}} \equiv \bar{z}$, and hence the Pareto Type II distribution is exactly the Pareto Type I distribution, but shifted so its support begins at zero.

In this section, the CA-CFAR and the OS-CFAR detectors are presented for both Pareto Type I and II distributions with independent samples along range. Variations that remove the need to estimate the scale parameter are also presented for the Pareto Type I model using sufficient statistics and for the Pareto Type II model using a Bayesian approach. To put this into a mathematical framework, consider the function, g, used to measure the level of interference and the CUT represented by the statistic z_k with the CRP described by the intensity samples, $z_k \in \mathbf{\Omega}_k$. To test the hypothesis that z_k contains interference only (denoted as H_0) against the alternative that it contains a target embedded within the interference (H_1), the following decision rule can be written,

$$z_k \underset{H_0}{\overset{H_1}{\gtrless}} \eta g(z_1, z_2, \ldots, z_M) \tag{7.13}$$

where the threshold multiplier, η, is determined from the desired P_{fa}, and the objective of each detector below is to select g such that the P_{fa} does not vary with unknown clutter parameters, thus acquiring the CFAR property. The notation used in (7.13) means that H_0 is rejected (i.e. a target is present) when z_k is greater than the right hand side of (7.13).

In the following subsections, the robustness of these detectors are described over a variety of realistic conditions [30]. This includes data that does not match the Pareto distribution and contains spatial correlation between samples in the CRP and thermal noise. Each of these has the potential to affect the model assumption and cause mismatch in the detection performance. Regarding the thermal noise, it has been shown in Chapter 2 that the estimated shape parameter can be considered an 'effective shape' which accounts for the noise and does not affect the distribution fit.

7.3.1 Type I transformed CFAR

The transformed CA-CFAR scheme in exponential clutter is known as the geometric mean (GM) detector when applied to Pareto clutter [24]. For the Pareto Type I distribution, it has a decision rule given by

$$\bar{z}_k \underset{H_0}{\overset{H_1}{\gtrless}} b_{\mathrm{p}}^{1-M\eta} \prod_{m \in \mathbf{\Omega}_k} \bar{z}_m^{\eta} \tag{7.14}$$

where η can be determined through the P_{fa} expression

$$P_{\mathrm{fa}}(\eta) = (1 + \eta)^{-M}. \tag{7.15}$$

Note that the decision rule in (7.14) requires a priori knowledge of the Pareto scale parameter.

If there are strong clutter returns in the CRP due to interfering targets or clutter spikes, then an OS-CFAR will offer better detection performance. The transformed Pareto OS-CFAR was derived in [25], with the special cases of the minimum and maximum OS examined in [24]. If the intensity samples ordered from minimum to maximum are represented by $\bar{z}_{(m)}$, then the OS-CFAR detector is given as

$$\bar{z}_k \underset{H_0}{\overset{H_1}{\gtrless}} \beta^{1-\eta}\bar{z}_{(J)}^{\eta} \tag{7.16}$$

where the index $1 \leq J \leq M$ defines the range sample where the interference power is maximum and hence $M - J$ strong returns are rejected from the CRP. With this detector, the threshold multiplier η can be set by inversion of

$$P_{\text{fa}}(\eta) = \frac{M!}{(M-J)!}\frac{\Gamma(\eta + M - J + 1)}{\Gamma(\eta + M + 1)}. \tag{7.17}$$

An issue with practical implementation of (7.14) and (7.16) is the required a priori knowledge of b_{p}. If the latter can be assumed to be known, or estimated accurately from the clutter, then these decision rules will achieve the desired P_{fa} reasonably well.

7.3.2 Type I complete sufficient statistics

For a generic clutter measurement function, $g(\cdot)$, that is assumed to be scale-invariant in the sense that $g(\lambda z_1, \ldots, \lambda z_M) = \lambda g(z_1, \ldots, z_M)$, it is shown in [31] that the general form of the transformed decision rule is

$$\bar{z}_k \underset{H_0}{\overset{H_1}{\gtrless}} b_{\text{p}} \exp\left(\eta g\left(\left[\ln\left(\frac{\bar{z}_1}{b_{\text{p}}}\right), \ldots, \ln\left(\frac{\bar{z}_M}{b_{\text{p}}}\right)\right]\right)\right) \tag{7.18}$$

where the threshold multiplier η is based upon the choice of $g(\cdot)$ and set via the corresponding P_{fa} relationship for the pre-transformed CFAR detector in exponentially distributed clutter. It can be shown that the sample minimum is a complete sufficient statistic for the Pareto scale parameter, thus justifying the replacement for b_{p}. Based upon this, an alternative detector can be defined as [31]

$$\bar{z}_k \underset{H_0}{\overset{H_1}{\gtrless}} \bar{z}_{(1)} \exp\left(\eta g\left(\left[\ln\left(\frac{\bar{z}_1}{\bar{z}_{(1)}}\right), \ldots, \ln\left(\frac{\bar{z}_M}{\bar{z}_{(1)}}\right)\right]\right)\right). \tag{7.19}$$

In [31], it was shown how the P_{fa} and threshold multiplier relationship can be modified to account for the application of the minimum to (7.18). In the case where $g(\cdot)$ is a sum, (7.19) reduces to

$$\bar{z}_k \underset{H_0}{\overset{H_1}{\gtrless}} \bar{z}_{(1)}^{1-M\eta} \prod_{m\in\Omega_k} \bar{z}_m^{\eta} \tag{7.20}$$

with its P_{fa} given by

$$P_{\mathrm{fa}}(\eta) = \frac{M}{M+1}(1+\eta)^{-M}. \tag{7.21}$$

Similarly, when $g(\cdot)$ is the Jth OS, the decision rule becomes

$$\bar{z}_k \underset{H_0}{\overset{H_1}{\gtrless}} \bar{z}_{(1)}^{1-\eta}\bar{z}_{(J)}^{\eta}, \tag{7.22}$$

with threshold multiplier set via inversion of

$$P_{\mathrm{fa}}(\eta) = \frac{M!}{(M+1)(M-J)!}\frac{\Gamma(\eta+M-J+1)}{\Gamma(\eta+M)}. \tag{7.23}$$

Both the decision rules (7.20) and (7.22) are CFAR with respect to both the Pareto shape and scale parameters.

7.3.3 Type II transformed CFAR

In order to develop detectors for operation in Pareto Type II clutter, the transformation approach can also be used [29]. In this case, the detector is given by

$$\ln\left(1+\frac{z_k}{b_{\mathrm{p}}}\right) \underset{H_0}{\overset{H_1}{\gtrless}} \eta g\left(\left[\ln\left(1+\frac{z_1}{b_{\mathrm{p}}}\right), \ldots, \ln\left(1+\frac{z_M}{b_{\mathrm{p}}}\right)\right]\right) \tag{7.24}$$

where the function, $g(\cdot)$, is required to be scale-invariant as for the Pareto Type I case. Also the P_{fa} and threshold multiplier relationship is preserved from the original exponential decision rule used for the Pareto Type I scenario.

As an example, consider the case where $g(\cdot)$ is a sum. The decision rule (7.24) then becomes

$$z_k \underset{H_0}{\overset{H_1}{\gtrless}} b_{\mathrm{p}}\left(\prod_{m\in\mathbf{\Omega}_k}\left(1+\frac{z_m}{b_{\mathrm{p}}}\right)^{\eta} - 1\right), \tag{7.25}$$

with the threshold multiplier η determined by numerical inversion of the same P_{fa} relationship as the Pareto Type I GM detector in (7.15).

A second decision rule can be based upon the selection of $g(\cdot)$ as the Jth OS of the CRP. Since the logarithm function is increasing and its argument in (7.24) begins at unity, it follows that the Jth OS of $\ln\left(1+z_m/b_{\mathrm{p}}\right)$ is exactly $\ln\left(1+z_{(J)}/b_{\mathrm{p}}\right)$ and so the detector (7.24) reduces to

$$z_k \underset{H_0}{\overset{H_1}{\gtrless}} b_{\mathrm{p}}\left(\left(1+\frac{z_{(J)}}{b_{\mathrm{p}}}\right)^{\eta} - 1\right) \tag{7.26}$$

where η is determined by numerical inversion of (7.17).

Note that these detectors are CFAR decision rules with respect to the shape parameter but still require a priori knowledge of the Pareto Type II scale parameter. It turns out that it is not possible to eliminate the b_p dependency via an application of a complete sufficient statistic, as in the Pareto Type I case.

7.3.4 Type II Bayesian CFAR

To achieve a CFAR detector for the Pareto Type II model, an approach based on Bayesian statistics was developed by Weinberg *et al.* [32]. The idea is to construct the Bayesian predictive distribution of the CUT, z_k, given the samples in the CRP. The expression for the P_{fa} is given by

$$P_{\text{fa}}(\eta) = \int_{\eta}^{\infty} P_{z_k|z_1,\ldots,z_M}(z_k|z_1, z_2, \ldots, z_M)dz_k \tag{7.27}$$

where $P_{z_k|z_1,\ldots,z_M}(z_k|z_1, z_2, \ldots, z_M)$ is the PDF of the predictive distribution which can be derived from the likelihood function and the assumed prior distribution for the unknown parameters. For the Pareto Type II model, the Bayesian predictive PDF is given by

$$\begin{aligned}P_{z_k|z_1,\ldots,z_M}(z_k|z_1, z_2, \ldots, z_M) = \int_0^{\infty}\int_0^{\infty} & P_z(z_k)P_{\nu_p,b_p|z_1,\ldots,z_M}(\nu_p, b_p|z_1, \ldots, z_M) \\ & \times P_{\nu_p,b_p}(\nu_p, b_p)d\nu_p db_p\end{aligned} \tag{7.28}$$

where the Pareto distribution is given by

$$P_z(z) = \frac{zb_p^{\nu_p}}{(b_p + z)^{1+\nu_p}}. \tag{7.29}$$

The joint PDF of the CRP is

$$P_{\nu_p,b_p|z_1,\ldots,z_M}(\nu_p, b_p|z_1, \ldots, z_M) = \frac{\nu_p^M b^{\nu_p M}}{\prod\limits_{m\in\boldsymbol{\Omega}_k} (z_m + b_p)^{\nu_p+1}} \tag{7.30}$$

and the Jeffreys prior, $P(\nu_p, b_p)$, can be determined by evaluating the square root of the determinant of the Fisher information matrix. For the Pareto Type II model, this is given by,

$$\begin{aligned}P_{\nu_p,b_p}(\nu_p, b_p) &= \frac{1}{\sqrt{\nu_p(\nu_p+1)^2(\nu_p+2)}b_p} \\ &\approx \frac{1}{\nu_p b_p}.\end{aligned} \tag{7.31}$$

After some manipulation, the P_{fa} can be written as

$$P_{\mathrm{fa}}(\eta) = \frac{\displaystyle\int_0^{\infty} \frac{\phi_{\mathrm{p}}^{M-1}}{\prod_{m\in\mathbf{\Omega}_k}(\phi_{\mathrm{p}} z_m + 1)} \left(\log(\eta\phi + 1) + \sum_{m\in\mathbf{\Omega}_k} \log(\phi_{\mathrm{p}} z_m + 1) \right)^{-M} d\phi_{\mathrm{p}}}{\displaystyle\int_0^{\infty} \frac{\phi_{\mathrm{p}}^{M-1}}{\prod_{m\in\mathbf{\Omega}_k}(\phi_{\mathrm{p}} z_m + 1)} \left(\sum_{m\in\mathbf{\Omega}_k} \log(\phi_{\mathrm{p}} z_m + 1) \right)^{-M} d\phi_{\mathrm{p}}} \tag{7.32}$$

where $\phi_{\mathrm{p}} = 1/b_{\mathrm{p}}$. Typically, for a given $P_{\mathrm{fa}}(\eta)$, this equation must be inverted to find the threshold, η. However, since P_{fa} is a decreasing function of η, when $z_k > \eta$, the hypothesis H_0 (no target present) can be rejected if $P_{\mathrm{fa}}(z_k) < P_{\mathrm{fa}}(\eta)$. As a result, the hypothesis test can be written as

$$\begin{aligned} -\int_0^{\infty} \frac{\phi_{\mathrm{p}}^{M-1}}{\prod_{m\in\mathbf{\Omega}_k}(\phi_{\mathrm{p}} z_m + 1)} & \left(\left(\log(\eta\phi + 1) + \sum_{m\in\mathbf{\Omega}_k} \log(\phi_{\mathrm{p}} z_m + 1) \right)^{-M} \right. \\ & \left. + P_{\mathrm{fa}}(\eta) \left(\sum_{m\in\mathbf{\Omega}_k} \log(\phi_{\mathrm{p}} z_m + 1) \right)^{-M} \right) d\phi_{\mathrm{p}} \underset{H_0}{\overset{H_1}{\gtrless}} 0. \end{aligned} \tag{7.33}$$

The performance of this detector was studied in [32], where a CFAR loss was found relative to a fixed threshold detector. While this was shown to decrease as the length of the CRP increased, the performance was always less than the transformed detector in (7.24).

7.3.5 Detection results

In the following section, a number of the CFAR detectors are compared using two data sets from the Ingara X-band data set. The data is horizontally polarised with a Douglas sea state between 4 and 5, the grazing angle is between 30.5° and 35.5°, the CNR is 14 dB and the look angle is from the up-wind direction as it contains the largest concentration of sea spikes and the greatest variability in backscatter. Figure 7.10 shows the variability of the data in the range/time intensity domain over a processing interval of 128 pulses (0.22 s). The two data sets are taken from grazing bins 1–200 and 601–800 with the CCDF shown in Figure 7.11 with a Pareto Type II model fit and the measured spatial correlation shown in Figure 7.12. The first data set fits the Pareto model well and has a small spatial decorrelation length of 1.87 m. However, for the second data set, the model fit overshoots in the tail region of the CCDF and there is a much greater spatial decorrelation length of 3.13 m. The shape and scale of the amplitude distribution for the first data set are 7.5 and 6.5 and for the second, 3.6 and 2.6.

To determine the detection performance, a Monte Carlo simulation has been used with Swerling 1 synthetic targets injected into the data sets at random range bins. The

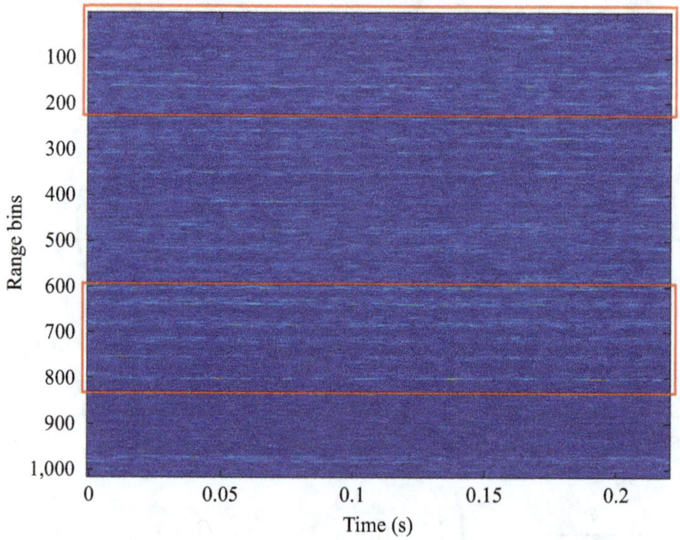

Figure 7.10 Range/time intensity images highlighting two data sets from the Ingara X-band data set [30]. © Commonwealth of Australia 2017

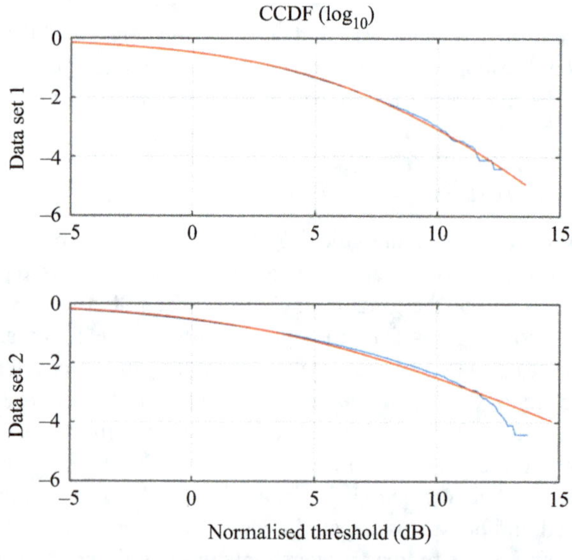

Figure 7.11 CCDF plot for the two data sets: (—) data, (—) Pareto distribution fit. The first data set matches the Pareto model extremely closely, while the second mismatches in the tail [30]. © Commonwealth of Australia 2017

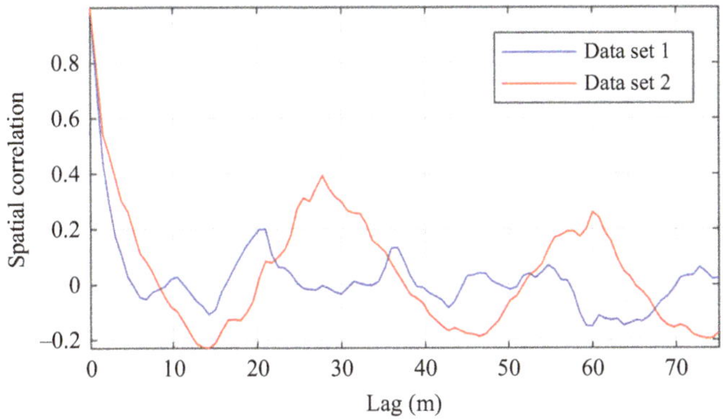

Figure 7.12 Spatial correlation of the two data sets [30]. © Commonwealth of Australia 2017

Table 7.1 Measured false alarm rates for the Pareto detection schemes (log_{10}). Desired $log_{10}(P_{\mathrm{fa}})$ is -3

	Data set 1	**Data set 2**
GM Type I/2	−2.95	−3.48
GM Type I SS	−2.77	−3.42
OS Type I/2	−2.97	−3.25
OS Type I SS	−2.95	−3.29

signal-to-interference ratio (SIR) is then varied and the probability of detection, P_{d}, is determined by the number of times the data exceeds the threshold. For reference, the fixed threshold is also determined for the desired P_{fa} as it allows a measure of the CFAR loss at given P_{d}. For the CFAR schemes, the desired $P_{\mathrm{fa}} = 10^{-3}$, the CRP length is $M = 32$ with $G_{\mathrm{c}} = 2$ guard bands either side of the CUT.

Table 7.1 shows the measured false alarm results with a desired $\log_{10}(P_{\mathrm{fa}})$ of -3. For the first data set, the actual false alarm rate is very close to the desired value, while the second data set shows smaller values. The GM results are also slightly worse than the OS. The detection results for the two data blocks are then shown in Figures 7.13 and 7.14 with the corresponding CFAR loss shown in Table 7.2 for a P_{d} of 0.5. For the first data set, the top plots show the GM results with the actual and ideal results on the left and right, respectively. The bottom plots show the equivalent results for the OS with $J = M - 2$. All these results are very similar with the GM performing slightly better by around 1 dB and the CFAR loss also approximately 1 dB.

The GM results for the second data set have the same shape, but there is now a 3 dB loss in detection performance when compared to the first data set. The improvement

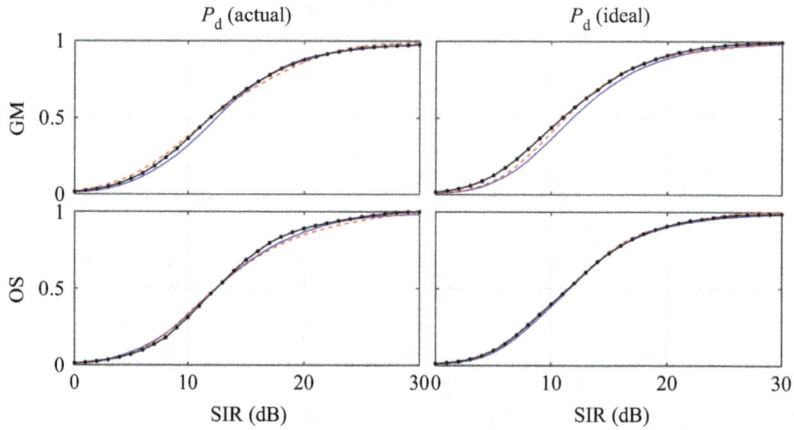

Figure 7.13 *Detection performance for data set 1. Geometric mean (top) and ordered statistic with* $J = M - 2$ *(bottom): (—) Type I, (- -) Type I SS, (-·-) Type II [30]. © Commonwealth of Australia 2017*

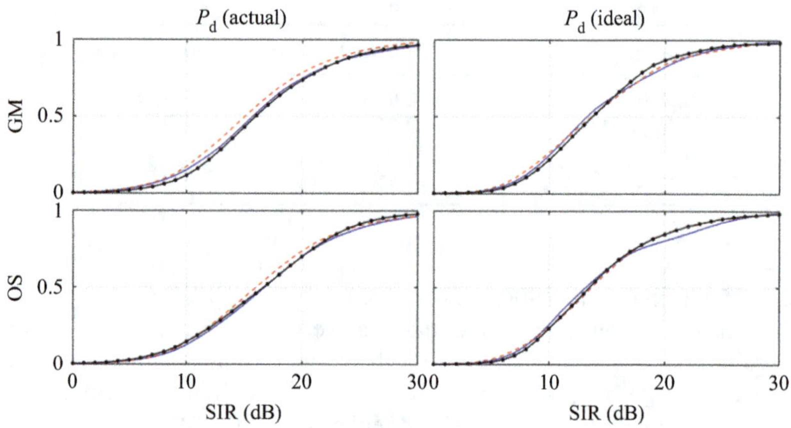

Figure 7.14 *Detection performance for data set 2. Geometric mean (top) and ordered statistic with* $J = M - 2$ *(bottom): (—) Type I, (- -) Type I SS, (-·-) Type II [30]. © Commonwealth of Australia 2017*

of 1 dB in the GM result is consistent, but the CFAR loss varies between 1.4 and 3.6 dB with the GM results slightly better due to the lower level of spatial correlation. The algorithms presented in this section expect M independent scatterers, which is reduced when spatial correlation is present and hence the CFAR schemes are not as effective. The slight mismatch in the tail of the distribution observed in Figure 7.11 is not likely to be significant given the desired P_{fa} of 10^{-3}.

Table 7.2 CFAR loss in dBs for the Pareto detection schemes. Desired P_{fa} is 10^{-3} and P_d is 0.5

	Data set 1	**Data set 2**
GM Type I	0.7	2.5
GM Type I SS	0.8	1.4
GM Type II	1.0	2.3
OS Type I	0.9	3.6
OS Type I SS	1.0	2.4
OS Type II	1.0	3.2

7.3.6 Summary

This section has studied the performance of both Type I and II Pareto CFAR detection schemes, including the original formulations of the GM and ordered statistic detectors, the modified Type I versions assuming complete sufficient statistics and a Bayesian Type II detector. Analysis of the detection performance using the Ingara X-band data set revealed good robustness to noise, but a mismatch when spatial correlation is present in the CRP. This mismatch in the model assumption caused the measured false alarm to be less than the desired value and hence the detection performance is sub-optimal. However, in the presence of discrete spikes, using a Pareto decision rule with the GM or OS-CFAR schemes should provide a significant improvement in the CFAR performance. The only extra processing step is the estimation of the scale parameter required to offset the data for the Type I sufficient statistic CFAR scheme or as an input to the test statistic with the Type II formulation.

7.4 Coherent detection

Traditionally, airborne maritime surveillance radars have used non-coherent processing to detect small targets with the radar platform operating at low altitudes. Coherent processing was not considered to offer sufficient performance gain to warrant the additional signal processing complexity and cost [33]. However, in recent years there has been growing interest in operating airborne maritime surveillance platforms at higher altitudes. When observing the sea clutter at steeper grazing angles, the returns become less spiky but have increased reflectivity, resulting in degraded performance for non-coherent detection schemes [33]. While coherent processing has the potential to give some small improvement for detecting slow moving targets, the improvement can be considerable for targets whose radial velocity places their Doppler shift outside the clutter Doppler spectrum.

A good overview of the many coherent detection methods in the literature can be found in [34]. They can be broadly categorised as pre- and post-Doppler with a number that have been applied to sea clutter [35,36]. Well-known detectors such as the generalised likelihood ratio test (GLRT) [37], adaptive matched filter (AMF) [38] and normalised adaptive matched filter (NAMF) [39] have the CFAR property in homogeneous Gaussian environments. However, the optimal coherent detector for K-distributed clutter [40] is not practically implementable and sub-optimal approaches must be used.

One of the particular problems with using coherent detection schemes for targets in sea clutter is the significant complexity required to maintain a CFAR over all Doppler frequencies. The work in [41,42] evaluated a number of pre- and post-Doppler space–time adaptive processing (STAP) algorithms on real experimental data from Fraunhofer FHRs PAMIR system. The study looked carefully at how the desired and actual false alarm rates varied across the frequency domain [42] and found that a post-Doppler version of the NAMF had superior CFAR performance when the detection statistic was calculated using a very limited number of local Doppler bins to estimate the background clutter power. This is due to the clutter heterogeneity which limited the number of suitable training data snapshots.

7.4.1 Multichannel radar

Airborne multichannel radar systems have proven successful at improving detection performance in land-based scenarios. Displaced phase centre array (DPCA) is one of the earliest techniques for moving target indication [1]. It uses two along-track antennas with the transmit antenna switched from pulse-to-pulse in such a way that the position of the second antenna at the second pulse is at the same position as the first antenna during the first pulse. This requires the distance between the two spatial channels to be equal to the radar platform velocity times the pulse repetition interval, $d = v_p T_r$. Echoes received from stationary targets should be the same and if their signals are subtracted, the difference signal will reveal any moving targets. However, with this design, the potential for channel mismatch is very high and a variation known as adaptive DPCA algorithm [43] was developed to address the mismatch under non-homogeneous ground clutter conditions [44].

A summary of different slow-time STAP algorithms is presented in [45], where four classes of sub-optimal algorithms are described. These include the pre- and post-Doppler element space algorithms as well as formulations in the spatial frequency domain, known as beamspace. In most airborne radars, there are typically more pulses than spatial channels and post-Doppler element space techniques are most commonly used [46,47]. For a stationary scatterer, the relationship between Doppler and angle is given by

$$f = \frac{2v_p}{\lambda} \sin\theta_t \tag{7.34}$$

where λ is the radar wavelength and θ_t is the spatial steering angle. Figure 7.15 is a famous figure used to represent the two-dimensional power spectrum of the received signal at a single range bin and includes clutter, a direct-path jammer and a slow and

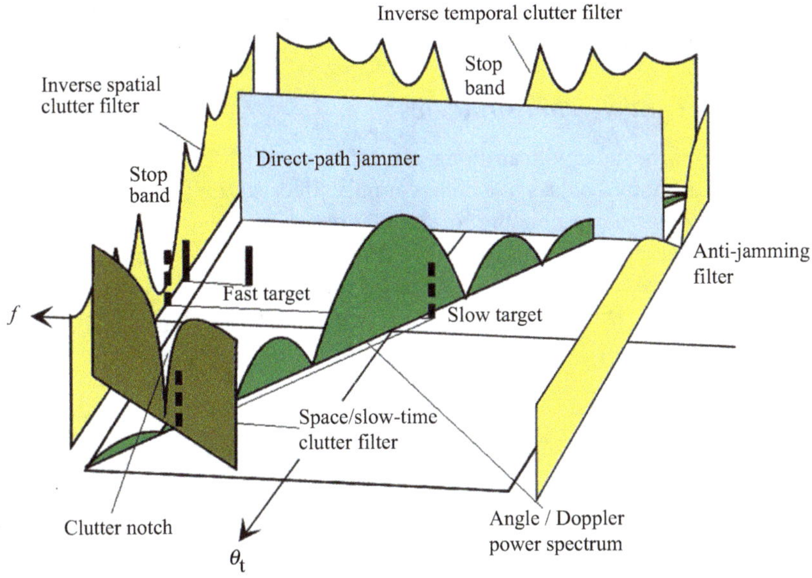

Figure 7.15 Principle of space–time adaptive processing

fast moving target. The clutter signal is determined by (7.34), with its position given by the slope of the clutter ridge,

$$\tilde{\beta} = \frac{2v_{\mathrm{p}}T_{\mathrm{r}}}{d}. \tag{7.35}$$

The shape of the clutter signal will also be modulated by the two-way antenna pattern and in the maritime environment, there will be broadening in Doppler due to the motion of the sea clutter. In Figure 7.15, the other major signal components are the slow and fast moving targets which are assumed to remain fixed in angle and Doppler and the direct-path jammer signal which covers the entire Doppler band at a fixed angle. To suppress the clutter using only temporal processing, an inverse temporal filter along the frequency axis could be used. This is depicted at the back of the plot, where the clutter notch is determined by the mainlobe of the projected ground clutter signal. Likewise, if only spatial processing were used to cancel the clutter or direct-path jammer, an inverse filter along the spatial angle axis could be used. If either of these inverse filters are used, they will form a broad stop-band and blind the radar in these intervals. The alternative is a space/slow-time filter that operates over the whole angle/Doppler plane forming a narrow ridge along the trajectory of the interference spectrum so that even slow targets will fall into the pass band. Both the clutter and direct-path jammer signals can be cancelled with the same filter or separately if preferred. All these approaches require knowledge of the clutter covariance matrix, which is not available in practice. Instead, an estimate is usually obtained from a set of training data comprising the returns from range gates surrounding the CUT.

This relies on the assumption that the clutter is spatially statistically homogeneous and will result in mismatch in many realistic scenarios.

7.4.2 Coherent detection strategies

Consider an airborne radar with an along-track side looking linear array comprising L antennas transmitting N pulses in a coherent pulse interval (CPI). The radar collects K fast-time samples corresponding to K range gates, yielding an $L \times N \times K$ data cube [48]. The target detection problem is then formulated as a hypothesis test by scanning for the presence of a target in each range gate over a grid of angles and Doppler frequencies. If the complex signal returns are given by $\mathbf{h}$, then the null and alternative hypotheses, H_0 and H_1, are defined by

$$H_0 : \mathbf{h} = \mathbf{c} + \boldsymbol{\xi} \tag{7.36}$$

$$H_1 : \mathbf{h} = \mathbf{c} + \boldsymbol{\xi} + A\mathbf{s} \tag{7.37}$$

where $\mathbf{c}$, $\boldsymbol{\xi}$ and $\mathbf{s}$ are the clutter, thermal noise and signal vectors and A is the complex target amplitude. Sea clutter can be modelled as a compound Gaussian process, $\mathbf{c} = \sqrt{\tau}\mathbf{x}$, with slowly varying texture intensity τ and fast fluctuating speckle $\mathbf{x}$, modelled as a zero mean complex Gaussian process. The $L \times N$ spatio-temporal steering matrix is given by

$$\mathbf{s} = \mathbf{s}_\theta \mathbf{s}_\mathrm{t}^T, \tag{7.38}$$

where the spatial and temporal steering vectors, $\mathbf{s}_\theta$ and $\mathbf{s}_\mathrm{t}$, are defined as

$$\begin{aligned} \mathbf{s}_\theta &= \frac{1}{\sqrt{L}}\left[1\ e^{j2\pi f_s}\ \ldots\ e^{j2\pi(L-1)f_s}\right]^T, \\ \mathbf{s}_\mathrm{t} &= \frac{1}{\sqrt{N}}\left[1\ e^{j2\pi f}\ \ldots\ e^{j2\pi(N-1)f}\right]^T. \end{aligned} \tag{7.39}$$

Here the target spatial and Doppler frequencies are given by

$$f_s = \frac{d}{\lambda}\sin\theta_\mathrm{t}, \quad f = \frac{2v_0}{c_0} \tag{7.40}$$

where v_0 is the radial velocity and c_0 is the speed of light.

7.4.3 Pre-Doppler whitening

Consider a CPI of sea clutter and noise in a single range gate, k. If the texture, τ, is assumed to be constant over the CPI and the returns are Gaussian, then detectors that whiten the spectrum can be considered optimum. This requires knowledge of the interference covariance matrix for the CUT, which is generally unknown and must be estimated, usually using data from surrounding range cells. A parallel to this is the estimation of the mean level of the CUT by a CA-CFAR using surrounding range cells.

There is a considerable body of literature describing different methods for estimating the covariance matrix (e.g. [49]). The methods that rely on secondary data from surrounding range cells are generally dependent on specific models for the clutter. In particular, many are predicated on the clutter being described as a spherically invariant random process. This essentially assumes that a compound Gaussian process can be modelled as two mutually independent processes comprising the local texture intensity, τ, and a normalised covariance matrix for the speckle, $\mathbf{R}$, so that

$$\left\langle \tilde{\mathbf{h}}\tilde{\mathbf{h}}^{H} \right\rangle = \langle \tau \rangle \left\langle \tilde{\mathbf{x}}\tilde{\mathbf{x}}^{H} \right\rangle + p_{\mathrm{n}}\mathbf{I} = \langle \tau \rangle \, \mathbf{R} + p_{\mathrm{n}}\mathbf{I} \tag{7.41}$$

where p_{n} is the noise power, $\mathbf{I}$ is the identity matrix, $\tilde{\mathbf{h}}$ is a vector formed by stacking the columns of $\mathbf{h}$. The $LN \times 1$ steering vectors can be similarly stacked using a Kronecker product, $\tilde{\mathbf{s}} = \mathbf{s}_{\theta} \otimes \mathbf{s}_{\mathrm{t}}$.

Let us assume that $\tau = 1$, meaning that the clutter is statistically homogeneous, following the multivariate (complex normal) distribution and resulting in a Rayleigh distribution for the clutter envelope. Under this assumption, three practically implementable coherent detectors include the GLRT [37], the AMF [38] and the NAMF [39]. The GLRT was derived by Kelly assuming homogeneous complex Gaussian clutter,

$$\frac{\left| \tilde{\mathbf{s}}^{H} \widehat{\mathbf{R}}^{-1} \tilde{\mathbf{h}} \right|^{2}}{\left(\tilde{\mathbf{s}}^{H} \widehat{\mathbf{R}}^{-1} \tilde{\mathbf{s}} \right) \left(1 + \frac{1}{K} \tilde{\mathbf{h}}^{H} \widehat{\mathbf{R}}^{-1} \tilde{\mathbf{h}} \right)} \underset{H_0}{\overset{H_1}{\gtrless}} \gamma \tag{7.42}$$

while the AMF is a commonly used sub-optimal form of the GLRT and has the following detection statistic,

$$\frac{\left| \tilde{\mathbf{s}}^{H} \widehat{\mathbf{R}}^{-1} \tilde{\mathbf{h}} \right|^{2}}{\tilde{\mathbf{s}}^{H} \widehat{\mathbf{R}}^{-1} \tilde{\mathbf{s}}} \underset{H_0}{\overset{H_1}{\gtrless}} \gamma. \tag{7.43}$$

When both the local texture and the clutter covariance matrix are unknown, pre-Doppler whitening can be undertaken using the NAMF,

$$\frac{\left| \tilde{\mathbf{s}}^{H} \widehat{\mathbf{R}}^{-1} \tilde{\mathbf{h}} \right|^{2}}{\left(\tilde{\mathbf{s}}^{H} \widehat{\mathbf{R}}^{-1} \tilde{\mathbf{s}} \right) \left(\tilde{\mathbf{h}}^{H} \widehat{\mathbf{R}}^{-1} \tilde{\mathbf{h}} \right)} \underset{H_0}{\overset{H_1}{\gtrless}} \gamma. \tag{7.44}$$

There are many approaches for estimating the covariance of the interference. The sample covariance matrix (SCM) averages the covariance over K range bins surrounding the CUT with guard bands used to reduce target contamination (self-nulling) due to range-extended targets. If the CRP is defined as $\boldsymbol{\Omega}_k = [k - M/2 - G_c, \ldots, k - G_c + 1, k + G_c + 1, \ldots, k + M/2 + G_c]$, then the SCM is given by

$$\widehat{\mathbf{R}}_{\text{SCM}}(k) = \frac{1}{K} \sum_{n \in \boldsymbol{\Omega}_k} \tilde{\mathbf{h}}_n \tilde{\mathbf{h}}_n^H. \tag{7.45}$$

This expression can be written in terms of the clutter and noise contributions,

$$\widehat{\mathbf{R}}_{\text{SCM}}(k) = \frac{1}{K} \sum_{n \in \boldsymbol{\Omega}_k} \left(\sqrt{\tau_n} \tilde{\mathbf{x}}_n + \tilde{\boldsymbol{\xi}}_n \right) \left(\sqrt{\tau_n} \tilde{\mathbf{x}}_n + \tilde{\boldsymbol{\xi}}_n \right)^H \tag{7.46}$$

$$\approx \frac{1}{K} \sum_{n \in \boldsymbol{\Omega}_k} \tau_n \tilde{\mathbf{x}}_n \tilde{\mathbf{x}}_n^H + \frac{1}{K} \sum_{n \in \boldsymbol{\Omega}_k} \tilde{\boldsymbol{\xi}}_n \tilde{\boldsymbol{\xi}}_n^H. \tag{7.47}$$

This shows that even if the clutter covariance estimate is in error, the thermal noise level should be estimated correctly with $\hat{p}_{\text{n}} \mathbf{I} \approx (1/K) \sum_{n=1}^{K} \tilde{\boldsymbol{\xi}}_n \tilde{\boldsymbol{\xi}}_n^H$. This is not the case for the normalised sample covariance matrix (NSCM) estimator, which attempts to normalise the covariance using an inner product of the data to estimate the local texture, τ_k,

$$\widehat{\mathbf{R}}_{\text{NSCM}}(k) = \frac{1}{K} \sum_{n \in \boldsymbol{\Omega}_k} \frac{\tilde{\mathbf{h}}_n \tilde{\mathbf{h}}_n^H}{\tilde{\mathbf{h}}_n^H \tilde{\mathbf{h}}_n} \tag{7.48}$$

$$= \frac{1}{K} \sum_{n \in \boldsymbol{\Omega}_k} \frac{\left(\sqrt{\tau_n} \tilde{\mathbf{x}}_n + \tilde{\boldsymbol{\xi}}_n \right) \left(\sqrt{\tau_n} \tilde{\mathbf{x}}_n + \tilde{\boldsymbol{\xi}}_n \right)^H}{\left(\sqrt{\tau_n} \tilde{\mathbf{x}}_n + \tilde{\boldsymbol{\xi}}_n \right)^H \left(\sqrt{\tau_n} \tilde{\mathbf{x}}_n + \tilde{\boldsymbol{\xi}}_n \right)}. \tag{7.49}$$

In spiky clutter when the local CNR ratio, τ_k / p_{n}, varies rapidly in range, the NSCM estimator may give a poor estimate of the noise level in the CUT. However, if the clutter characteristics are locally correlated in range, a smaller value of K may be desirable when estimating the covariance matrix, so that the CRP is more representative of the CUT. If K is too small, problems may arise from numerical instability when inverting the matrix. In that event, diagonal loading, η_{DL}, may be used so that

$$\widehat{\mathbf{R}}'(k) = \widehat{\mathbf{R}}(k) + \eta_{\text{DL}} \mathbf{I}. \tag{7.50}$$

The level of diagonal loading is discussed in [34,50,51], where a typical value of about 1% of the estimated mean power is desirable, $\eta_{\text{DL}} = 0.01 \text{Tr} \left[\widehat{\mathbf{R}}(k) \right] / (NL)$, where Tr[·] is the trace of a matrix.

Unfortunately, as described in Chapter 2, the Doppler spectrum of sea clutter fluctuates in width randomly over range and time and may exhibit a mean Doppler shift that is correlated with the local texture, τ. This means that attempts to estimate the covariance matrix of the CUT from surrounding cells may be prone to error and the detectors are unlikely to achieve CFAR performance. Depending on the variation in performance that is acceptable, further steps may be needed to adjust the detection thresholds as a function of Doppler frequency. This may require, for example, an independent estimation of the average CNR ratio and interference PDF in each Doppler bin.

7.4.4 Post-Doppler detection

The simplest form of Doppler processing involves filtering the returns at each range to match to individual Doppler and spatial frequencies. This is typically implemented by a two-dimensional discrete Fourier transform. For the returns at a given range bin, k,

$$h'_k(n,l) = \sum_{p=1}^{L}\sum_{q=1}^{N} h_k(m,l)w_{\mathrm{D}}(m,l)\exp\left(-j2\pi(p-1)\frac{n}{N}\right)\exp\left(j\frac{2\pi d}{\lambda}(q-1)\sin(l\Delta\theta)\right) \tag{7.51}$$

where there are N temporal and L spatial filters. The temporal filters are each centred on a frequency, $f = nf_{\mathrm{r}}/N$ with $n = -N/2, \ldots, N/2 - 1$ and the spatial filters, $\theta = l\Delta\theta$ for $l = -L/2, \ldots, L/2$ with angular resolution, $\Delta\theta$. In each bin, the Doppler spectrum can be obtained with

$$H_k(n,l) = \left|h'_k(n,l)\right|^2 \tag{7.52}$$

and intensity value thresholded with

$$H_k(n,l) \underset{H_0}{\overset{H_1}{\gtrless}} \gamma'(n,l) \tag{7.53}$$

where $\gamma'(n,l)$ is the post-Doppler threshold. Setting the correct value of $\gamma'(n,l)$ requires an estimate of the interference statistics at each frequency and spatial location. This is the same problem addressed in Section 7.2 for non-coherent detection methods and the same methods can be applied. For example, the interference in each frequency and spatial bin could be estimated by running a CA-CFAR over adjacent range bins and the CNR ratio and the amplitude statistics can be estimated using data from surrounding cells.

7.4.5 Post-Doppler whitening

The pre-Doppler NAMF requires the repeated estimation and inversion of the covariance matrix for each CUT which is a significant processing burden. In addition, if small values of K are used, there is potential for numerical instability. Alternatively, there are a number of post-Doppler frequency-domain whitening approaches that can be considered [52,53]. One method is to estimate an average power spectral density for the CUT from surrounding range cells and use that to whiten the spectrum,

$$\left(\frac{H_k(n,l)}{\widehat{H}_k(n,l)}\right) \Big/ \left(\frac{1}{N}\sum_{p=1}^{N}\frac{H_k(p,l)}{\widehat{H}_k(p,l)}\right) \underset{H_0}{\overset{H_1}{\gtrless}} \gamma'(n,l) \tag{7.54}$$

where $\widehat{H}_k(n,l)$ is the estimated average power spectral density that flattens the spectrum of the CUT with the denominator containing a normalising term that estimates the value of the local clutter intensity in the CUT. The estimate of the average power spectral density can be obtained from surrounding range cells by

$$\widehat{H}_k(n,l) = \frac{1}{K}\sum_{n\in\Omega_k} H_m(n,l). \tag{7.55}$$

Unfortunately, any estimate of the mean spectrum in the CUT may also be inaccurate due to the range-varying nature of the clutter Doppler spectra. Also, self-eclipsing by targets is a potential weakness of this method.

7.4.6 Model-based detection

Due to the range-varying nature of the Doppler spectra, estimating the power spectrum or covariance matrix directly from surrounding range cells can lead to errors. As with a CA-CFAR in Section 7.2, if the clutter is highly correlated over local range samples, there may be some advantage in using only a small number of range cells, K, when estimating parameters for a CUT. However, due to the small sample size, this can also lead to poor estimates and numerical instability when inverting covariance matrices. If small values of K are used or if only small sample sizes are available, one approach to improve the estimation accuracy is to introduce prior knowledge into the estimator. In the examples given here, this prior knowledge typically involves assumptions on the shape of the clutter power spectrum.

The method proposed by Román *et al.* [54] assumes that clutter can be described as an autoregressive (AR) process. The application of an AR model to detection in sea clutter is described in [36]. If the clutter can be represented by an AR process with an order P_{AR} that is lower than the CPI length N, then it should be possible to obtain good estimates of the covariance matrix using only P_{AR} samples. Once the AR coefficients are estimated from the data, better estimates of the covariance matrix may then be achieved using overlapping groups of P_{AR} samples. Further averaging over adjacent range samples may still be necessary, but the hope is that sufficient accuracy may be achieved with a smaller number of secondary range cells, K. The lower order covariance matrix estimated in this way can then be repeatedly applied to give a more robust detection statistic using the NAMF [54]. Alternatively, the AR coefficients derived from the data can be used to estimate the clutter spectral density and used in a post-Doppler whitening filter.

The evolving Doppler spectrum model, described in Chapter 2, is based on a Gaussian-shaped spectrum that is characterised by a mean power τ_k, a mean Doppler shift $m_{\mathrm{f}}(k)$ and a spectrum width $w(k)$,

$$G(f) = \frac{\tau_k}{\sqrt{2\pi}w(k)}\exp\left(-\frac{(f-m_{\mathrm{f}}(k))^2}{2w(k)^2}\right). \tag{7.56}$$

Rather than estimating the covariance matrix or power spectral density directly from the data, it is possible to derive an estimate from knowledge of the parameter values for this model [36]. For example, estimates of $m_{\mathrm{f}}(k)$ and $w(k)$ can be used to determine the local power spectral density or covariance matrix, and then averaged over surrounding range cells to predict values for the CUT. Alternatively, estimates of the basic spectrum model parameters τ_k, $m_f(k)$ and $w(k)$ can be averaged over adjacent range cells and then used to determine the local power spectral density or covariance matrix.

The first important measure of performance in a detector is its ability to control false alarms. This may be more difficult when secondary data from surrounding range cells is not representative of the CUT or when only a small number of range cells can be used. In the analysis of model-based detectors in sea clutter [36], it was found that detectors based on low-order AR models produced the smallest variation in the probability of false alarms, P_{fa}, as a function of Doppler frequency. Both post-Doppler whitening and a post-Doppler detector using the Gaussian-shaped spectrum model also performed well. The main reason for this improved control of false alarms is that these detectors were able to use smaller numbers of range samples for training data. In a practical radar application, it would be necessary to assess the variation in P_{fa} across frequency. In many of the detectors described here, it may be necessary to adjust the detection threshold as a function of frequency, rather than relying on ideal clutter cancellation as would be expected from an optimum whitening filter. This could be achieved by running an independent CFAR detection system in each Doppler bin, using the various techniques described in Section 7.2.

The primary aim of the radar will be to detect targets with maximum sensitivity, while also maintaining an acceptable false alarm rate. In [36], the detection performances of a number of different coherent detectors were compared, after correcting the detection thresholds at each Doppler frequency to ensure true CFAR performance. Under these conditions, most detectors performed in a similar manner. The choice of detector in a practical radar application might then be driven more by the relative processing complexity. For example, post-Doppler whitening and Gaussian shaped spectrum models may be preferred to time domain whitening models that require repeated matrix inversions.

7.4.7 Doppler spectrum aliasing

Doppler processing can provide significant detection gains for fast moving targets when compared with non-coherent processing. For slow targets, the gain will depend on the dwell time and the Doppler resolution relative to the clutter Doppler spectrum width. High Doppler resolution can be used to improve the signal-to-clutter-ratio of slow moving targets but will require a long dwell time that may not be available for radars using a fast scan rate.

An alternative approach to filtering or whitening the clutter spectrum is to spread the clutter energy over the Doppler space. This can be achieved by under-sampling the coherent radar returns or reducing the pulse repetition frequency (PRF), so that the clutter Doppler spectrum is aliased. Provided that the target has a relatively narrow spectrum, this should remain unaffected with a lower PRF. If the PRF is reduced sufficiently, the clutter spectrum will become approximately uniformly distributed over the relatively narrow unambiguous Doppler space. Following a Fourier analysis of the aliased spectrum, the CNR in each Doppler bin will then be the same as in the time domain signal. Moreover, the amplitude statistics over range will also be the same for the original time domain data as well as the data in each Doppler bin after filtering. The target's energy is assumed to be confined to a single Doppler bin and its relative power will be increased by the length of the coherent CPI. This has the potential for a useful improvement in performance when compared to non-coherent integration over the same number of pulses or when compared to coherent processing when the target and most of the clutter spectrum fall into a single Doppler bin.

The extent to which the aliased clutter will be uniformly spread in Doppler will depend on the sampling frequency and the width of the Doppler spectrum. Figure 7.16 shows the power spectral density of a Gaussian shaped clutter spectrum as a function of the spectrum standard deviation, w, normalised by the PRF, f_r. It can be seen qualitatively that if $w \geq 0.4f_r$, then the resulting aliased spectrum will be approximately uniform. With typical mean values of w in the region of 30–70 Hz for X-band, this implies that a PRF of about 75–150 Hz might be required to achieve uniformity. For side-looking antennas, there might be spectrum broadening due to platform motion and in that case proportionally higher values of PRF may be acceptable. After aliasing, the degree of amplitude variation over frequency that may be acceptable will depend on the performance required. If a single fixed threshold is used over all Doppler bins, then an amplitude ripple will result in a variation in the probability of false alarm over Doppler. This can be controlled to achieve an acceptable average false alarm rate by setting a higher threshold, with a resulting loss in detection performance. It may be noted that there will also be a variation in the amplitude statistics over range and across Doppler bins if the whitening is incomplete and that will cause an additional variation in the false alarm rate.

Achieving a PRF of around 100 Hz may be very difficult with a conventional scanning radar, due to constraints on the scan rate and the required number of pulses in a dwell. However, if an electronically scanned antenna is available, it is possible

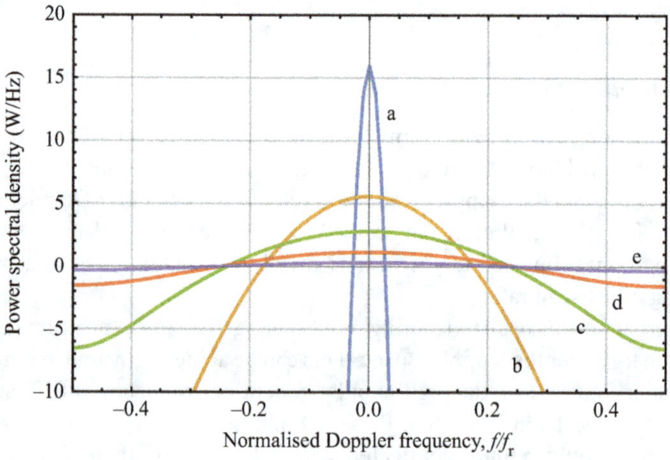

Figure 7.16 Aliased Gaussian-shaped spectrum for different normalised spectrum standard-deviations, w/f_r: (a) 0.01, (b) 0.1, (c) 0.2, (d) 0.3, (e) 0.4

to achieve such a PRF while maintaining an acceptable average scanning rate in azimuth. One possible scheme has been proposed using an electronic-scanned antenna [55–57]. It is assumed that the antenna is mechanically scanned in azimuth at a fixed rate, while at the same time from pulse-to-pulse, the beam pointing direction can be steered electronically. Using beam steering from pulse-to-pulse, coherent returns can be collected from a given direction at a desired lower PRF.

7.5 Detection in heterogeneous environments

The dynamic nature of sea clutter can lead to significant degradation in the performance of traditional coherent target detectors such as the GLRT, AMF and NAMF. These detectors are designed for complex Gaussian distributed clutter and rely on the availability of homogeneous training data. However, these conditions can be violated in the maritime environment, where the clutter statistics slowly vary over time and range and can be non-Gaussian.

The issue of heterogeneity has been extensively studied for land clutter and many approaches have been proposed to reduce its impact [58]. The majority of these focus on pre-screening the training data or by using prior knowledge of the clutter environment [59]. The direct data domain (D^3)-STAP [60] and its robust implementation, RD^3-STAP [61], are approaches that use only a single range bin to cancel the clutter without relying on secondary training data. They work by highlighting any target signal contributions in the range gate under test, while reducing any other sources of interference in the scene [60,61]. However, there is assumed knowledge of the target direction of arrival and Doppler frequency, which would not be known in a real-world scenario and hence the application of RD^3-STAP is limited.

An alternative philosophy that eliminates the need for training data was proposed in [62]. It is referred to as the single data set (SDS) approach as it relies only on data extracted from the CUT to mitigate the problem of statistical variability. The SDS approach was generalised into a fixed scale (FS) hybrid [63,64] that averages the SDS covariance estimate with the SCM, hence utilising both the test and training data sets to improve the detection performance. These two detectors were applied to simulated single-channel sea clutter in [65] and to real multichannel sea clutter in [66]. In both studies, the SDS and FS hybrid approaches were found to offer significant performance improvement over the traditional approaches. Two other hybrid variations have been proposed in [63,64]. The first is known as the variable scale hybrid and uses the generalised inner product (GIP) to pre-screen the training data to better deal with clutter heterogeneity. The second variation uses the entire test and training data sets, with a normalisation of the texture fluctuations resulting in good performance in homogeneous environments and superior performance in heterogeneous environments. The SDS and hybrid algorithms are now described in more detail with results from both simulated and real data sets.

7.5.1 The single data set detector

Traditional coherent detectors can perform poorly in heterogeneous environments and limit the availability of suitable training data [36]. Indeed even in Gaussian clutter, the violation of the heterogeneity assumption has been shown to cause significant degradation in performance when the SCM estimate is used. This led to the development of the SDS approach, where detectors only utilise data from the CUT, thus reducing or even eliminating the need for training data [62,63]. Consider a multichannel radar with a CPI of N pulses and L spatial channels. Suppose there are K_{T} snapshots from the CUT generated using a two-dimensional sliding window of size $P_{\mathrm{T}} \times Q_{\mathrm{T}}$ over the data slice of the CUT, $\mathbf{h}_k$,

$$\mathbf{h}_k = \begin{bmatrix} h(0,0) & h(0,1) & \cdots & h(0,Q_{\mathrm{T}}-1) & \cdots & h(0,L-1) \\ h(1,0) & h(1,1) & \cdots & h(1,Q_{\mathrm{T}}) & \cdots & h(1,L-1) \\ \vdots & \vdots & \ddots & \vdots & \ddots & \vdots \\ h(P_{\mathrm{T}}-1,0) & h(P_{\mathrm{T}}-1,1) & \cdots & h(P_{\mathrm{T}}-1,Q_{\mathrm{T}}-1) & \cdots & h(P_{\mathrm{T}}-1,L-1) \\ \vdots & \vdots & \ddots & \vdots & \ddots & \vdots \\ h(N-1,0) & h(N-1,1) & \cdots & h(N-1,Q_{\mathrm{T}}-1) & \cdots & h(N-1,L-1) \end{bmatrix} \tag{7.57}$$

Each window is vectorised and arranged into an $P_{\mathrm{T}}Q_{\mathrm{T}} \times K_T$ matrix, $\bar{\mathbf{h}}_k$, where $K_{\mathrm{T}} = K_{\mathrm{T_t}}K_{\mathrm{T_s}}$ and the temporal and spatial window lengths, $K_{\mathrm{T_t}} = N - P_{\mathrm{T}} + 1$ and $K_{\mathrm{T_s}} = L - Q_{\mathrm{T}} + 1$. The statistical expressions for the two SDS detectors mirror the forms of the AMF and the GLRT and are given by

$$\frac{\left|\tilde{\mathbf{s}}^H \widehat{\mathbf{Q}}_{\mathrm{T}}^{-1}(k)\tilde{\mathbf{g}}_k\right|^2}{\tilde{\mathbf{s}}^H \widehat{\mathbf{Q}}_{\mathrm{T}}^{-1}(k)\tilde{\mathbf{s}}} \underset{H_0}{\overset{H_1}{\gtrless}} \gamma \tag{7.58}$$

and

$$\frac{\left|\tilde{\mathbf{s}}^H \widehat{\mathbf{Q}}_{\mathrm{T}}^{-1}(k)\tilde{\mathbf{g}}_k\right|^2}{\left(\tilde{\mathbf{s}}^H \widehat{\mathbf{Q}}_{\mathrm{T}}^{-1}(k)\tilde{\mathbf{s}}\right)\left(1 + \frac{1}{K_T-1}\tilde{\mathbf{g}}_k^H \widehat{\mathbf{Q}}_{\mathrm{T}}^{-1}(k)\tilde{\mathbf{g}}_k\right)} \underset{H_0}{\overset{H_1}{\gtrless}} \gamma \tag{7.59}$$

where the data 'mean' vector of size $P_{\mathrm{T}}Q_{\mathrm{T}} \times 1$ is given by

$$\tilde{\mathbf{g}}_k = \frac{1}{|\tilde{\mathbf{s}}_{K_{\mathrm{T}}}|}\bar{\mathbf{h}}_k \tilde{\mathbf{s}}_{K_{\mathrm{T}}}^*, \tag{7.60}$$

and $\tilde{\mathbf{s}}_{K_{\mathrm{T}}}$ is the vectorised $K_{\mathrm{T}} \times 1$ target steering vector of the same form as (7.38). Effectively, the SDS detector estimates the data mean vector, $\tilde{\mathbf{g}}_k$, and then obtains a covariance matrix estimate, $\widehat{\mathbf{Q}}_{\mathrm{T}}(k)$ of size $P_{\mathrm{T}}Q_{\mathrm{T}} \times P_{\mathrm{T}}Q_{\mathrm{T}}$, from the subspace that is orthogonal to $\tilde{\mathbf{g}}_k$. That is,

$$\widehat{\mathbf{Q}}_{\mathrm{T}}(k) = \frac{1}{K_{\mathrm{T}} - 1}\left(\bar{\mathbf{h}}_k \bar{\mathbf{h}}_k^H - \tilde{\mathbf{g}}_k \tilde{\mathbf{g}}_k{}^H\right). \tag{7.61}$$

In this way, the SDS algorithm can be seen as a subspace detector. Calculation of the covariance inverse can also be implemented by exploiting the matrix inversion lemma [63]. This enables the algorithm to run with the same computational load as the NAMF statistic in (7.44).

7.5.2 Fixed scale hybrid

When the environment contains a mix of both heterogeneous and homogeneous scattering, the test and training data sets can be combined into an extended data set to form an improved hybrid detection scheme [63,67]. The resulting combined estimate of the covariance matrix is referred to as the fixed scale (FS) hybrid,

$$\hat{\boldsymbol{\Sigma}}_{\mathrm{FS}}(k) = \frac{(K_{\mathrm{T}} - 1)\widehat{\mathbf{Q}}(k) + K\widehat{\mathbf{R}}_{\mathrm{SCM}}(k)}{K_{\mathrm{T}} + K - 1}. \tag{7.62}$$

Note that the training data is now partitioned in the same way as the test data so that $\widehat{\mathbf{R}}_{\mathrm{SCM}}$ is reduced in size to $P_{\mathrm{T}}Q_{\mathrm{T}} \times P_{\mathrm{T}}Q_{\mathrm{T}}$. Furthermore, both the AMF and NAMF statistics of (7.43) and (7.44) can be modified to use this reduced size SCM estimate with the data vector, $\mathbf{g}_{\mathrm{T}}$.

7.5.3 Variable scale hybrid

While the FS hybrid detector can lead to significant performance gains in homogeneous environments, the clutter will generally exhibit some degree of heterogeneity in practical scenarios. Therefore, a variable scale hybrid detector based on the GIP [68–70] was proposed in [63] where the training data is screened to determine the level of homogeneity. The GIP is essentially a hypothesis test that the training range gate, $\tilde{\mathbf{h}}_k$, belongs to the same multivariate normal distribution as the test data. Observations belonging to the tail of the distribution are deemed to be outliers and are therefore rejected as being heterogeneous. Given the estimate of the covariance matrix, $\widehat{\mathbf{Q}}(k)$, the GIP test is given by

$$P_{\mathrm{GIP}}(k) = \tilde{\mathbf{h}}_k^H \widehat{\mathbf{Q}}^{-1}(k)\tilde{\mathbf{h}}_k \tag{7.63}$$

where the filtered samples can be represented by the standard F-distribution [63]. Then a two-sided hypothesis test with confidence level, ϵ_{GIP}, would accept the null hypothesis that $\tilde{\mathbf{h}}_k$ belongs to the same complex normal distribution as the test data if $\vartheta_{\mathrm{L}} \leq P_{\mathrm{GIP}}(k) \leq \vartheta_{\mathrm{U}}$, where the lower and upper thresholds are [63]

$$\begin{aligned} P(P_{\mathrm{GIP}}(k) \leq \vartheta_{\mathrm{L}}) &= 1 - \beta_{\mathrm{inc}}\left(\frac{1}{\vartheta_{\mathrm{L}} + 1}, P_{\mathrm{T}}Q_{\mathrm{T}}, K_{\mathrm{T}} - 1\right), \\ P(P_{\mathrm{GIP}}(k) \geq \vartheta_{\mathrm{U}}) &= \beta_{\mathrm{inc}}\left(\frac{1}{\vartheta_{\mathrm{U}} + 1}, P_{\mathrm{T}}Q_{\mathrm{T}}, K_{\mathrm{T}} - 1\right) \end{aligned} \tag{7.64}$$

and $\beta_{\mathrm{inc}}(\cdot)$ is the incomplete beta function. Finally, the covariance matrix estimate, $\widehat{\boldsymbol{\Sigma}}_{\mathrm{GIP}}$, is obtained from the combination of the test data and the training snapshots that are accepted by the hypothesis test.

7.5.4 Texture estimation and normalisation hybrid

Unlike the GIP-based hybrid approach that rejects some of the data, the texture estimation and normalisation (TN) hybrid capitalises on the entire available training data to improve performance [64]. Also, in contrast to the FS hybrid algorithm, which can suffer from degradation in severely heterogeneous environments, this detector is designed to restore homogeneity in maritime scenarios by normalising the texture fluctuations.

Assume that the clutter for range bin k is K-distributed with texture, τ_k, and follows the complex normal distribution $\mathcal{CN}(\mathbf{0}, \boldsymbol{\Sigma}_k)$ where $\boldsymbol{\Sigma}_k = \tau_k \mathbf{R}$. Similarly, the test data used for estimating the covariance also follows the complex normal $\mathcal{CN}(\mathbf{0}, \mathbf{R}_{\text{av}})$ with $\mathbf{R}_{\text{av}} = \tau_{\text{av}} \mathbf{R}$. In both cases, the textures τ_{av} and τ_k are drawn from a gamma distribution with $\widehat{\mathbf{Q}} = \tau_{\text{av}} \widehat{\mathbf{R}}_{\text{av}}$ and $\widehat{\boldsymbol{\Sigma}}_k = \tau_k \widehat{\mathbf{R}}_k$. Then

$$\begin{aligned} \text{Tr}\left[\widehat{\mathbf{Q}}^{-1}\widehat{\boldsymbol{\Sigma}}_k\right] &= \frac{\tau_k}{\tau_{\text{av}}}\text{Tr}\left[\widehat{\mathbf{R}}_{\text{av}}^{-1}\widehat{\mathbf{R}}_k\right] \\ &= \frac{\tau_k}{\tau_{\text{av}}}\text{Tr}\left[\widehat{\mathbf{R}}_{\text{av}}^{-1}\widehat{\mathbf{R}}_k\right] \\ &= P_{\text{T}} Q_{\text{T}} \frac{\tau_k}{\tau_{\text{av}}}. \end{aligned} \tag{7.65}$$

Now define $\zeta_{\text{TN}}(k) = 1/(P_{\text{T}} Q_{\text{T}}) \text{Tr}\left[\widehat{\mathbf{Q}}^{-1}\widehat{\boldsymbol{\Sigma}}_k\right]$ to re-scale the covariance matrix estimate of the kth range gate, giving

$$\tilde{\boldsymbol{\Sigma}}_k = \frac{1}{\zeta_{\text{TN}}(k)}\widehat{\boldsymbol{\Sigma}}_k = \tau_{\text{av}}\widehat{\mathbf{R}}_k, \tag{7.66}$$

and hence the training range covariance matrix should become homogeneous with that of the test data. Finally, the covariance estimate of size $P_{\text{T}} Q_{\text{T}} \times P_{\text{T}} Q_{\text{T}}$ is constructed as

$$\widehat{\boldsymbol{\Sigma}}_{\text{TN}} = \frac{1}{2K_{\text{T}} - 1}\left((K_{\text{T}} - 1)\widehat{\mathbf{Q}} + \sum_{k=1}^{K_{\text{T}}} \tilde{\boldsymbol{\Sigma}}_k\right). \tag{7.67}$$

7.5.5 Detection results

Two data sets are used to evaluate the performance of the coherent detection schemes. The first data set comes from a side looking L-band radar simulation with a 4-channel linear array. The clutter is modelled with a K-distributed amplitude distribution and a uniform Doppler spectrum. To change the heterogeneity, the shape parameter, ν, is varied with a small value leading to greater variation in the texture and hence an increased level of heterogeneity. Thermal noise is also added to the clutter realisation to achieve a CNR of 20 dB. For the experiments here, the key parameters include a carrier frequency of 1.33 GHz, a bandwidth of 140 MHz giving a range resolution of 1.07 m, a PRF of 1,500 Hz and a two-way 3 dB azimuth beamwidth of 16°.

The second data set was collected by the Ingara L-band airborne radar, which was developed by the Australian Defence Science and Technology Group. This radar uses a side looking polarimetric antenna which transmits with a single linear polarisation and then receives with two 4-channel linear arrays. The data set used here was collected off the coast near Townsville in Queensland, Australia and is from the horizontal polarisation with a Douglas sea state between 3 and 4. It was collected with a 30° grazing angle in the cross-wind direction. An analysis of the amplitude statistics found that the CNR of the data was 10.1 dB and the underlying K-distribution shape parameter of the sea clutter was 22.1, making it reasonably homogeneous. Figure 7.17 shows an example angle/Doppler clutter spectra using the Ingara data set, highlighting both the clutter ridge and the location of an injected point target that is close to the clutter. Note that there is aliasing present due to the inter-element spacing, $d > \lambda/2$. A summary of the key parameters for both data sets is given in Table 7.3.

Results are now presented using both simulated and real data to highlight the effectiveness of the SDS and hybrid detectors at mitigating the clutter heterogeneity and fully capitalising on the training data. The detection schemes use a CPI of 135 pulses and include the AMF, NAMF, the SDS from (7.58) and three hybrid schemes,

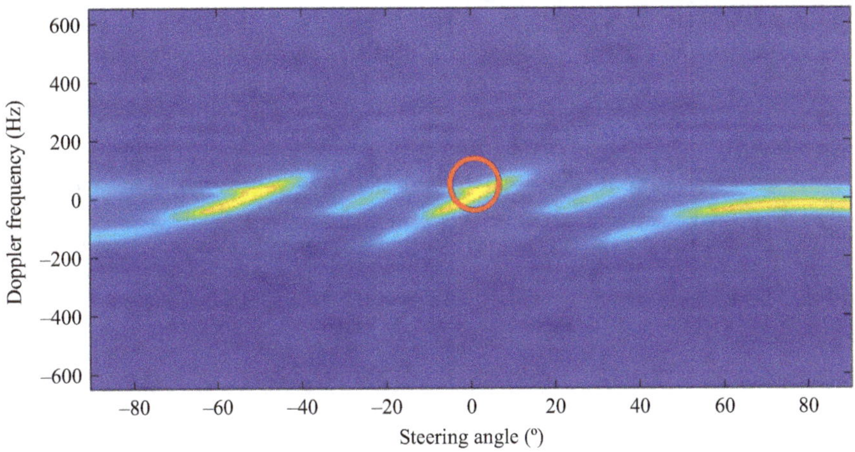

Figure 7.17 Optimal clutter spectrum of the Ingara L-band data set showing the location of an injected target at 0° azimuth and 50 Hz Doppler [64].

Table 7.3 Data characteristics for the sea clutter data sets

Parameter	**Simulated**	**Ingara**
Centre frequency (GHz)	1.33	1.33
Bandwidth (MHz)	140	140
Pulse repetition frequency (Hz)	1,500	1,500
Number of spatial channels	4	4
Inter-element spacing (m)	0.113 ($\lambda/2$)	0.261 (1.16λ)
Aircraft speed (m/s)	100	88
Azimuth one-way 3 dB beamwidth	16°	13°
Clutter to noise ratio (dB)	20	10.1

including the fixed scale (Hyb-FS), GIP-based variable scale (Hyb-VS) and the texture estimation and normalisation (Hyb-TN). Also, for a fairer comparison, the AMF and NAMF statistics of (7.43) and (7.44) are modified to employ $\widehat{\mathbf{R}}_{\mathrm{SCM}}$ of size $P_{\mathrm{T}}Q_{\mathrm{T}} \times P_{\mathrm{T}}Q_{\mathrm{T}}$ with the data vector $\tilde{\mathbf{g}}_{\mathrm{T}}$.

The probability of detection, P_{d}, results were obtained for a $P_{\mathrm{fa}} = 10^{-3}$, by injecting a Swerling type 1 point target into different range bins. The target Doppler is set as 50 Hz and the azimuth to 0°. For all detectors, sliding windows of size $P_{\mathrm{T}} = 8$ and $Q_{\mathrm{T}} = 4$ are used, which results in $K_{\mathrm{T}} = 135 - 8 + 1 = 128$ snapshots for the SDS matrix $\tilde{\mathbf{h}}_{\mathrm{T}}$, and $K = K_{\mathrm{T}}$. All the results reported later were obtained using 10,000 Monte Carlo runs.

For the simulated data, the first two results in Figures 7.18 and 7.19 show the P_{d} as a function of the SIR for the homogeneous and heterogeneous cases, respectively. The former case has exponential statistics for the intensity, while the latter is K-distributed with a shape parameter of $\nu = 1$. For the homogeneous case, there is little difference observed for the AMF and hybrid algorithms, while the NAMF and SDS algorithms show reduced performance by approximately 4 dB at a P_{d} of 0.5. In the case of the SDS, this is due to the loss of degrees of freedom in using the mean test vector, $\tilde{\mathbf{g}}_{\mathrm{T}}$, and covariance matrix estimate, $\widehat{\mathbf{Q}}$. For the NAMF however, the test statistic is more sensitive due to the use of the mean test vector. For the heterogeneous case in Figure 7.19, the difference in the algorithms is more distinct with the Hyb-TN having the best performance, while the others show worse results by different amounts. The AMF is particularly bad as it does nothing to account for variations in the texture.

The performance as a function of the target Doppler frequency is now shown with the target SIR fixed at −20 dB. Figure 7.20 shows the homogeneous case with

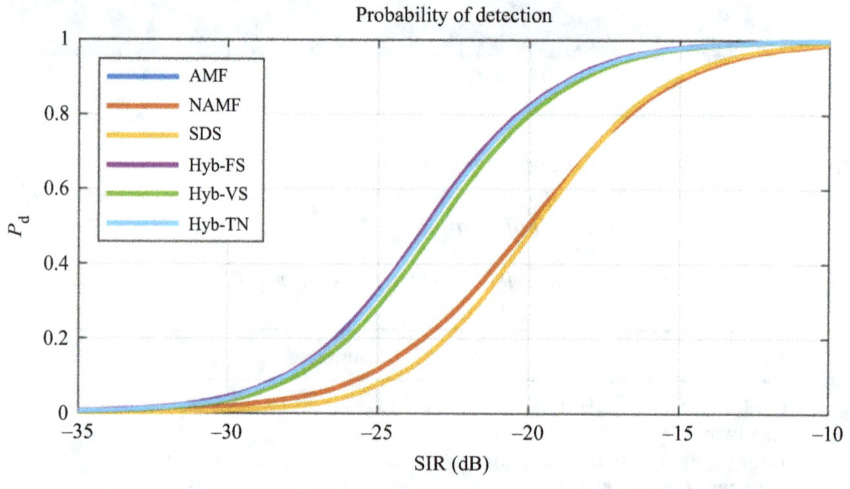

Figure 7.18 Probability of detection for $P_{\mathrm{fa}} = 10^{-3}$ obtained for an homogeneous environment. The target azimuth and Doppler frequency were set to 0° and 50 Hz, respectively, indicating a slow target scenario [64].

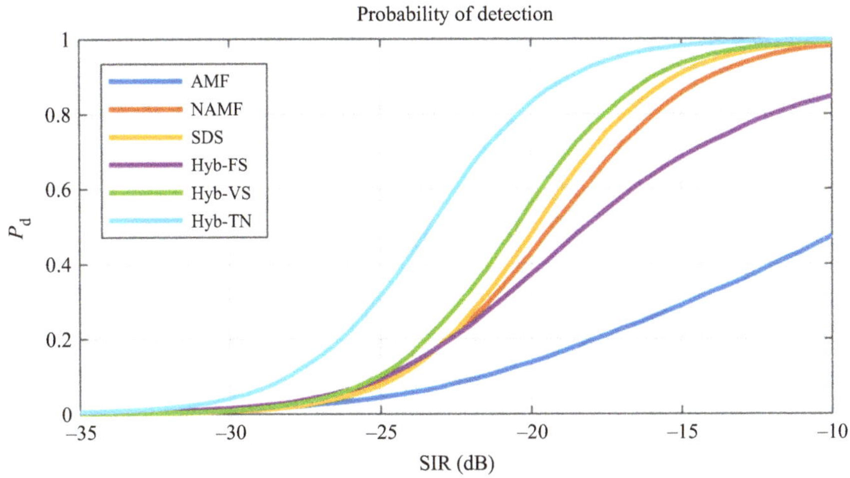

Figure 7.19 Probability of detection for $P_{fa} = 10^{-3}$ obtained for an heterogeneous environment with $\nu = 1$. The target azimuth and Doppler frequency were set to 0° and 50 Hz, respectively, indicating a slow target scenario [64].

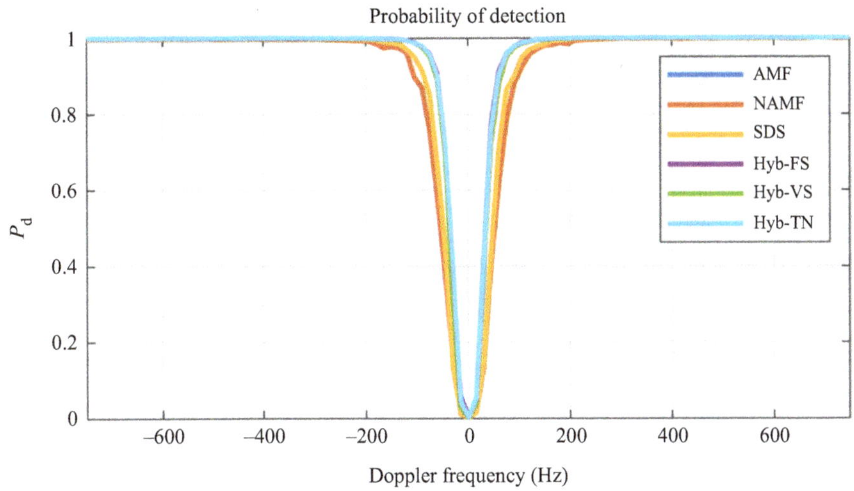

Figure 7.20 Probability of detection for $P_{fa} = 10^{-3}$ obtained for homogeneous data with a fixed SIR of −20 dB and an azimuth of 0° [64].

the best performance in the endo-clutter region observed for the AMF and all three hybrid algorithms. For the heterogeneous case in Figure 7.21, it is again clear that the Hyb-TN offers the best performance. In this example, the AMF and Hyb-FS cannot match the performance in the exo-clutter region due to the heterogeneity of the clutter.

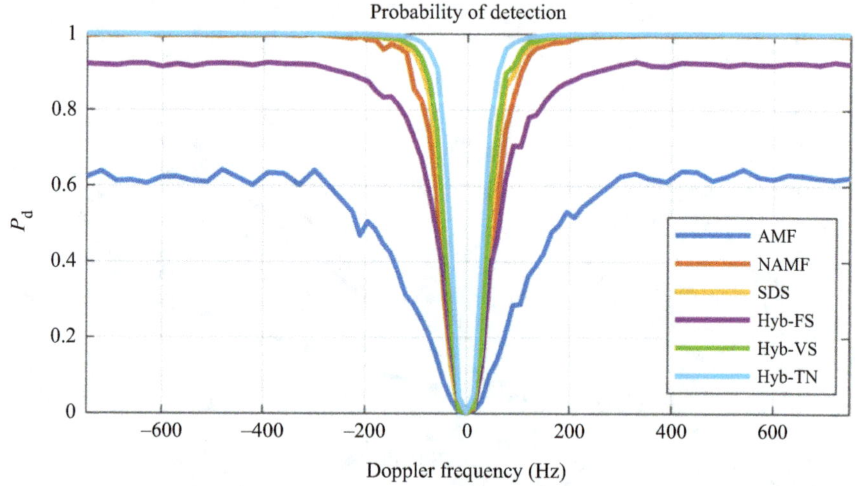

Figure 7.21 Probability of detection for $P_{fa} = 10^{-3}$ obtained for an heterogeneous environment with $\nu = 1$, a fixed SIR of −20 dB and an azimuth of 0° [64]. © Commonwealth of Australia 2020

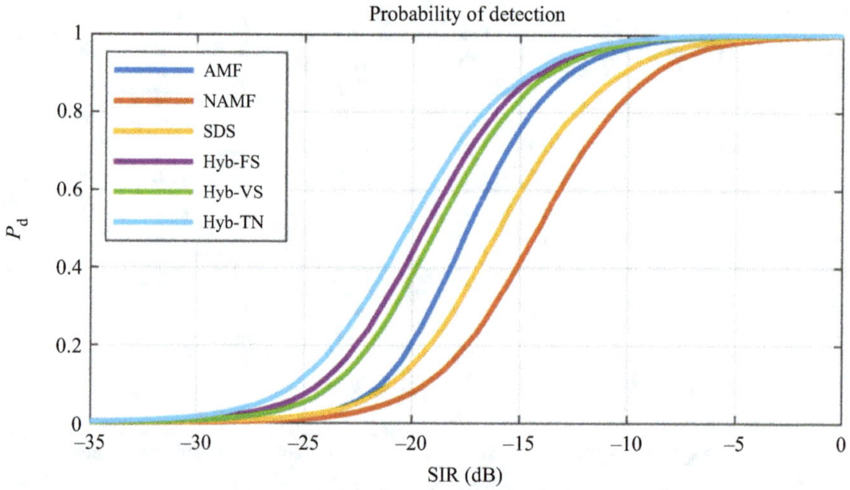

Figure 7.22 Probability of detection for $P_{fa} = 10^{-3}$ obtained for the Ingara L-band data set. The target azimuth and Doppler frequency were set to 0° and 50 Hz respectively indicating a slow target scenario [64]. © Commonwealth of Australia 2020

The SDS algorithm, however, is completely immune to the clutter heterogeneity as it does not use any training data.

A similar analysis is now performed using the Ingara L-band data. The first result in Figure 7.22 shows the performance against the target SIR for a frequency of 50

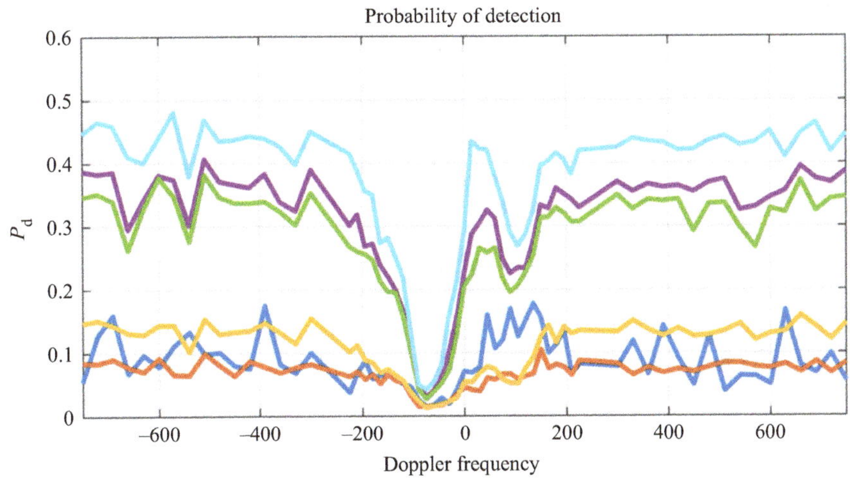

Figure 7.23 Probability of detection for $P_{fa} = 10^{-3}$ obtained for the Ingara L-band data set with a fixed SIR of −20 dB and an azimuth of 0°. The line colours match those from Figure 7.22 [64]. © Commonwealth of Australia 2020

Hz and an azimuth angle of 0°. The Hyb-TN again shows the best performance with the other hybrid approaches within 3 dB at a P_d of 0.5. The SDS, AMF and NAMF show performance that lies part way between the two extremes shown in Figures 7.18 and 7.19. The final plot in Figure 7.23 shows the variation in Doppler for an SIR of −20 dB. These results are similar to Figure 7.22 with the best performance attributed to the Hyb-TN and the worst for the AMF and NAMF.

References

[1] M. I. Skolnik, *Radar handbook*, 3rd ed. McGraw-Hill, New York, 2008.

[2] J. J. Ousborne, D. Griffith, and R. W. Yuan, "A periscope detection radar," *Johns Hopkins APL Technical Digest*, vol. 18, no. 1, pp. 125–133, 1997.

[3] K. D. Ward, R. J. A. Tough, and S. Watts, *Sea clutter: Scattering, the K-distribution and radar performance*, 2nd ed. The Institute of Engineering Technology, London, 2013.

[4] L. Rosenberg, "Sea-spike detection in high grazing angle X-band sea-clutter," *IEEE Transactions on Geoscience and Remote Sensing*, vol. 51, no. 8, pp. 4556–4562, 2013.

[5] S. Bocquet, "Parameter estimation for Pareto and K distributed clutter with noise," *IET Radar Sonar and Navigation*, vol. 9, no. 1, pp. 104–113, 2015.

[6] A. Balleri, A. Nehorai, and J. Wang, "Maximum likelihood estimation for compound-Gaussian clutter with inverse gamma texture," *IEEE Transactions on Aerospace and Electronic Systems*, vol. 43, no. 2, pp. 775–779, 2007.

[7] M. Farshchian and F. L. Posner, "The Pareto distribution for low grazing angle and high resolution X-band sea clutter," *IEEE Radar Conference*, 2010, pp. 789–793.

[8] G. V. Weinberg, "Assessing Pareto fit to high-resolution high-grazing-angle sea clutter," *IET Electronic Letters*, vol. 47, no. 8, pp. 516–517, 2011.

[9] G. Weinberg, *Radar detection theory of sliding window processes*. CRC Press, London, 2017.

[10] M. M. Finn and R. S. Johnson, "Adaptive detection mode with threshold control as a function of spatially sampled clutter-level estimates," *RCS Review*, vol. 30, pp. 414–465, 1968.

[11] V. Y. F. Li and K. M. Miller, "Target detection in radar: Current status and future possibilities," *The Journal of Navigation*, vol. 50, pp. 303–313, 1997.

[12] S. Watts, "Radar detection prediction in sea clutter using the compound K-distribution model," *IEE Proceedings F (Communications, Radar and Signal Processing)*, vol. 132, no. 7, pp. 613–620, 1985.

[13] M. McDonald and S. Lycett, "A comparison of fast and slow scan radars," *International Radar Conference*, 2004, pp. 1–6.

[14] S. Watts and A. G. Stove, "Non-coherent integration strategies in scanning radar for the detection of targets in sea clutter and noise," *International Radar Conference*, no. 10P-SP-229, 2004, pp. 1–6.

[15] P. Reuillon, F. Parenty, and F. Perret, "Scan-to-scan sea spikes filtering for radar," *European Radar Conference*, 2010, pp. 272–275.

[16] G. B. Goldstein, "False alarm regulation in log-normal and Weibull clutter," *IEEE Transactions on Aerospace and Electronic Systems*, vol. AES-7, pp. 942–950, 1971.

[17] R. J. A. Tough, K. D. Ward, and P. W. Shepherd, "The modelling and exploitation of spatial correlation in spiky sea clutter," *European Radar Conference*, 2005, pp. 5–8.

[18] S. Watts, "The Searchwater radar on Nimrod MR2," *International Radar Conference*, 2017, pp. 1–6.

[19] S. Watts, "Constant false alarm rate adaptive radar control, UK Patent GB 2449884 B," Nov. 2011.

[20] L. Rosenberg, D. J. Crisp, and N. J. Stacy, "Analysis of the KK-distribution with medium grazing angle sea-clutter," *IET Proceedings of Radar Sonar and Navigation*, vol. 4, no. 2, pp. 209–222, 2010.

[21] L. Rosenberg and S. Bocquet, "The Pareto distribution for high grazing angle sea-clutter," *IEEE Proceedings on Geoscience and Remote Sensing*, 2013, pp. 4209–4212.

[22] D. Middleton, "New physical-statistical methods and model for clutter and reverberation: The KA-distribution and related probability structures," *IEEE Journal of Oceanic Engineering*, vol. 24, no. 3, pp. 261–284, 1999.

[23] Y. Dong, "Distribution of X-band high resolution and high grazing angle sea clutter," Defence Science Technology Organisation, Research Report DSTO-RR-0316, 2006.

[24] G. V. Weinberg, "Constant false alarm rate detectors for Pareto clutter models," *IET Radar Sonar and Navigation*, vol. 7, no. 2, pp. 153–163, 2013.

[25] G. V. Weinberg, "Constant false alarm rate detection in Pareto distributed clutter: further results and optimality issues," *Contemporary Engineering Sciences*, vol. 7, no. 6, pp. 231–261, 2014.

[26] G. V. Weinberg, "Examination of classical detection schemes for targets in Pareto distributed clutter: Do classical CFAR detectors exist, as in the Gaussian case?" *Multidimensional Systems and Signal Processing*, vol. 26, no. 3, pp. 599–617, 2015.

[27] G. V. Weinberg, "The constant false alarm rate property in transformed noncoherent detection processes," *Digital Signal Processing*, vol. 51, no. C, pp. 1–9, 2016.

[28] G. V. Weinberg, "General transformation approach for constant false alarm rate detector development," *Digital Signal Processing*, vol. 30, pp. 15–26, 2014.

[29] G. V. Weinberg, L. Bateman, and P. Hayden, "Constant false alarm rate detection in Pareto type II clutter," *Digital Signal Processing*, vol. 68, pp. 193–198, 2017.

[30] L. Rosenberg and G. V. Weinberg, "Performance analysis of Pareto CFAR detectors," *International Radar Conference*, 2017, pp. 1–6.

[31] G. V. Weinberg, "On the construction of CFAR decision rules via transformations," *IEEE Transactions on Geoscience and Remote Sensing*, vol. 55, no. 2, pp. 1140–1146, 2017.

[32] G. V. Weinberg, S. D. Howard, and C. Tran, "Bayesian framework for detector development in Pareto distributed clutter," *IET Radar Sonar and Navigation*, vol. 13, pp. 1548–1555, 2019.

[33] S. Watts and L. Rosenberg, "A comparison of coherent and non-coherent radar detection performance in radar sea clutter," *International Conference on Radar Systems*, 2017, pp. 1–6.

[34] A. De Maio and M. S. Greco, Eds., *Modern radar detection theory*. SciTech, London, 2016.

[35] L. Rosenberg and S. Watts, "Model based coherent detection in medium grazing angle sea-clutter," *IEEE Radar Conference*, 2016, pp. 1–6.

[36] L. Rosenberg and S. Watts, "Coherent detection in medium grazing angle sea-clutter," *IET Radar Sonar and Navigation*, vol. 11, no. 9, pp. 1340–1348, 2017.

[37] E. J. Kelly, "An adaptive detection algorithm," *IEEE Transactions on Aerospace and Electronics Systems*, vol. AES-22, no. 1, pp. 115–127, 1986.

[38] W. Chen and I. Reed, "A new CFAR detection test for radar," *Digital Signal Processing*, vol. 1, no. 4, pp. 198–214, 1991.

[39] S. Kraut and L. L. Scharf, "The CFAR adaptive subspace detector is a scale-invariant GLRT," *IEEE Transactions on Signal Processing*, vol. 47, no. 9, pp. 2538–2541, 1999.

[40] F. Gini, M. S. Greco, A. Farina, and P. Lombardo, "Optimum and mismatched detection against K-distributed plus Gaussian clutter," *IEEE Transactions on Aerospace and Electronic Systems*, vol. 34, no. 3, pp. 860–876, 1998.

[41] V. Gracheva and J. Ender, "Multichannel analysis and suppression of sea clutter for airborne microwave systems," *IEEE Transactions on Geoscience and Remote Sensing*, vol. 54, no. 4, pp. 2385–2399, 2016.

[42] K. McDonald, M. and D. Cerutti-Maori, "Coherent radar processing in sea clutter environments part 2: Adaptive normalised matched filter versus adaptive matched filter performance," *IEEE Transactions on Aerospace and Electronic Systems*, vol. 52, no. 4, pp. 1818–1833, 2016.

[43] D. J. Coe and R. G. White, "Experimental moving target detection results from a three-beam airborne SAR," *European SAR Conference*, 1996, pp. 419–422.

[44] Z. Zhiqin and H. Shunji, "A research of moving targets detection and imaging of SAR," *European SAR Conference*, 1996, pp. 427–430.

[45] J. Ward, "Space-time adaptive processing for airborne radar," Lincoln Labs. MIT, Technical Report 1015, 1994.

[46] R. Klemm, *Principles of space-time adaptive processing*, 3rd ed. The Institution of Engineering and Technology, London, 2006, vol. 21.

[47] J. R. Guerci, *Space-time adaptive processing for radar*, 2nd ed. Artech House, Norwood, 2003.

[48] W. L. Melvin, "A STAP overview," *IEEE Aerospace and Electronic Systems Magazine*, vol. 19, no. 1, pp. 19–35, 2004.

[49] M. Greco, P. Stinco, F. Gini, and M. Rangaswamy, "Impact of sea clutter nonstationarity on disturbance covariance matrix estimation and CFAR detector performance," *IEEE Transaction on Aerospace and Electronic Systems*, vol. 46, no. 3, pp. 1502–1513, 2010.

[50] J. Li, P. Stoica, and Z. Wang, "On robust Capon beamforming and diagonal loading," *IEEE Transactions on Image Processing*, vol. 51, no. 7, pp. 1702–1715, 2003.

[51] N. Ma and J. T. Goh, "Efficient method to determine diagonal loading value," *IEEE International Conference on Acoustics, Speech, and Signal Processing*, vol. 5, 2003, pp. 341–344.

[52] S. Watts and L. Rosenberg, "Coherent radar performance in sea clutter," *IEEE Radar Conference*, Johannesburg, 2015, pp. 103–108.

[53] E. Conte, M. Lops, and G. Ricci, "Adaptive detection schemes in compound-Gaussian clutter," *IEEE Transaction on Aerospace and Electronic Systems*, vol. 34, no. 4, pp. 1058–1069, 1998.

[54] J. R. Román, M. Rangaswamy, D. W. Davis, Q. Zhang, B. Himed, and J. H. Michels, "Parametric adaptive matched filter for airborne radar applications," *IEEE Transactions on Aerospace and Electronic Systems*, vol. 36, no. 2, pp. 677–692, 2000.

[55] F. Gordon and C. Mountford, "A novel small target mode for the Leonardo Seaspray and Osprey E-scan radars," *NATO SET-239*, vol. MP-SET-239-16, 2016, pp. 1–6.

[56] B. J. Wilcox, S. Stevenson, and C. P. Mountford, "Coherent sea clutter modelling," *International Radar Conference*, 2017, pp. 1–6.

[57] R. L. Sinclair, "Radar surveillance system, UK Patent GB 2500931A," 2012.

[58] M. Rangaswamy, "Statistical analysis of the nonhomogeneity detector for non-Gaussian interference backgrounds," *IEEE Transactions on Signal Processing*, vol. 53, no. 6, pp. 2101–2111, 2005.

[59] P. Antonik, H. Schuman, W. Melvin, and M. Wicks, "Implementation of knowledge-based control for space-time adaptive processing," *IEEE Radar Conference*, Oct. 1997, pp. 478–482.

[60] T. Sarkar, H. Wang, S. Park, *et al.*, "A deterministic least-squares approach to space-time adaptive processing (STAP)," *IEEE Transactions on Antennas and Propagation*, vol. 49, no. 1, pp. 91–103, 2001.

[61] D. Cristallini and W. Bürger, "A robust direct data domain approach for STAP," *IEEE Transactions on Signal Processing*, vol. 60, no. 3, pp. 1283–1294, 2012.

[62] E. Aboutanios and B. Mulgrew, "A STAP algorithm for radar target detection in heterogeneous environments," *Proc. IEEE/SP 13th Workshop on Statistical Signal Processing*, 2005, pp. 966–971.

[63] E. Aboutanios and B. Mulgrew, "Hybrid detection approach for STAP in heterogeneous clutter," *IEEE Transactions on Aerospace and Electronic Systems*, vol. 46, no. 3, pp. 1021–1033, 2010.

[64] E. Aboutanios and L. Rosenberg, "Hybrid detection approaches using the single data set algorithm," *IEEE Radar Conference*, 2020, pp. 1–6.

[65] E. Aboutanios and L. Rosenberg, "Single snapshot coherent detection in sea clutter," *IEEE Radar Conference*, 2019, pp. 1–6.

[66] E. Aboutanios and L. Rosenberg, "Multichannel target detection for maritime radar," *2019 53rd Asilomar Conference on Signals, Systems, and Computers*, 2019, pp. 113–117.

[67] E. Aboutanios and B. Mulgrew, "A hybrid STAP approach for radar target detection in heterogeneous environments," *14th European Signal Processing Conference*, 4–8 Sep. 2006, pp. 1–5.

[68] W. Melvin and M. C. Wicks, "Improving practical space-time adaptive radar," *IEEE National Radar Conference*, 1997, pp. 48–53.

[69] P. Chen, W. Melvin, and M. Wicks, "Screening among multivariate normal data," *Journal of Multivariate Analysis*, vol. 69, pp. 10–29, 1999.

[70] M. Rangaswamy, J. Michels, and B. Himed, "Statistical analysis of the non-homogeneity detector for STAP applications," *Digital Signal Processing: A Review Journal*, vol. 14, no. 3, pp. 253–257, 2004.

Chapter 8
New approaches for target detection

Reliable and robust radar target detection is a challenging task in the maritime environment. This is particularly true for aircraft operating at higher altitudes where the grazing angle is higher and backscatter from the sea becomes stronger. Bright returns or sea spikes also add to the difficulty as they share many similarities with true targets and can be difficult to distinguish. While coherent detection in the maritime domain is difficult due to heterogeneous sea clutter, it is increasingly sought after due to the potential gain that can be achieved. Doppler filtering may be non-adaptive, with a constant false alarm rate (CFAR) scheme run along each frequency bin, or the processing could be adaptive and whiten the clutter prior to detection. Such coherent techniques are described in Chapter 7.

In this chapter, a number of new approaches for target detection are presented. These exploit knowledge from other areas of signal processing and potentially offer some significant advantages in maritime target detection. The first approach in Section 8.1 explores how the time/frequency domain can be exploited through the use of wavelets. This approach is designed to improve target detection in the range/time domain by a careful selection of different signal components (sub-bands) in the wavelet decomposition. In Section 8.2, a number of different detection schemes that exploit sparsity in the time/frequency domain are described. These rely on knowledge of the target model with appropriate 'dictionaries' defined such as tuned Q-wavelet transforms (TQWTs) or short-time Fourier transforms (STFTs). A related approach is to learn the dictionary from the data in order to distinguish targets from the background sea clutter. Section 8.3 introduces a dictionary learning (DL) detection approach that is able to learn the characteristics of sea clutter. Section 8.4 then looks at track before detect (TBD) as another paradigm for target detection. This is a powerful way to take advantage of targets that move through the scene and does not rely on thresholding to first identify the target location. There are other techniques not covered here that have been proposed. Some of these include background subtraction [1] for capturing the dynamics of a scene to detect anomalous behaviour and the use of machine learning approaches for exploiting knowledge of what the underlying clutter and/or target look like [2,3].

8.1 Time/frequency target separation

There have been a number of studies into the use of wavelet transforms (WT) for detecting targets in Gaussian noise. Ehara *et al.* [4] proposed two methods to enhance the target return prior to a detection scheme. The first is based on the idea that given an optimal scale, the wavelet function can approximate the target response with significant gains in the signal-to-noise ratio (SNR). The second method extends this idea by integrating a small range of wavelet coefficients around the optimal scale. This provides more robustness and was shown to improve the SNR even further. Wang *et al.* [5] then proposed the use of a wavelet-based function to design the transmitted waveform in a radar system. If the target was matched to the wavelet function, it should appear stronger in the discrete wavelet transform (DWT) domain while the noise will be small and spread out. By suppressing these noise values with a threshold (de-noising), the target SNR should substantially increase. However, it was found that matching the wavelet waveform was difficult unless the target was large and had sufficient radar cross section. Zhang *et al.* [6] also used WTs for radar target detection and found that removing some of the high-frequency 'detail' wavelet coefficients was effective in reducing the noise of radar echoes and thus improving the detection performance. With this approach, the DWT sub-bands suffer from reduced detail (resolution), so the authors applied an independent component analysis of the wavelet coefficients to improve the performance. However, with only a single simulation and limited experimental details provided in [6], it is unclear whether this approach is effective for a broader range of signals. Another wavelet-based approach was proposed by Davidson *et al.* [7], who used a continuous wavelet transform (CWT) to identify the dominant scatterers in sea clutter. The authors then applied a 'persistence' statistic to detect slow moving targets from the surrounding backscatter. Wavelets have also found their way into radar image processing. Jangal *et al.* [8] used a DWT and subsequent selective reconstruction (by ignoring some sub-bands) to process range/Doppler images of a high-frequency surface wave radar and suppress the clutter-plus-noise (interference) in the images.

Related to these studies, Torres [9] investigated radar detection by comparing three popular and widely used wavelet functions: Morlet, Mexican hat and Daubechies-9. Each wavelet was applied to the problem of finding a simulated chirp in the presence of white Gaussian noise, with Daubechies-9 offering the best detection performance. Selecting a suitable scale in a WT implementation is also important to better localise a signal of interest [4,10,11]. Ball and Tolley [11] suggested an algorithm to automatically determine the optimal scale parameter. This involved searching the CWT coefficients to find a peak and then performing a de-noising procedure around the optimum scale. They achieved an average SNR improvement of up to 10 dB when compared to a conventional matched filter.

In contrast to these existing wavelet-based schemes, the use of two-dimensional (2D) stationary wavelet transforms (SWT) has been investigated in [12], where a single-level 2D SWT was applied to the range/time domain with the filtered outputs individually reconstructed to produce four different range/time images. Analysis of the amplitude distribution of these reconstructed sub-bands revealed that some were less spiky than others and offered the potential for improved detection performance. Compared to the unfiltered data, a good improvement was shown for moving targets with a slight improvement for stationary targets. A limitation of this work was the lack of a method to select which sub-band to use when the target velocity is unknown. This was addressed in subsequent studies [13,14], where similar or better performance was achieved with a one-dimensional SWT applied along slow time. This work is described further in the following sections.

8.1.1 Sub-band isolation and reconstruction

WTs have been used for signal processing in various applications to obtain compact representations of non-stationary signals. They have the ability to represent signals at multiple scales, or 'resolutions', with an adaptive time window, thus offering a different partitioning of the time–frequency plane compared to techniques such as STFTs [10,15–18]. Multi-resolution analysis separates the input data into different sets of detail and approximation (scaling function) coefficients, called 'sub-bands'. Each of these coefficients corresponds to a unique region of the time–frequency plane. It is reasonable to expect interference and target returns would produce different distributions of coefficients, thus allowing them to be separated in the coefficient domain. The SWT was used in [13] due to its translation-invariance property [19–21]. This is a non-decimated transform and has been used successfully in applications that may be sensitive to artefacts arising from the more popular, decimated version of the WT [19,22,23].

The SWT signal decomposition can be implemented as cascaded two-channel filter banks, with each filter bank consisting of a pair of complementary low- and high-pass filters. The filter bank at level $q \geq 0$ accepts the approximation sub-band at level $q-1$ as an input and produces detail (D_q) and approximation (A_q) sub-bands at the outputs of the high- and low-pass filters, respectively. The data input is commonly taken to be the A_0 sub-band, with subsequent levels:

$$A_q(n) = \sum_m h_a(n-m)A_{q-1}(m), \tag{8.1}$$

$$D_q(n) = \sum_m g_a(n-m)A_{q-1}(m), \tag{8.2}$$

where $h_a(\cdot)$ and $g_a(\cdot)$ are the analysis impulse responses of the low- and high-pass filters. The process can be reversed to perform reconstruction by

$$A_q(n) = \sum_m h_s(n-m)A_{q+1}(m) + \sum_m g_s(n-m)D_{q+1}(m) \tag{8.3}$$

where $h_s(\cdot)$ and $g_s(\cdot)$ are the low- and high-pass synthesis filters, respectively.

The set of filters $\{h_a(\cdot), g_a(\cdot), h_s(\cdot), g_s(\cdot)\}$ must be designed together to achieve perfect (error-free) reconstruction and the literature describes numerous sets of wavelet filters, each designed to satisfy different properties [9,10,24,25]. For the work presented in [13,14], the Daubechies-4 wavelet was chosen based on earlier work [5,8,9,26] and provides short finite impulse response filters in the corresponding two-channel filter bank.

The components of the interference and target (if present) generally have different distributions of SWT coefficients across sub-bands. This implies that one or more sub-bands have a stronger signal-to-interference ratio (SIR) than the original data. This observation led to the idea of isolating different SWT sub-bands for target detection. The 'sub-band isolation and reconstruction' procedure for the first level decomposition is illustrated in Figure 8.1 in which the desired sub-bands are reconstructed, while the remainder are set to zero. The four filters are represented by their equivalent frequency responses, $G_a(\cdot)$, $H_a(\cdot)$, $G_s(\cdot)$ and $H_s(\cdot)$. In the frequency domain, these can be interpreted as band-pass filters with one or more pass-bands. Examples of the power spectral density (PSD) using the Ingara X-band data set are shown in Figure 8.2 with the SWT applied along slow time. The reconstruction from the level-1 approximate sub-band, $\tilde{A}_1$, only retains the low-frequency part of the spectrum, while the reconstruction from the detail sub-band, $\tilde{D}_1$, only contains the high-frequency part of the spectrum. The bottom result shows the $\tilde{D}_{23}$-reconstructed sub-band which combines the D_2 and D_3 sub-band components.

In computing the SWT, the total number of levels is usually a user-chosen parameter with a finite number of pulses placing a practical ceiling on the value of q. For the SWT decomposition, the filter impulse responses are up-sampled by 2 at every level of the decomposition. This causes rapid growth in the length of the filters as the level increases, eventually reaching the length of the data series. Further decomposition ceases to produce meaningful sub-bands. More precisely, if the coherent processing interval (CPI) contains N pulses and the length of the analysis filters $h_a(\cdot)$, $g_a(\cdot)$ is M_a, then the maximum level K_a must satisfy

$$K_a < \log_2\left(\frac{N}{M_a}\right). \tag{8.4}$$

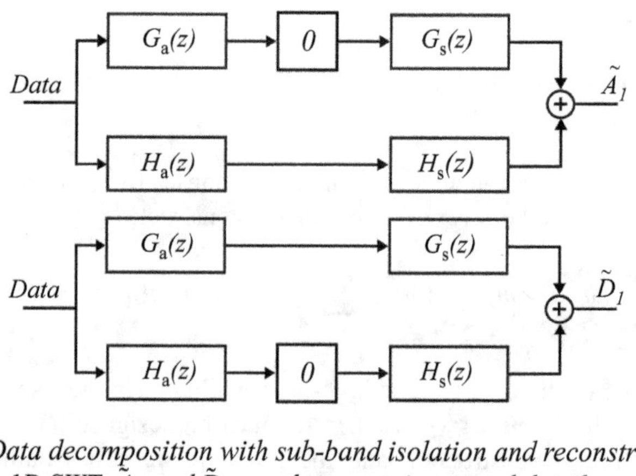

Figure 8.1 Data decomposition with sub-band isolation and reconstruction using a 1D SWT: $\tilde{A}_1$ and $\tilde{D}_1$ are the approximate and detail reconstructed sub-bands of the original data [13]. © IEEE 2017

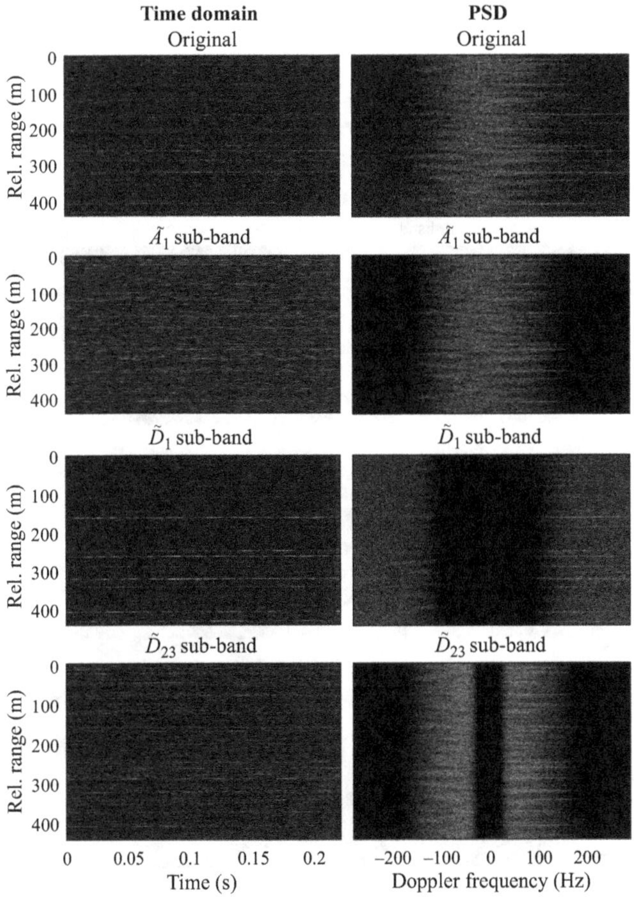

Figure 8.2 Time and PSD domain representations of the Ingara X-band data set after sub-band isolation and reconstruction [13]. © IEEE 2017

For example, using length 8 Daubechies-4 filters with a CPI of 128 pulses implies that the SWT is limited to 4 levels or lower.

8.1.2 Sub-band selection

Each reconstructed sub-band is now analysed to investigate the impact of the SWT decomposition when targets are present. Synthetic non-fluctuating point targets with an SIR of 10 dB are injected into the horizontally polarised data. Figure 8.3 shows the 3-level decomposition of the data with two point targets (stationary and moving) at relative ranges of 190 and 175 m, respectively. The stationary target is always preserved in the reconstructed approximate sub-bands, $\tilde{A}_q$, and appears brighter at lower resolutions. The moving target has a higher Doppler frequency and is maintained in the reconstructed detail sub-band, $\tilde{D}_1$. In this case, decomposing the A_1 sub-band

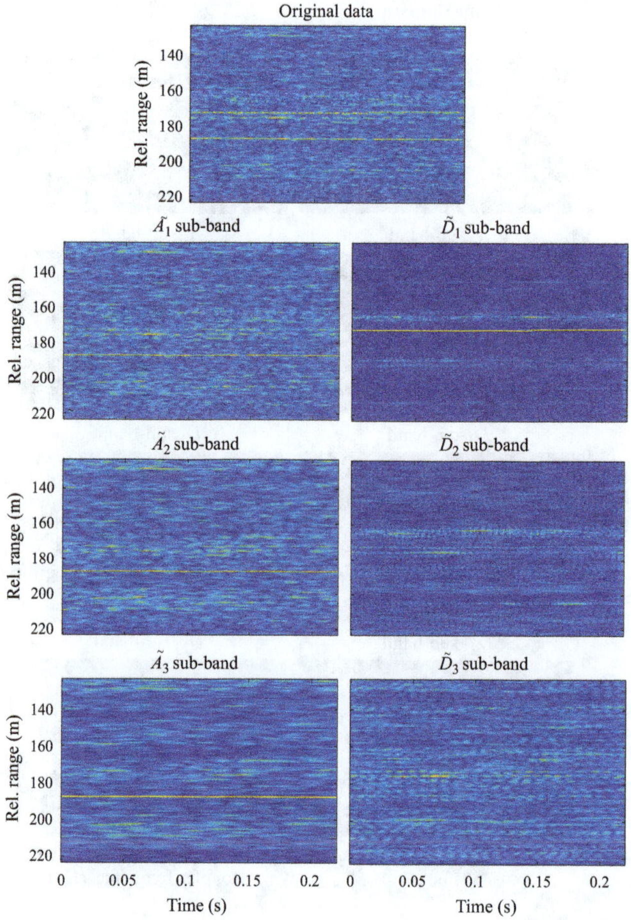

Figure 8.3 Range/time intensity images with stationary and moving point targets injected into the Ingara X-band data set. The stationary target is maintained in the reconstructed $\tilde{A}_q$ *sub-bands, while the moving target is maintained in the first level reconstructed* $\tilde{D}_1$ *sub-band [13].* *© IEEE 2017*

further will not give any improvement and consequently, the reconstructed $\tilde{D}_2$ and $\tilde{D}_3$ sub-bands show less information about the two targets.

After extensive experimentation by Duk *et al.* [13], a group of reconstructed sub-bands were selected for further analysis and to test the detection performance. A moving target outside the endo-clutter region will always be present in the $\tilde{D}_1$ reconstructed sub-band, while a stationary target will always be located in $\tilde{A}_1$. Detecting a slowly moving target is more difficult as it may be present in either the $\tilde{A}_2$ or $\tilde{D}_2$ reconstructed sub-bands or even have a Doppler frequency that lies partway between the frequency extent of these sub-bands. If this happens, it can potentially reduce the detection performance. Therefore to ensure good detection performance in all cases, combinations of D_1, D_2 and D_3 sub-bands were also selected. The spectra of the final group of reconstructed sub-bands are shown in Figure 8.4 and include $\tilde{D}_1, \tilde{D}_2, \tilde{D}_{12}, \tilde{D}_{23}$ and $\tilde{A}_3$. Unfortunately, the target velocity is not known ahead of time and a method is required to select which sub-band to use to ensure good detection performance.

Entropy is a measure of information that has been widely used to measure system disorder in statistical analysis [27] and the global or average uncertainty of random samples [28–30]. It was proposed in [13] as a means of determining which reconstructed sub-band contains the most information about a target. The motivation for using entropy is that the target is persistent over time while the interference returns are more random. To determine the entropy, $H_e(\cdot)$, for the kth range bin, the probability density function (PDF) of the data is determined using all the slow-time samples in a given CPI. An entropy indicator for the reconstructed sub-band, X, can then be defined as the absolute value of the entropy value with zero mean,

$$E(X_k) = |H_e(X_k) - \langle H_e \rangle|, \tag{8.5}$$

where $\langle H_e \rangle$ is the mean entropy over all range bins. To demonstrate the entropy variation when a target is present, Figure 8.5 shows the entropy metric for a number of reconstructed sub-bands as the SIR varies. Three different target velocities, 0, 1.1 and 2.6 m/s are shown with each result baselined by first determining the maximum entropy over an interference only region and then offsetting the measured entropy value. This ensures that the entropy value for each reconstructed sub-band is equal

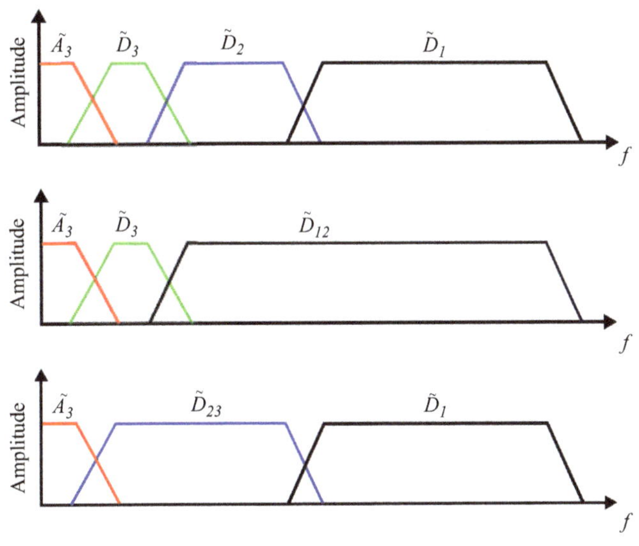

Figure 8.4 Spectra of the multi-resolution reconstructed sub-bands [13]. *© IEEE 2017*

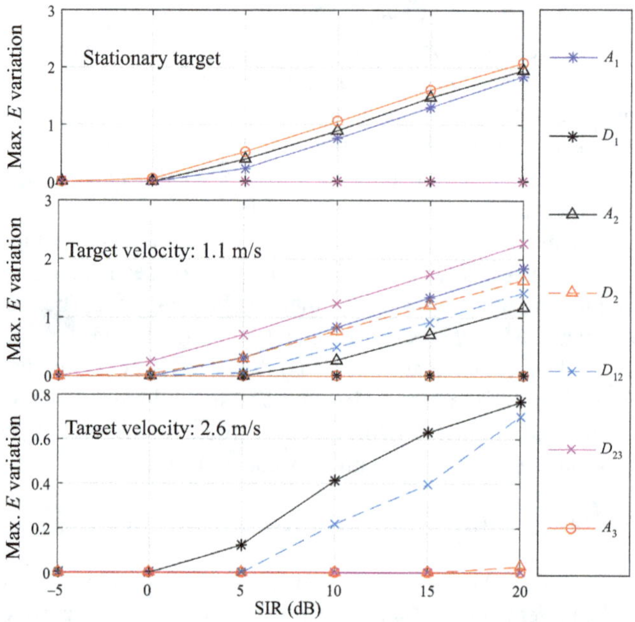

Figure 8.5 Maximum entropy variation with a Swerling 0 target and varying SIR [13]. © IEEE 2017

when no target is present. When the target is stationary, the $\tilde{A}_3$ sub-band has the highest maximum entropy followed by $\tilde{A}_2$ and $\tilde{A}_1$. For the faster moving target, the entropy metric is maximum for the $\tilde{D}_1$ reconstructed sub-band. Finally, when the target moves at 1.1 m/s, the combination, $\tilde{D}_{23}$, reconstructed sub-band is best.

Rather than decomposing and isolating every sub-band from the chosen set, a more efficient method was proposed in [13], where the entropy variations of the sub-bands are compared after each decomposition level. This idea is based on an image-processing application for selecting the minimum decomposition level of a natural image [31]. Figure 8.6 shows a sub-band selection scheme, where the maximum entropy is $\gamma_X = \text{Max}[E(X_k)]$. To implement this scheme, the first step is to compute the maximum entropy of the $\tilde{A}_1$ and $\tilde{D}_1$ reconstructed sub-bands after a single level of decomposition. If the maximum entropy of $\tilde{D}_1$ is greater than $\tilde{A}_1$, then $\tilde{D}_1$ is selected to perform the detection because the target is no longer maintained in $\tilde{A}_1$. However, if the maximum entropy of $\tilde{D}_1$ is less than $\tilde{A}_1$, the SWT for the next level is computed and compared to the maximum entropy of $\tilde{D}_2$ and $\tilde{A}_2$. This time, if $\tilde{D}_2$ is larger than $\tilde{A}_2$, the maximum of the three reconstructed sub-bands is selected: $\tilde{D}_2, \tilde{D}_{12}$ or $\tilde{D}_{23}$. If $\tilde{A}_2$ has the larger entropy, then a further decomposition to $\tilde{A}_3$ is performed. For most cases, this method reduces the computation cost by avoiding the decomposition and reconstruction of sub-bands which are not necessary.

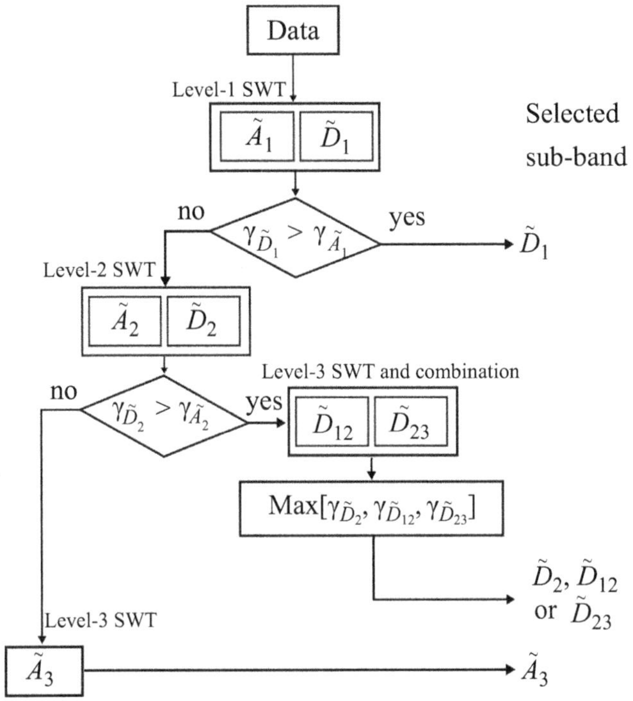

Figure 8.6 Sub-band selection scheme, where γ_X is the maximum entropy of each reconstructed sub-band used to indicate the sub-band with the most information about the target [13]. © IEEE 2017

8.1.3 Detection performance

To assess the performance of the entropy-SWT (eSWT) detector, a Monte Carlo simulation was performed using the Ingara X-band data set with a CPI of 128 pulses and 200 range bins. The SWT approach is a coherent scheme designed to improve target detection in the range/time domain. If the filtered output was transformed into the frequency domain and assuming that the correct sub-band was chosen, it would offer identical performance to a coherent detection scheme. To compare performance, the correct SWT reconstructed sub-band is shown as well as a cell-averaging CFAR (CA-CFAR) detector with 2 guard bins and 32 range bins in the clutter range profile. If the correct sub-band was selected, the performance of the eSWT scheme should match the correct reconstructed sub-bands. Targets with a Swerling 1 target fluctuation and radial velocities of 0, 1.1 and 2.6 m/s are used, corresponding to $\tilde{A}_3, \tilde{D}_{23}$ and $\tilde{D}_1$, respectively. For the following comparisons, the SIR has been compared with a probability of detection, $P_\mathrm{d} = 0.5$ and a desired probability of false alarm, $P_\mathrm{fa} = 10^{-3}$.

Figure 8.7 shows the detection performance (first column) and reconstructed sub-band selections (second column). The CA-CFAR detection result is shown in blue, the eSWT result in dark green and the correct reconstructed sub-band detection results are shown in red, magenta and black for the three target velocities. For the stationary

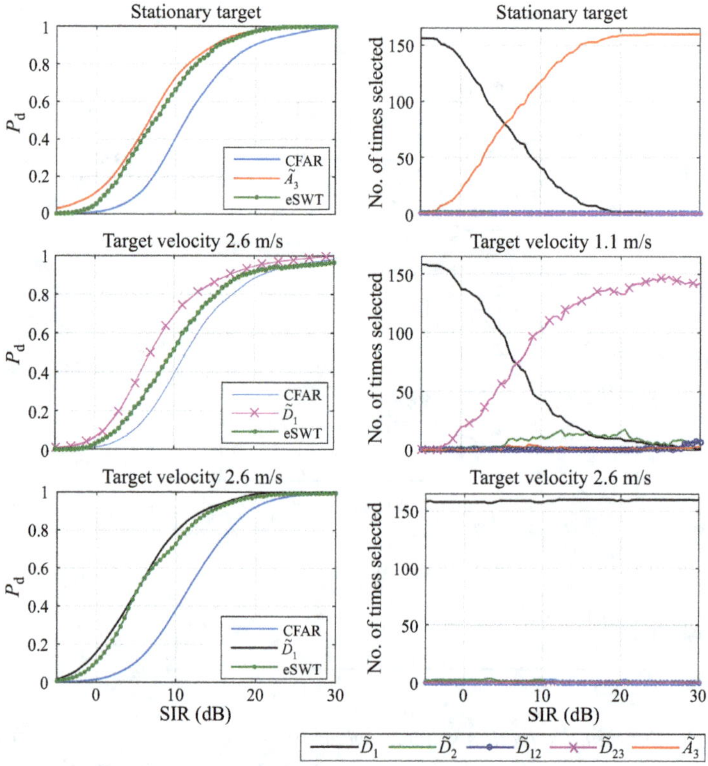

Figure 8.7 The probability of detection for a Swerling 1 target (left column) and the number of times which the reconstructed sub-bands are selected (right column) [13]. © IEEE 2017

target, the detection in the $\tilde{A}_3$ sub-band has the highest detection performance and is about 6 dB greater than the CA-CFAR result. The eSWT performance is not as good as it has no knowledge of the target velocity. While the majority of the indicator results are incorrect when the SIR goes below 5 dB, there is a 5 dB improvement over the CA-CFAR result. For the slow moving target, the $\tilde{D}_{23}$ performance is 3 dB greater than the CA-CFAR, while the eSWT has an improvement of only 2.5 dB. This is because the indicator is often confused between the $\tilde{D}_2$ and $\tilde{D}_{23}$ reconstructed sub-bands. However, since both of these contain information about the target, there is only a minor impact in the final detection result. For the faster moving target, the indicator points to $\tilde{D}_1$ at a rate of almost 100%, and the detection result is 7 dB greater than the CA-CFAR.

8.1.4 Summary and future directions

This section has reported on the application of SWTs for target detection in the maritime domain. The process of sub-band isolation and reconstruction was used to highlight different features of the sea clutter and better distinguish targets. A scheme for selecting the correct sub-band was presented based on entropy, and a computationally efficient scheme was presented based on the maximum entropy at different levels of the SWT decomposition. Using this scheme, an improvement between 3 and 7 dB in the required SIR was demonstrated over a CA-CFAR. Ideas for future work include understanding the impact of different mother wavelets and alternative indication schemes for determining the best choice of reconstructed sub-bands. Application to a real system would require further refining of the sub-band selection scheme to ensure robustness in different environments.

8.2 Sparse signal separation

Compressive sensing is a signal processing approach widely used for exploiting sparsity. For an imaging radar, it has been used for image formation [32], radio-frequency interference suppression [33] and the isolation of moving targets in radar imagery [34]. In maritime detection, sparse signal separation has been used to improve the separation of target signals from unwanted interference, thereby enhancing performance when used in conjunction with a detection algorithm [35,36]. The underlying assumption is that the transformed signals can be represented by a limited number of basis functions, or 'atoms', within a larger group, referred to as a 'dictionary'.

Consider the following formulations [35], where the input data is represented by $\mathbf{h}$ and the target component by $\mathbf{s}$. The single component case focuses on separating the target signal from the interference or residual $\mathbf{r}_s$,

$$\mathbf{h} = \mathbf{s} + \mathbf{r}_s. \tag{8.6}$$

To achieve this, a dictionary, $\mathbf{\Phi}_s$, is used with the basis pursuit de-noising (BPD) optimisation. This algorithm determines the target weight vector, $\mathbf{w}_s$, by minimising the squared error of the input signal subject to a sparsity constraint [37],

$$\underset{\mathbf{w}_s}{\arg\min}\ \ \|\mathbf{h} - \mathbf{\Phi}_s\mathbf{w}_s\|_2^2 + \lambda_s\|\mathbf{w}_s\|_1 \tag{8.7}$$

where the ℓ_2-norm, $\|.\|_2$, and ℓ_1-norm, $\|.\|_1$, are the fidelity and penalty terms, respectively, and $\lambda_s \geq 0$ is the penalty parameter. The target component is then given by $\mathbf{s} = \mathbf{\Phi}_s\mathbf{w}_s$ and the residual by, $\mathbf{r}_s = \mathbf{h} - \mathbf{s}$. While this formulation is written as a matrix multiplication, it is typically implemented using a fast Fourier transform based algorithm that greatly reduces the computational complexity. Also, for certain choices of dictionary (i.e. Fourier transforms or wavelets), there is no need to compute and store the large matrix $\mathbf{\Phi}_s$ when computing (8.7).

The dual component algorithms are commonly referred to as morphological component analysis (MCA) and are an energy preserving signal separation technique. This approach uses two dictionaries, $\mathbf{\Phi}_s$ and $\mathbf{\Phi}_q$, that represent both the target, $\mathbf{s}$, and interference signals, $\mathbf{q}_I$,

$$\mathbf{h} = \mathbf{s} + \mathbf{q}_I. \tag{8.8}$$

The target and interference weight vectors $\mathbf{w}_s$ and $\mathbf{w}_q$ are then found by solving the following minimisation:

$$\underset{\mathbf{w}_q,\mathbf{w}_s}{\arg\min}\ \ \lambda_s||\mathbf{w}_q||_1 + (1-\lambda_s)||\mathbf{w}_s||_1 \tag{8.9}$$

such that $\mathbf{h} = \mathbf{\Phi}_s\mathbf{w}_s + \mathbf{\Phi}_q\mathbf{w}_q$.

A third variation uses different dictionaries for both the target and clutter with the noise captured in the residual component.

$$\mathbf{h} = \mathbf{s} + \mathbf{c} + \mathbf{r}_s. \tag{8.10}$$

The interference weight and dictionary are then replaced by equivalent clutter signals $\mathbf{w}_c$ and $\mathbf{\Phi}_c$,

$$\underset{\mathbf{w}_c,\mathbf{w}_s}{\arg\min} \|\mathbf{h} - \mathbf{\Phi}_c\mathbf{w}_c - \mathbf{\Phi}_s\mathbf{w}_s\|_2^2 + \lambda_s\|\mathbf{w}_c\|_1 + (1-\lambda_s)\|\mathbf{w}_s\|_1 \tag{8.11}$$

where the individual components are determined by $\mathbf{s} = \mathbf{\Phi}_s\mathbf{w}_s$ and $\mathbf{c} = \mathbf{\Phi}_c\mathbf{w}_c$ with the residual, $\mathbf{r}_s = \mathbf{h} - \mathbf{c} - \mathbf{s}$.

To solve these optimisation algorithms, there are a number of potential approaches, including the iterative shrinking thresholding algorithm (ISTA), fast ISTA and the iterative split augmented Lagrangian shrinkage algorithm (SALSA) [38]. For the work in [39], SALSA has been used as a robust and quick solution method [37,38,40–42]. It is also an example of the alternating direction method of multipliers optimisation [43].

8.2.1 Tuned Q wavelet transform

Resonance based signal separation using the TQWTs was first used in [37] to perform MCA. It was then applied to radar data in [44] with dual TQWTs for classifying different signals based on their oscillatory nature. The TQWT is characterised by a triplet of parameters [45]:

- Q_0 is the centre frequency divided by the bandwidth, $Q_0 = f_{RF}/B$, and determines the 'oscillatory' nature of the wavelet. A high value of Q_0 implies a highly oscillatory signal. In the extreme limit, $Q_0 \to \infty$ results in a pure sine wave (zero bandwidth).
- r_0 controls the redundancy of the wavelet. If the received signal, $\mathbf{h}$, has N samples, then the number of coefficients is $r_0 N$. For wavelets limited in time support, r_0 must be greater than 1.
- J_0 specifies the number of successive two-channel filter banks to apply in the computation.

The dual TQWT was applied to maritime radar in [46] and showed great promise in extracting targets. To further understand the choice of the TQWT parameters for discriminating targets, Ng *et al.* [47] investigated the effect of different Q-factors on the target separation. This was implemented using a single-TQWT dictionary with the data decomposed into a target component and residue. When applied to real data, it was found that a low Q-transform led to better detection for stationary targets while a high Q-transform led to better detection of moving targets. Figure 8.8 shows an example using the NetRAD data set, where a target was extracted using $(Q_0, r_0, J_0) = (1, 2, 3)$ and $(Q_0, r_0, J_0) = (8, 8, 30)$. Due to the target motion, the higher Q_0 value is better able to extract the target.

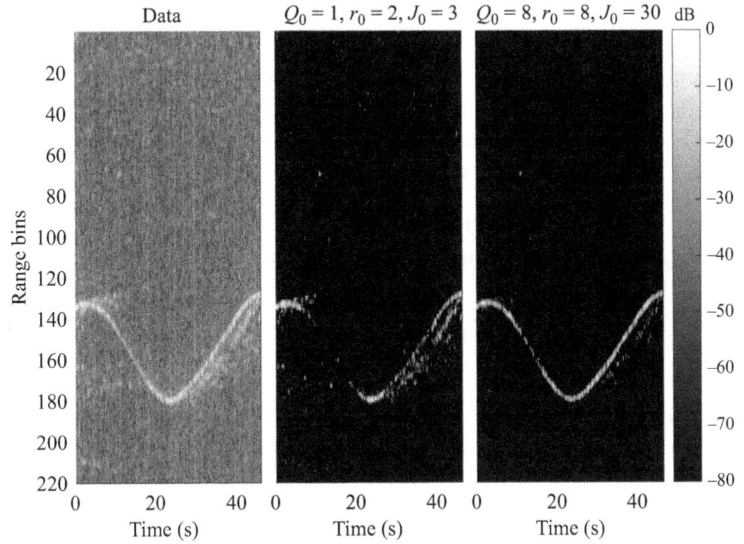

Figure 8.8 Signal separation example using the NetRAD data set with a real target present. Original data shown on left, BPD outputs shown in the middle and right for different TQWT parameter sets [47]. © *IEEE 2018*

8.2.2 Short-time Fourier transform

One of the issues with the application of TQWTs is that they are tuned to different resonant frequencies. This means that a wavelet with a low-resonance dictionary suitable for stationary or slow moving targets cannot extract targets which require a high-resonance dictionary suitable for faster moving targets. An alternative method is to use an invertible STFT as a dictionary. MCA using STFTs was first applied to image processing to decompose an image into several components that are assumed to be sparse in different transform domains [48]. It was later applied to extract targets in sea clutter by Farshchian [49], using a two-component decomposition. The advantage of this approach is that it allows signals to overlap in time and frequency, yet is able to separate the components based on their relative bandwidth. This is especially relevant for target detection, as the radar return from a small target on the sea surface has a relatively small number of long-lived scattering centres that move at nearly constant velocities.

The STFT dictionary is implemented with a half-cosine window to ensure perfect reconstruction. It decomposes the input signal into a set of Fourier coefficients by sliding an analysis window through the data. There are four parameters required to define the two dictionaries:

- N_s, N_q are the lengths of each filter. The 'target' component length, N_s, should result in a narrowband filter in the frequency domain with a relatively long-time domain analysis window, while the 'interference' component window length N_q

should result in a wideband frequency response as a result of a shorter analysis window, $N_q < N_s$.

- L_s, L_q are the overlap factors for the target and interference dictionaries, respectively. Typical values are 2 or 4 which correspond to 50% and 75% overlap of the STFT window. The overlap factor for the interference component is typically matched to the target component, $L_q = L_s$.

Note that for the dual dictionary with residual in (8.11), the interference component is replaced by the clutter component.

To determine the best choice of the STFT target filter length, consider the 3 dB width of the (magnitude squared) frequency response after Fourier transforming a half-cosine windowed rectangle function. Figure 8.9(a) shows that as the window length gets longer, the 3 dB width gets smaller [35]. Assuming that the target signal spans the entire CPI length, N (i.e. with no range walk), its frequency domain 3 dB width is approximately equal to the frequency resolution, $f_r/(N-1)$. For this reason, the best window length is always the total CPI length, $N_s = N$, and the STFT

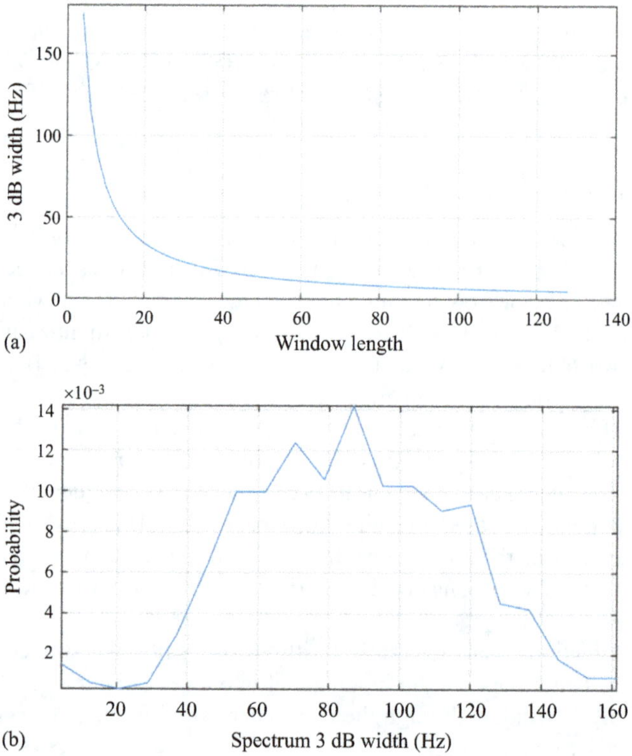

Figure 8.9 (a) Variation of the 3 dB spectral width with window length, (b) PDF of the 3 dB clutter spectrum width [35]. © Commonwealth of Australia 2020

dictionary reduces to a windowed discrete Fourier transform. For the dual component algorithms, the window length of the interference/clutter component should ideally match the 3 dB width of the clutter spectrum. Figure 8.9(b) shows the PDF of the Ingara X-band data set Doppler spectrum 3 dB width, with a mean of 93.2 Hz that matches quite closely to a window length of eight pulses in Figure 8.9(a). Regarding the overlap factor, the study in [35] found almost identical results with varying the overlap factor and hence the extra processing effort required with greater overlap (and hence redundancy) is not required to improve performance.

An example of the dual component with residual is shown in Figure 8.10 using the Ingara X-band sea clutter data set. Three synthetic targets with SIRs of 10 dB are injected at range bins 100, 200 and 300 with Doppler frequencies of −100, 0 and 100 Hz, respectively. For this example, a CPI of 64 pulses was used with $(N_s, N_q) = (64, 8)$, $L_s = L_q = 2$ and a penalty parameter of $\lambda_s = 0.18$. The targets are clearly extracted with good definition of the clutter component and the noise is captured in the residue. There are also faint target remnants that have leaked into the clutter component and faint clutter remnants in the residual component.

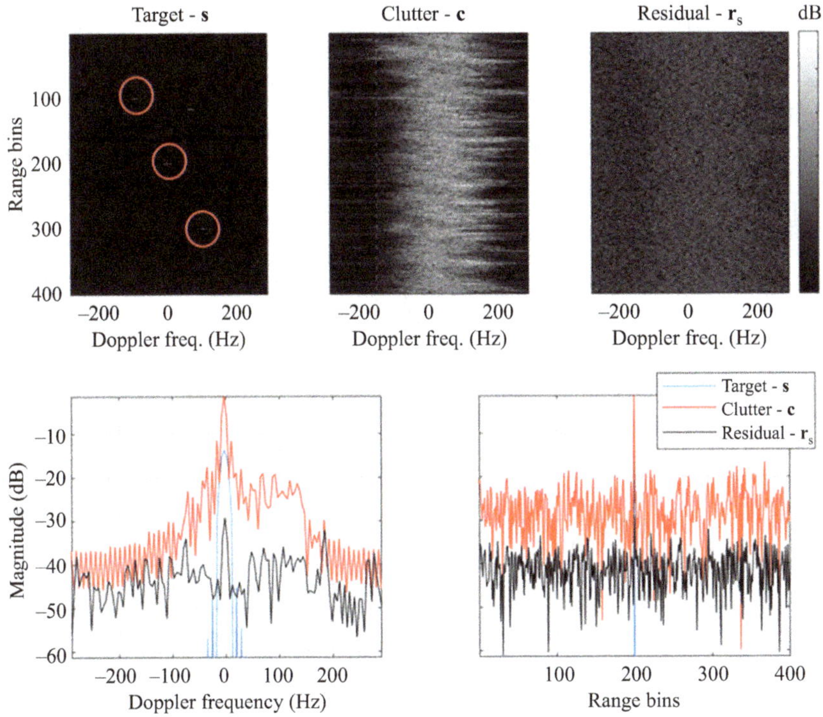

Figure 8.10 Dual component decomposition with residual. The target component is represented by **s**, *the clutter by* **c** *and the residue by* $\mathbf{r}_s$. *Target slices are shown for the stationary target location at range bin 200 [35].*

There have been a few studies applying sparse signal separation using STFTs to maritime target detection. In [39], an algorithm based on the energy ratio of the target and interference was proposed, while in [35], the performance of the three sparse signal decompositions were compared. Further details about the different detection schemes are given next.

8.2.3 Detection approaches

To utilise the sparse signal separation algorithms in a practical detection scheme, four processing stages are required as shown in Figure 8.11, where γ is the threshold chosen to give a desired probability of false alarm, P_{fa}, ζ is the test statistic and H_0 and H_1 are the null and alternative hypotheses, respectively. The first stage determines the penalty parameter using a block of target free training data. This is then used in the optimisation algorithm and solved using SALSA. Finally, a detection statistic is formed and compared with the appropriate threshold.

The example in Figure 8.10 uses a carefully selected penalty parameter to extract the targets from the background interference. Choosing the correct penalty parameter is therefore key for any detection scheme to work effectively. To investigate this issue, Anitori *et al.* [50] proposed an adaptive scheme based on the complex approximate message passing algorithm. The scheme was designed for radars that use stepped frequency waveforms and solve the BPD problem using partial Fourier matrices as the sparse dictionary. By assuming that the data is corrupted by additive Gaussian noise, the algorithm estimates the noise variance to determine the correct penalty parameter to use. Unfortunately, the scheme was found to be unsuitable for non-Gaussian sea clutter [41].

To understand the impact of the penalty parameter, the energy ratio between the target and the interference components can be used to determine the separation effectiveness,

$$\alpha_s(\lambda_s) = \frac{||\mathbf{s}||_2}{||\mathbf{h} - \mathbf{s}||_2} \tag{8.12}$$

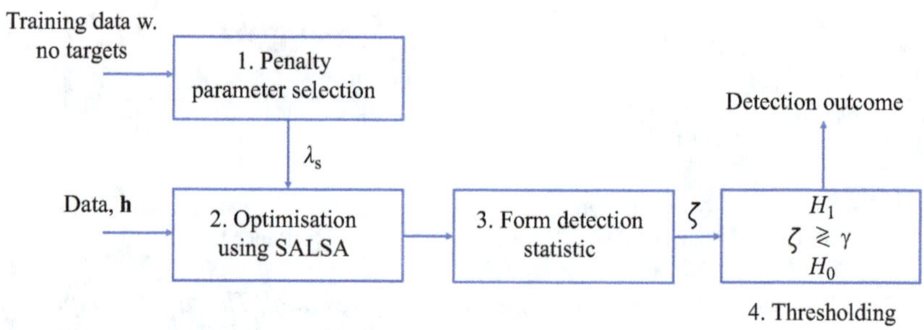

Figure 8.11 Sparse signal separation detection approach [39]. © Commonwealth of Australia 2018

where $\alpha_s(\cdot)$ is a scalar determined for each range bin. Consider the three separation cases with the STFT used as the dictionary, a CPI of 64 pulses, $(N_s, N_q) = (64, 8)$ and $L_s = L_q = 2$. Figure 8.12(a) shows the energy ratio for the Ingara X-band data set as the range and penalty parameter, λ_s, varies. The single component result has larger values with low penalty parameters as they contain more interference in the target component. For the dual component results, the opposite is observed with larger parameters indicating that more interference is present. Figure 8.12(b) shows slices along the first ten range bins for each algorithm. For the single component result, the decay in the energy ratio is now clearer. This result is expected though, as when there is only interference present, the target component should ideally be completely removed and increasing the penalty parameter simply causes more energy to be shifted to the residual component. For the dual component algorithms, there is a distinct pattern for each range bin with a significant change of slope between 0.1 and 0.2. When the penalty parameter is low, the decompositions are ill-matched to their respective intended components (target or interference), which leads to low-energy ratios. However, when the penalty parameter increases, the rebalancing of coefficients between the target and interference simultaneously improve, which leads to the sudden jump in the energy ratio. For good separation, the energy ratio should be large, but not so large that the interference leaks into the target signal.

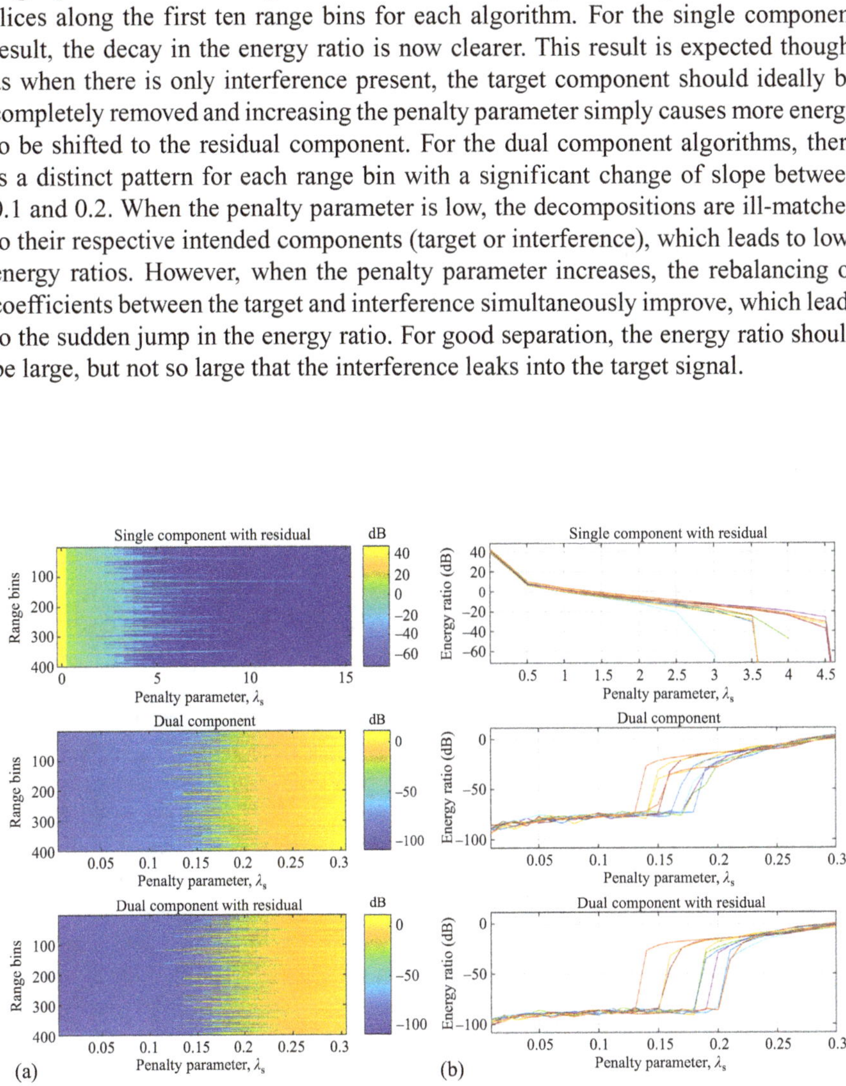

Figure 8.12 (a) Variation of the energy ratio with penalty parameter and range, (b) energy ratio as a function of penalty parameter for the first ten range bins [39]. © Commonwealth of Australia 2018

The inflection point appears to offer the greatest separation between the target and interference and can be found by finding the maximum of the first derivative

$$\lambda_s = \arg\max_{\lambda_s} \alpha_s'(\lambda_s) \tag{8.13}$$

which can be solved empirically using the first-order difference for each range bin in the target free data set. A number of different detection statistics have been developed for the maritime environment.

- The first statistic is based on the work in [36,39] and utilises the energy ratio, $\alpha_s(\lambda_s)$. This requires λ_s and γ to be chosen together such that the desired P_{fa} is achieved. One way to achieve this is to select γ and then determine the value of λ_s that achieves the desired P_{fa} [39].
- The second statistic is the magnitude squared of the range/time elements of the target component, **s**.
- The third statistic is the magnitude squared of the range/Doppler elements of the target component, **s**.

For each detection scheme, the threshold can be determined by the analysis of the target free output. In a real-world implementation, target free 'training' data is required with similar statistics to the data being analysed. To obtain sufficient samples to determine the threshold, adjacent CPIs can be utilised. For example, 15 CPI blocks of $N = 64$ pulses give 7.7×10^6 samples in the range/time domain, or in the range/frequency domain where different thresholds are used for each frequency bin, there are 12,000 samples per bin. If a low P_{fa} is desired that cannot be achieved due to the finite number of training samples, the threshold can be determined by extrapolating a model of the amplitude distribution (see Chapter 7).

To implement the energy ratio detection statistic, it is straightforward to determine a single penalty parameter that achieves the desired P_{fa}. However, for the single component case, it is possible to exploit the sea clutter structure and choose different λ_s values for each range bin. Consider the top plot of Figure 8.13 which shows the penalty parameter across range bins for a fixed energy ratio of $\alpha_s = -20$ dB. As shown in Figure 8.12(a), the Ingara X-band data set has strong waves that move through the scene. Since the optimisation is performed independently for each range bin, it is possible to choose penalty parameters which relate to the strength of the signal backscatter. For example, range bins corresponding to the wave crests can have larger penalty parameters than those in the trough. One measure that clearly shows this fluctuation is the backscatter standard deviation, defined for the kth range bin as $\sigma_s(k)$. However, the standard deviation cannot directly be used to scale the penalty parameter, as any strong targets will cause the optimisation to weight the penalty term too strongly. To avoid penalising the target, the standard deviation is instead averaged over $K_s/2$ range cells on either side of the range cell of interest. The averaged standard deviation, $\tilde{\sigma}_s(k, K_s)$, is shown in the bottom plot of Figure 8.13 where $K_s = 4$

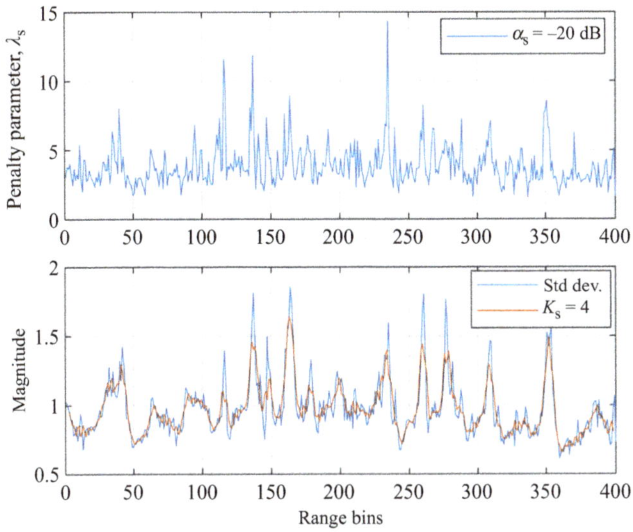

Figure 8.13 Top: penalty parameter for a fixed energy ratio (threshold) of $\alpha_s = -20$ dB. Bottom: range bin standard deviation, $\sigma_s(k)$, and the averaged standard deviation, $\tilde{\sigma}_s(k, K_s = 4)$ [39]. © Commonwealth of Australia 2018

provides a reasonable amount of smoothing, while preserving the fluctuation across range.

A method for selecting λ_s to achieve a fixed energy ratio (and the eventual P_{fa}) was presented in [36,39], where the adaptive penalty parameter for the kth range bin is given by

$$\lambda_s(k) = \lambda_0 + \eta_s \left(\tilde{\sigma}_s(k, K_s) - \text{Min}\left[\tilde{\sigma}_s(k, K_s)\right]\right), \tag{8.14}$$

where Min[·] is the minimum value, λ_0 is a constant offset for the penalty parameter, η_s is a multiplier which can be chosen to match the clutter background (i.e. amplitude fluctuations due to the swell) and the averaged standard deviation, $\tilde{\sigma}_s(k, K_s)$, is offset by its minimum value across all range bins. The effect of varying $\lambda_s(k)$ across range is to account for differences in the energy ratio. As discussed in [36,39], the choice of threshold does not greatly affect the detection result and an arbitrary threshold can be chosen before estimating the parameters λ_0 and η_s that achieve the desired P_{fa}. This is demonstrated in Figure 8.14 where the penalty parameter, $\lambda_s(k)$, for a desired threshold of $\alpha_s = -20$ dB is plotted against the scaled and averaged standard deviation of the data in that range bin, $\sigma_s(k) = \tilde{\sigma}_s(k, K_s) - \text{Min}[\tilde{\sigma}_s(k, K_s)]$. In this example, there are 12,000 range bins and to achieve the desired $P_{fa} = 10^{-3}$, a line must be fitted with

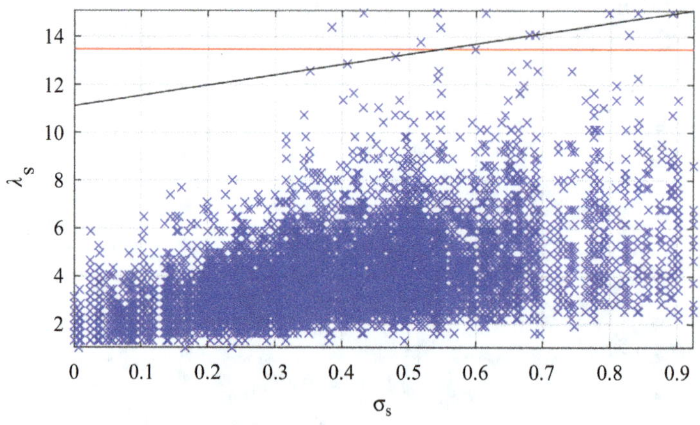

Figure 8.14 Variation of the penalty parameter for a fixed threshold of −20 dB using the single component algorithm with an STFT dictionary. Shown in red and black are the fixed and adaptive fits which achieve the desired P_{fa} *[39]. © Commonwealth of Australia 2018*

only 12 points above it. The fixed result shown in red has $\lambda_0 = 13.48$, while the adaptive fit in black has λ_0 and η_s values of 11.11 and 4.32, respectively.

For the time and frequency domain detection schemes, a suitable choice of penalty parameter can be determined by studying the probability of detection, P_d, using a Monte Carlo simulation. This can be achieved using 800 range bins from the Ingara X-band data set and synthetically injecting point targets with a Swerling 1 fluctuation. Figure 8.15 shows the detection performance of the time and frequency domain detectors as a function of the penalty parameter, λ_s. Three target frequencies are considered: 0, 115 and 230 Hz with the target SIR fixed at 5 dB for the time domain case. In the frequency domain, the detection performance is improved and the target SIR is reduced to 0 dB. For the single component time domain result, there is little difference in the performance for the three target frequencies and the detection performance increases with the penalty parameter. However, the opposite is true for the frequency domain where larger penalty parameters cause the interference data to decrease in value and spread, resulting in a longer tailed distribution. Hence to pick the appropriate P_{fa}, the threshold needs to be higher and the detection performance is reduced. Improved performance is observed for the faster target at smaller penalty parameters, while little difference is observed for the other two cases. Clearly, the choice of penalty parameter should be high for the time domain case and low for the frequency domain. With the dual component and dual component with residual (not shown), the target SIR is 5 and 0 dB for the time and frequency domain cases, respectively. Both results show a clear peak which varies with the target frequency. A good compromise for the time domain case is to use the 1% quantile of the energy ratio, 0.11, while for the frequency domain, the median of the energy ratio, 0.16, offers a good compromise for detection of both stationary and moving targets.

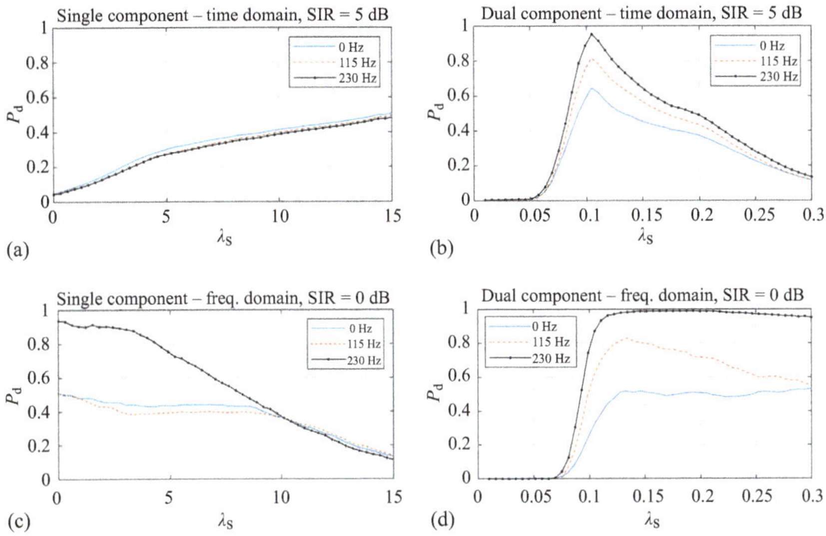

Figure 8.15 Probability of detection as a function of the penalty parameter for $P_{fa} = 10^{-3}$ and target Dopplers of the following: (—) 0 Hz, (- -) 115 Hz, (-·-) 230 Hz. (a) single component time domain, (b) dual component time domain, (c) single component frequency domain, (d) dual component frequency domain [39].

8.2.4 Detection performance

Two sets of results are now presented in Figures 8.16 and 8.17 showing the detection performance for the single and dual component algorithms with results for a stationary target shown on the top plot, while the moving target (Doppler of 230 Hz) result is shown on the bottom. For the following comparisons, the relative performance is compared by measuring the difference in the input SIR at a fixed P_d of 0.5. The baseline detector comprises a Fourier transform followed by a CA-CFAR (coherent CFAR) run along each frequency bin. For the examples here, the CFAR scheme uses 1 guard cell and 16 cells in the clutter range profile on either side of the cell under test (CUT).

For the single component results in Figure 8.16, only the frequency domain algorithm shows a slight improvement in performance of 0.5 dB greater than the coherent CFAR algorithm, while the energy ratio and time domain results show worse performance by 4 and 6 dB, respectively. For the moving target, both the frequency domain and coherent CFAR results show similar performance, with the time domain results being significantly worse by approximately 14 dB. For the dual component results in Figure 8.17, both of the frequency domain algorithms show an improvement of 1 dB over the coherent CFAR for the stationary target, while the time domain algorithms

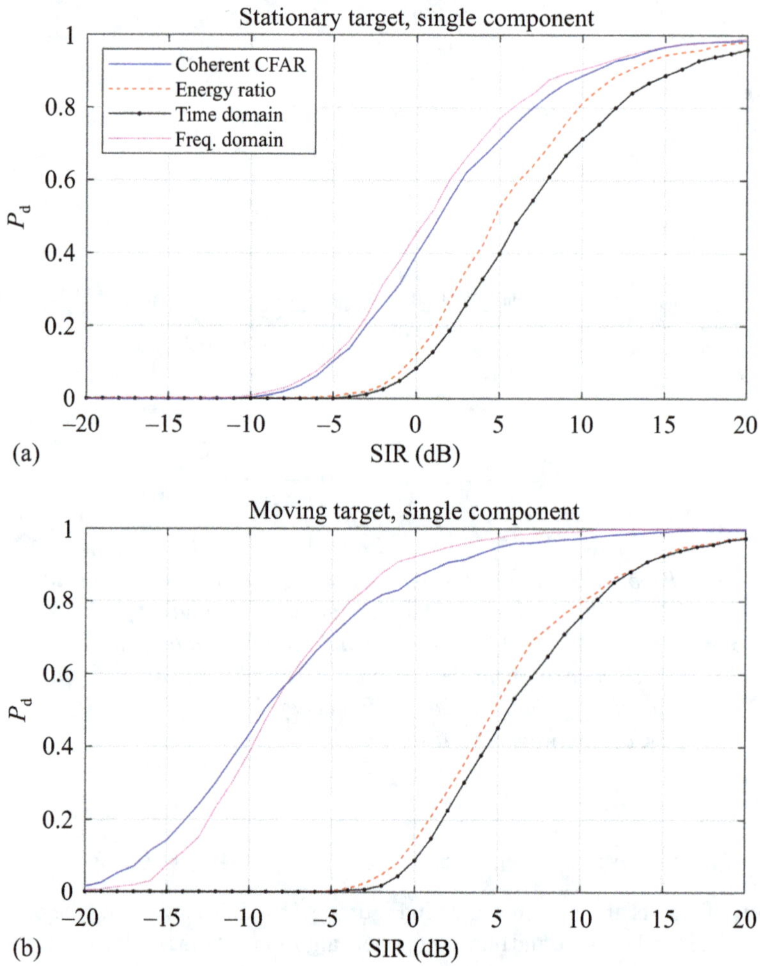

Figure 8.16 Probability of detection as a function of SIR with $P_{fa} = 10^{-3}$: (a) stationary and (b) moving target with Doppler of 230 Hz: (—) coherent CFAR, (- -) single component energy ratio, (-·-) single component time domain, (—) single component frequency domain [35]. © Commonwealth of Australia 2020

are slightly worse by 3 dB. This trend is repeated for the moving target with the dual component result showing improved performance by 8 dB over the coherent CFAR, with the two time domain algorithms showing worse performance by 1 dB. The final result in Figure 8.18 shows the detection performance of the frequency domain algorithms over all target frequencies. This clearly shows that both the dual component results offer the largest and most consistent results, while the single component and coherent CFAR show a loss of performance in the positive Doppler region. For the dual

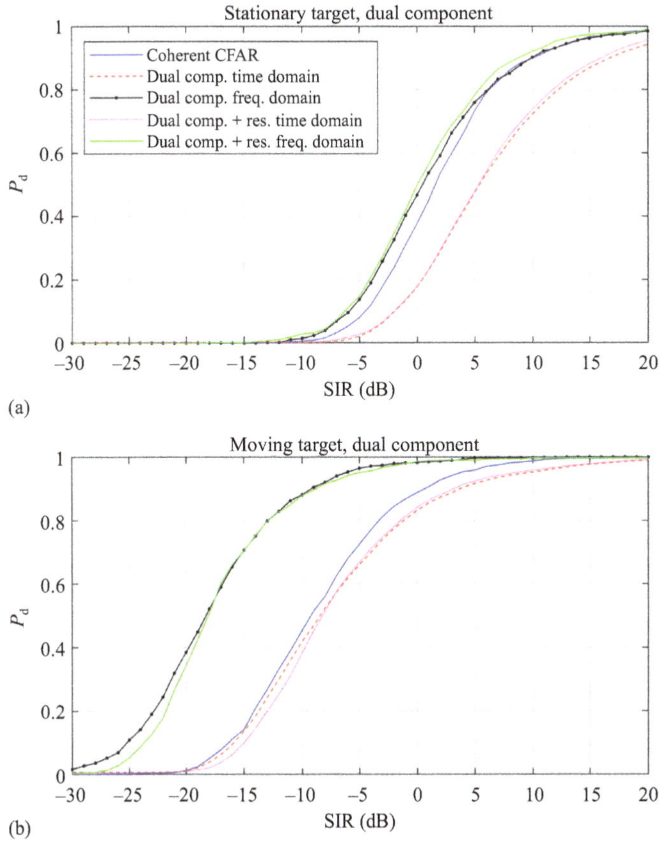

Figure 8.17 Probability of detection as a function of SIR with $P_{fa} = 10^{-3}$: (a) stationary and (b) moving target with Doppler of 230 Hz. (—) Coherent CFAR, (- -) dual component time domain, (-·-) dual component frequency domain, (—) dual component with residual time domain, (—) dual component with residual frequency domain [35].

component decompositions, there is not a lot to be gained with the dual component plus residual approach, concluding that the dual component approach offers the best performance result, particularly in the frequency domain region where the clutter is the most variable.

8.2.5 Summary and further developments

Sparse signal separation offers the potential for significant improvement over traditional coherent processing. However, the performance is very much determined by

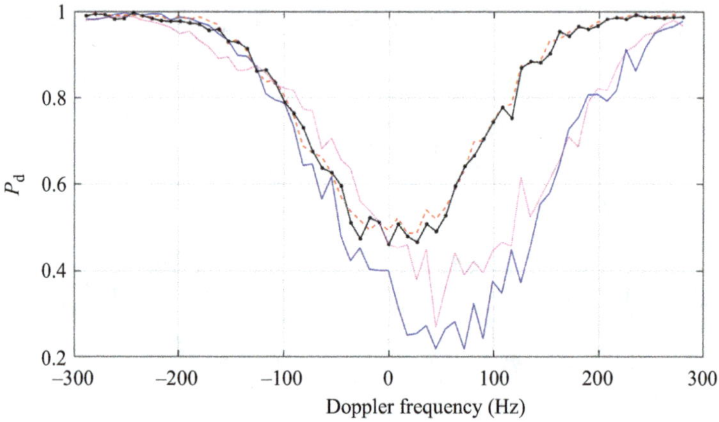

Figure 8.18 Probability of detection as a function target frequency for a fixed SIR = 0 dB and $P_{fa} = 10^{-3}$: (—) coherent CFAR, (—) single component, (- -) dual component, (-·-) dual component with residual [35]. © Commonwealth of Australia 2020

the choice of penalty parameter and the dictionary. Further work in [51] has proposed a multichannel version of the dual component algorithm utilising STFTs as the dictionaries. This work looked at a number of formulations with promising results. However, after publication it was found that the normalised adaptive matched filter (NAMF) comparison was incorrect and the multichannel MCA results actually performed worse than the NAMF. Another study [52] has looked at improved ways to determine the penalty parameter when the sea clutter is very spiky and proposed some alternative detection approaches that utilise the single data set algorithm from Chapter 7. There is also a large computational burden in solving the optimisation problem for many values of the penalty parameter. However, a graphical processing unit accelerated the implementation of the algorithms can be used, which runs quickly and could conceivably be running in the background of a radar processor.

8.3 Dictionary learning

Another application of compressive sensing is known as dictionary learning (DL). Rather than using knowledge of the target or clutter, the DL approach learns a dictionary from a wide range of training data and then uses that to estimate the desired component from the data. The advantage of this data-driven approach is that it does not rely on statistical modelling of the clutter, but rather on the quality of the training data used to form the dictionary and on the characteristics of the DL algorithm.

Sparse representations aim to represent the input target free signal, $\mathbf{c}$, as a linear combination of a few columns of a given dictionary matrix, $\boldsymbol{\Phi}_c$, with a coefficient vector, $\mathbf{w}_c$. This is achieved by the solution of a regularisation problem which enforces the sparsity of $\mathbf{c}$. The solution of such an under-determined problem has been tackled with many techniques that are often based on greedy and convex optimisation strategies. Popular greedy approaches such as orthogonal matching pursuit (OMP), block-OMP, batch-OMP, etc. [53–55] use iterative hard thresholding strategies and are computationally quick, but may not lead to a global optimum solution. Convex optimisation based strategies for sparse representations such as BPD [56], the least absolute shrinkage and selection operator (LASSO) [57] and least-angle regression (LARS) [58] use less restrictive constraints (ℓ_1- and ℓ_2-norms), while still achieving a sparse solution for $\mathbf{w}_c$ [59]. These methods are more computationally demanding, but also more robust against noise, and lead to a better reconstruction of $\mathbf{c}$.

The selection of the dictionary matrix is crucial for obtaining a sparse representation of a signal, i.e. a representation with a minimum number of non-zero coefficients. Depending on the application, a dictionary can be built with an arbitrary basis (such as Fourier, wavelets, etc.) [60] or by collecting empirical or synthetic signatures of the signal from a defined training set [61]. Classical DL algorithms such as the K-means singular value decomposition (KSVD) [62,63] operate in batches dealing with the entire training set in each iteration. Although quite successful, these methods are computationally demanding and not scalable to high-dimensional training sets. An efficient alternative is to use an online dictionary learning (ODL) algorithm [64], which converges quickly and can infer the dictionary from large or time varying training sets. A recent example is the drop off mini-batch online dictionary learning (DOMINODL) algorithm that was introduced in [65] and offers improved performance by exploiting training data that is correlated (i.e. with a similar sparse representation). In [66], these three DL approaches were used to learn the sea clutter 'signature' from a number of Ingara L-band data sets. This dictionary was then applied to estimate the clutter component in a different data set, hence revealing potential targets. This work is further summarised in this section.

8.3.1 Dictionary learning techniques

The first step in DL consists of building a training database $\mathcal{T} = \{\mathbf{c}(k)\}_{k=1}^{K_\mathcal{T}}$ where each of the $K_\mathcal{T}$ vectors contains N elements. It is assumed that every vector, $\mathbf{c}(k)$, is generated by a linear combination of K_{atom} atoms from the dictionary matrix $\boldsymbol{\Phi}_c \in \mathbb{R}^{N\times K_{\text{atom}}}$ and an associated sparse vector of coefficients, $\mathbf{w}_c(k)$. The key problem in DL is to find the dictionary, $\boldsymbol{\Phi}_c$, that gives the sparsest coefficient vector, $\mathbf{W}_c = \{\mathbf{w}_c(k)\}_{k=1}^{K_\mathcal{T}}$, for the input training database, $\mathcal{T}$. This can be found by solving the following optimisation:

$$\begin{aligned} &\underset{\boldsymbol{\Phi}_c,\{\mathbf{w}_c(k)\}_{k=1}^{K_\mathcal{T}}}{\operatorname{argmin}} \sum_{k=1}^{K_\mathcal{T}} \|\mathbf{w}_c(k)\|_0 \\ &\text{subject to } \|\mathbf{c}(k) - \boldsymbol{\Phi}_c\mathbf{w}_c(k)\|_2 \le \varepsilon_{\text{dev}} \end{aligned} \tag{8.15}$$

where ε_{dev} is the residual error or model deviation and can be selected using the entropy-thresholded strategy described in [65]. Batch DL methods such as KSVD deal with the entire training set in each iteration. They retain a guess for $\mathbf{\Phi}_c$ and $\mathbf{W}_c$ and iteratively update either $\mathbf{W}_c$ using basis/matched pursuit or $\mathbf{\Phi}_c$ using least squares or singular value decomposition [62]. ODL algorithms process one signal at a time and have a faster convergence with respect to classical DL methods. The ODL method in [64] was one of the first ODL algorithms. It achieves the sparse decomposition, $\mathbf{W}_c$, with a Cholesky-based implementation of the LARS-LASSO algorithm [67] and updates each column of $\mathbf{\Phi}_c$ sequentially using a block coordinate descent strategy.

DOMINODL was introduced in [65] and exploits the fact that some of the training data could be correlated and have a similar sparse representation. At each iteration in DOMINODL, computations are saved by considering only a small batch of previous elements that are correlated with a small batch of new ones. The two sets are defined as correlated if they have at least one common non-zero element in their sparse decomposition. At the end of each iteration, DOMINODL also drops off those previous training set elements that have not been defined as correlated after a specified number of iterations. The mini-batch drawing, combined with dropping off training elements and an entropy based criterion to control sparsity, results in a faster ODL algorithm that can be beneficial for real-time radar processing.

The second step in the DL-based approach is to estimate the clutter component from the received data and then subtract it to reveal potential targets. If the received data is $\mathbf{h} \in \mathbb{C}^{N\times K}$ where K is the number of range bins, then sparse reconstruction using the learned dictionary, $\mathbf{\Phi}_c$, can be applied to each range bin using the following optimisation:

$$\begin{aligned} \mathbf{w}_c(k) &= \underset{\mathbf{c}(k)}{\operatorname{argmin}} \ ||\mathbf{c}(k)||_1 \\ &\text{subject to } \ ||\mathbf{h}(k) - \mathbf{\Phi}_c\mathbf{w}_c(k)||_2 \le \varepsilon_{rec} \end{aligned} \tag{8.16}$$

with the output coefficients for K range bins, $\mathbf{W}_c = \{\mathbf{w}_c(k)\}_{k=1}^{K}$. The threshold, ε_{rec}, is again selected using the entropy thresholded strategy and the clutter contributions of $\mathbf{h}$ are obtained by $\mathbf{C} = \mathbf{\Phi}_c^T\mathbf{W}_c$, since $\mathbf{\Phi}_c$ has been learned from clutter only data. Finally, the clutter suppressed signal is obtained by subtracting $\mathbf{C}$ from $\mathbf{h}$.

8.3.2 Performance evaluation

To test these algorithms, a dictionary is created using target free runs of the Ingara L-band data set. For each run and spatial channel, range/Doppler intensity maps were formed by selecting 3,052 range bins and $N = 64$ pulses. From these range/Doppler maps, $K_{\mathcal{T}} = 5{,}036$, Doppler profiles were extracted across the different data sets and stacked together to form the training set $\mathcal{T} \in \mathbb{R}^{N\times K_{\mathcal{T}}}$. This number of training set examples ensures that the characteristics of the clutter data are sufficiently represented while maintaining a reasonable computational time. The training set, $\mathcal{T}$, is then randomly permuted and used to learn the clutter sparsifying dictionary by solving the optimisation in (8.15).

To demonstrate the relative performance of the three algorithms, an example from the Ingara L-band data set is shown in the top left plot of Figure 8.19. The data is horizontally polarised with a small vessel moving at 10 kt that is located at a range of 3.8 km, a Doppler of 170 Hz and is partially masked by the endo-clutter region between −200 and 200 Hz. The other plots show the processed data after the DL-based clutter suppression in (8.16) using KSVD, ODL and DOMINODL with $K_{\text{atom}} = 200$ atoms. These images all show that the proposed DL-based approaches strongly suppress the sea clutter, while maintaining the target return. It is important to note that each input Doppler profile needs to be normalised by its Euclidean norm in order for the sparse reconstruction in (8.16) to perform well. However, since the norm of each Doppler profile around the target range is considerably different with respect to the norm of the adjacent target free ranges, this would cause modulations across range and problems for any potential detection scheme. To avoid this, each Doppler profile was instead normalised by the mean norm of the neighbouring 400 range profiles, thus mitigating the aforementioned effect while still ensuring the algorithm performs properly. This type of normalisation is also employed for the range/Doppler maps that generate the training set for the DL algorithms.

To measure the relative quality, the gain in the SIR can be determined between the original data and each of the DL results. For this example, the KSVD, ODL and DOMINODL algorithms have values of 15.14, 14.65 and 10.53 dB, respectively. In terms of computational load, DOMINODL is the fastest at 2.3 times faster than ODL and more than 124 times faster than KSVD.

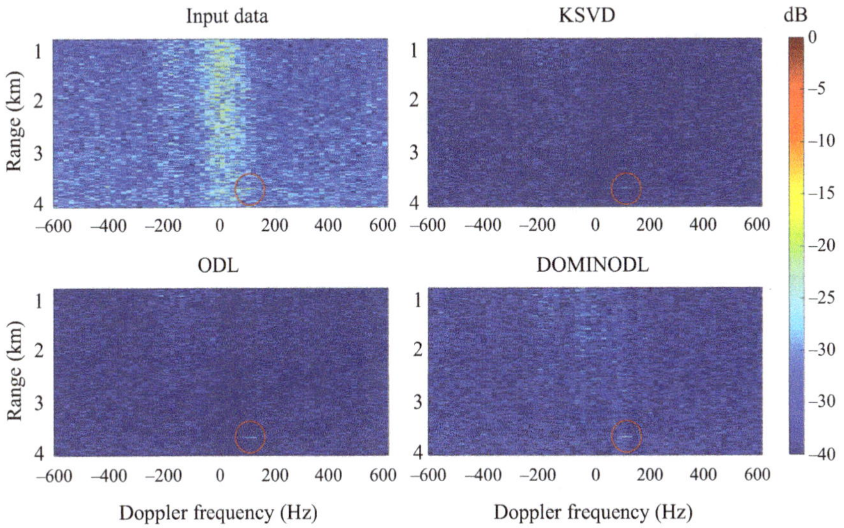

Figure 8.19 Range/Doppler maps for the test data (top left) and clutter suppressed data after DL-based clutter suppression using KSVD, ODL and DOMINODL with $K_{\text{atom}} = 200$ atoms [66]. © IEEE 2020

The learning performance of the DL algorithms are sensitive to the selection of the choice of iterations and the number of atoms in the dictionary, K_{atom} [65]. To ensure that a sufficient number of training set elements are available for both ODL and DOMINODL, while not making KSVD extremely slow, the number of iterations is set to 100. To understand the impact of the number of atoms in the learned dictionary, $\mathbf{\Phi}_c$, a 'similarity measure' can be defined that quantifies the closeness of the original training set, $\mathcal{T}$, with the reconstructed set, $\hat{\mathcal{T}}$. Consider the cross-correlation of the kth vector of the original training set, $\mathbf{c}(k)$, and its reconstruction, $\hat{\mathbf{c}}(k)$:

$$\rho_{\mathbf{c}(k),\hat{\mathbf{c}}(k)}(l) = \sum_{n=-\infty}^{\infty} \mathbf{c}(k,n)\hat{\mathbf{c}}(k,n+l). \tag{8.17}$$

The normalised cross-correlation is then defined as

$$\overline{\rho}_{\mathbf{c}(k),\hat{\mathbf{c}}(k)}(l) = \frac{\rho_{\mathbf{c}(k),\hat{\mathbf{c}}(k)}(l)}{\sqrt{\rho_{\mathbf{c}(k),\mathbf{c}(k)}(0)\rho_{\hat{\mathbf{c}}(k),\hat{\mathbf{c}}(k)}(0)}} \tag{8.18}$$

and for the vector $\mathbf{c}(k)$, a similarity measure can be defined as

$$q_{\text{sim}}(k) = \text{Max}\left[|\overline{\rho}_{\mathbf{c}(k),\hat{\mathbf{c}}(k)}(l)|\right], \tag{8.19}$$

where a value of $q_{\text{sim}}(k)$ closer to unity demonstrates greater similarity of the reconstructed data with the original training set. To assess the performance, the number of atoms, K_{atom}, is varied and $\{q_{\text{sim}}(k)\}_{k=1}^{K_{\mathcal{T}}}$ is determined for all vectors in $\mathcal{T}$.

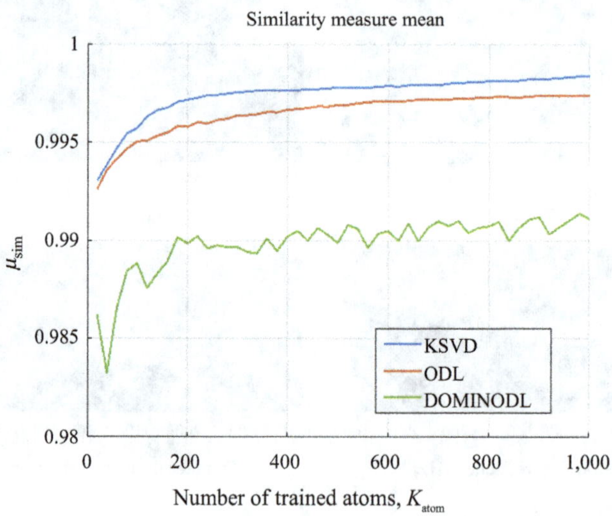

Figure 8.20 Similarity measure mean for the proposed DL algorithms [66]. © IEEE 2020

Figure 8.20 shows the mean value, μ_{sim}, of the similarity measure for each proposed DL algorithm. These are in general very close to unity, indicating that the learned dictionary, $\mathbf{\Phi}_{\mathrm{c}}$, has done well in reconstructing the training set $\mathcal{T}$ for all three approaches. KSVD is the most accurate, while DOMINODL exhibits slightly lower values with respect to KSVD and ODL. This is because KSVD uses all the training set elements, while ODL employs a convex optimisation approach (LARS) for the sparse representation step, thus giving an improved reconstruction in the presence of noise. DOMINODL uses even less training set elements than ODL (due to the mini-batch approach), and the sparse representation step carried out by the greedy batch-OMP algorithm is not as accurate as LARS. Therefore, to ensure a good reconstruction (with an over-complete dictionary, i.e. $K_{\mathcal{T}} > N$), while maintaining fast computational times, this result implies that $K_{\mathcal{T}} \approx 200$ should be sufficient to represent sea clutter in the DL schemes.

8.3.3 Detection performance

To assess the probability of detection, P_{d}, a Monte Carlo simulation is performed by injecting non-fluctuating (Swerling 0) point targets into 1,000 different range bins and then varying the SIR. For each algorithm, a CPI of $N = 64$ pulses is used with the detection threshold determined by a desired false alarm rate, $P_{\mathrm{fa}} = 10^{-3}$. Two standard detection algorithms are compared with the DL approaches. These include the coherent CFAR and the NAMF described in Chapter 7. The coherent CFAR algorithm takes a Fourier transform of the data, and then runs a CA-CFAR along each frequency bin. Different threshold multipliers are determined for each frequency bin to achieve the desired false alarm rate with the CFAR window using 32 range cells (16 on either side of the CUT) and 2 guard cells to prevent target self-nulling.

The algorithms were applied to the Ingara L-band data set with $K = 128$ range cells for the covariance estimate and 2 guard cells on either side of the CUT. When comparing performance, the SIR is measured at a P_{d} of 0.5. The first result in Figure 8.21 shows the P_{d} as a function of the input SIR with $K_{\mathrm{atom}} = 1{,}000$ atoms used to represent the dictionary. The top plot shows the results with a target in the endo-clutter region (0 Hz), while the bottom plot shows the results with a target in the exo-clutter region (445 Hz). For the former case, very similar results are observed for the three DL algorithms, with both the coherent CFAR and NAMF performing equally worse by 2 dB. In the exo-clutter region, the ODL algorithm outperforms the coherent CFAR and NAMF by 1 dB, with the KSVD and DOMINODL algorithms both performing slightly worse by less than 1 dB. Figure 8.22 then shows the P_{d} as a function of the target Doppler frequency for a fixed SIR of 0 dB. It is now clearer that the best performance over the entire endo-clutter region is achieved by the DOMINODL algorithm, with the ODL and KSVD also showing excellent performance. The NAMF and coherent CFAR both show similar results which are up to 30% worse at 0 Hz. However, there is a small region at the edge of the spectrum where the DL algorithms perform worse. In the exo-clutter region at this SIR, all the algorithms perform equally well.

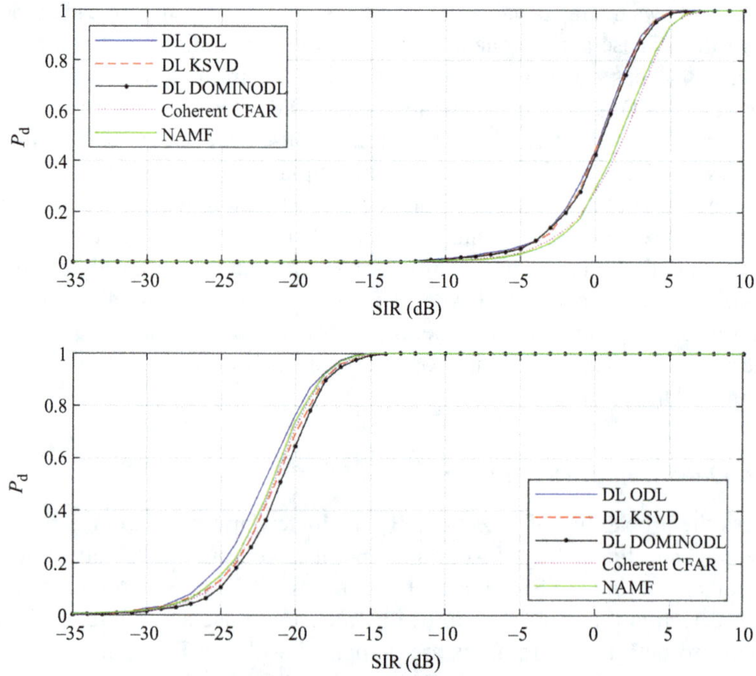

Figure 8.21 Probability of detection as the SIR varies, top plot: endo-clutter, bottom plot: exo-clutter. (—) ODL, (- -) KSVD, (-·-) DOMINODL, (· · ·) coherent CFAR, (—) NAMF [66]. © IEEE 2020

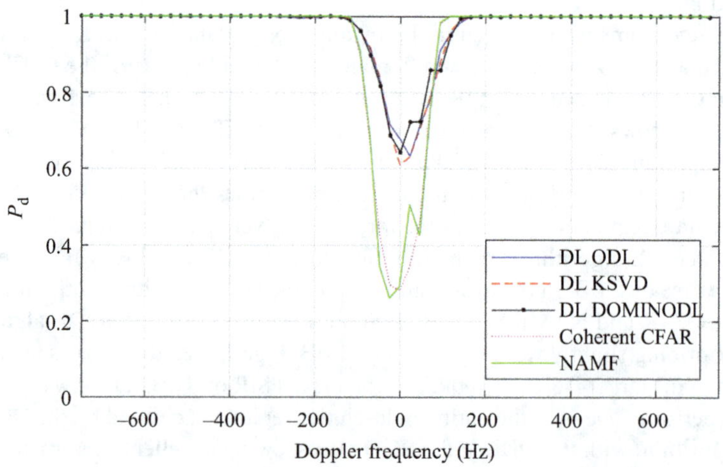

Figure 8.22 Probability of detection with variable frequency and SIR = 0 dB. (—) ODL, (- -) KSVD, (-·-) DOMINODL, (· · ·) coherent CFAR, (—) NAMF [66]. © IEEE 2020

8.3.4 Summary and further developments

In this section, a sea clutter suppression technique based on DL was presented and applied to the Ingara L-band data set. By conducting a statistical evaluation of the batch and ODL techniques, it was shown that they can efficiently learn a dictionary and sparsely decompose sea clutter signals. The learned dictionary can then be used to suppress the clutter contributions in a range/Doppler map, thereby enhancing the returns from any targets in the scene. All employed DL strategies achieved an improvement in the SIR of around 10 dB, with KSVD yielding a slightly better result and DOMINODL being the fastest. The detection analysis compared the performance of the DL algorithms with two coherent detection schemes. Results showed that the DOMINODL algorithm performed best in the endo-clutter region, while ODL performed marginally better in the exo-clutter region.

To apply this approach to a real system, approaches would need to be developed for building target free dictionaries that cover a wide range of possible environments. There are a number of avenues for future work, including a study of the algorithm robustness when different data sets are used for training the dictionary and then suppressing the clutter. It is also possible to extend this work to take advantage of multiple spatial channels by exploiting the angle/Doppler representation of the scene. Finally, DL could equally be applied to learning the signatures of targets in a scene and then applying the signal separation schemes developed in Section 8.2.

8.4 Track before detect

The traditional approach to target tracking in the maritime domain is to first detect the target using amplitude information and then estimate its positional and kinematic characteristics. Detection is typically performed using CFAR processing to deal with changes in the underlying mean of the clutter (see Chapter 7). The resulting point measurements are subsequently input to a conventional target tracking system to initialise new and maintain existing tracks. For targets with a low SIR, a common approach is to set the detection threshold very low and let the tracking algorithm deal with the high rate of false detections. However, this creates difficulties in track-to-measurement association and the use of thresholding will cause some information in the processing chain to be lost.

A potentially more powerful approach to tracking low-SIR targets is to use the raw output from the radar signal processing as a measurement. By removing the detection threshold, the TBD approach means that targets with lower SIRs can be detected and tracked [68, Ch.17]. The concept was originally proposed for a staring satellite-based imaging sensor [69] with its first application to a scanning pulse-Doppler airborne radar discussed in [70]. Both [69] and [70] use dynamic programming to solve the TBD problem. Subsequently, this was the method of choice in numerous publications [71–77]. Despite its popularity, dynamic programming is a fairly cumbersome framework for real-time implementation of TBD, mainly due to the impracticality of discretising the state space in realistic problems and its requirement for high-dimensional maximisation.

Two other TBD approaches include the Hough transform [78,79] which assumes constant velocity targets and the histogram probabilistic multi-hypothesis tracker (H-PMHT) [80]. For the latter approach, the raw output from radar signal processing is interpreted as a histogram of observations of an underlying random process. By a series of assumptions and approximations, the problem is simplified and solved in a computationally efficient manner using Kalman filters. Also, since the original H-PMHT algorithm is a track maintenance algorithm [81], it requires extra processing to initiate and terminate tracks [82].

The first recursive Bayesian estimator for a single target TBD was reported in [83] and solved using a particle filter [84, Ch.11]. Subsequently, a multiple target TBD particle filter was proposed in the context of radar processing [74] and image processing [85]. The recursive Bayesian framework for TBD soon become popular as the optimal Bayesian approach for a single target as it could be formulated analytically [86,87]. Extensions to multiple target recursive Bayesian TBD were numerous, particularly using the random finite set framework [88–91].

A comparison among H-PMHT, dynamic programming and a recursive Bayesian TBD algorithm has been reported in [92] in the context of a constant amplitude target in Rayleigh distributed clutter. The comparison favours the H-PMHT because it is fast and applicable to multiple targets. However, when using a high resolution radar in the context of airborne maritime surveillance, the assumption of a constant amplitude target in Rayleigh distributed clutter is inappropriate. In reality, the target amplitude is fluctuating and is typically modelled by one of the Swerling models [93]. Furthermore, the probability distribution of the sea clutter amplitude is characterised by long tails due to variations in the local mean intensity (the clutter texture) and the effect of sea spikes. This led to a study in [94], which considered a fluctuating target amplitude in K-distributed clutter and found that the H-PMHT-based TBD is difficult to calibrate and achieves worse performance than the Bernoulli-type recursive Bayesian TBD. The development of a the Bernoulli TBD filter for maritime radar was presented in [95] and is described in this section. It assumes a Swerling 1 fluctuating target, K-distributed clutter, thermal noise and a spatially varying texture.

8.4.1 Mathematical models

The Bernoulli TBD filter is built on a number of mathematical models. A discrete-time motion model is adopted with a constant sampling period, T_{sc}, determined by the radar scan rate. While the measurement at each scan is in the range/azimuth domain, it is convenient to consider the centre of each cell to be in the far field and represented in Cartesian coordinates. As a result, the state of the target for the mth scan is fully characterised with the state vector, $\mathbf{u}_m = [\mathrm{x}_m \; \dot{\mathrm{x}}_m \; \mathrm{y}_m \; \dot{\mathrm{y}}_m \; A_m^2]$, where $(\mathrm{x}_m, \mathrm{y}_m)$ are the Cartesian coordinates of the target, $(\dot{\mathrm{x}}_m, \dot{\mathrm{y}}_m)$ is the target velocity and A_m^2 is the energy or mean power of the target signal, which could be spread across more than one measurement cell.

The target state evolution satisfies the nearly constant velocity motion model [96] with the transition density given by

$$\vartheta_{m|m-1}(\mathbf{u}_m|\mathbf{u}_{m-1}) = \mathcal{N}(\boldsymbol{\Theta}\mathbf{u}_{m-1}, \boldsymbol{\Omega}) \tag{8.20}$$

where $\mathcal{N}(\boldsymbol{\Theta}\mathbf{u}_{m-1}, \boldsymbol{\Omega})$ denotes the Gaussian PDF with mean, $\boldsymbol{\Theta}\mathbf{u}_{m-1}$ and covariance matrix, $\boldsymbol{\Omega}$. The matrices $\boldsymbol{\Theta}$ and $\boldsymbol{\Omega}$ represent the transition matrix and the process noise covariance matrix, respectively. They are given by [96]

$$\boldsymbol{\Theta} = \begin{bmatrix} \boldsymbol{\vartheta} & 0 & 0 \\ 0 & \boldsymbol{\vartheta} & 0 \\ 0 & 0 & 1 \end{bmatrix}, \qquad \boldsymbol{\vartheta} = \begin{bmatrix} 1 & T_{\mathrm{sc}} \\ 0 & 1 \end{bmatrix}, \tag{8.21}$$

$$\boldsymbol{\Omega} = \begin{bmatrix} \boldsymbol{\omega} & 0 & 0 \\ 0 & \boldsymbol{\omega} & 0 \\ 0 & 0 & \omega_{\mathrm{i}} T_{\mathrm{sc}} \end{bmatrix}, \qquad \boldsymbol{\omega} = \omega_{\mathrm{s}} \begin{bmatrix} T_{\mathrm{sc}}^3/3 & T_{\mathrm{sc}}^2/2 \\ T_{\mathrm{sc}}^2/2 & T_{\mathrm{sc}} \end{bmatrix} \tag{8.22}$$

where ω_{s} and ω_{i} denote the process noise intensity in the spatial and amplitude domains, respectively. These standard expressions describe the discrete time dynamic model resulting from conversion of the continuous time model, where velocity and amplitude are random walks and position is the integral of velocity.

In order to indicate the presence of a target in TBD, it is helpful to introduce a binary random variable $\varepsilon_m \in \{0, 1\}$, where $\varepsilon_m = 1$ means that the target is present at scan m and $\varepsilon_m = 0$ indicates that the target is absent. The dynamics of ε_m are modelled by the first-order two-state Markov chain, whose transitional probability matrix is given by

$$\boldsymbol{\Pi} = \begin{bmatrix} 1 - P_{\mathrm{b}} & P_{\mathrm{b}} \\ 1 - P_{\mathrm{s}} & P_{\mathrm{s}} \end{bmatrix} \tag{8.23}$$

where $P_{\mathrm{b}} = P\{\varepsilon_{m+1} = 1|\varepsilon_m = 0\}$ and $P_{\mathrm{s}} = P\{\varepsilon_{m+1} = 1|\varepsilon_m = 1\}$ are referred to as the probability of target birth and target survival, respectively. If the target appears at $m - 1$, its state vector can be viewed as a random sample from the target birth PDF, $\tilde{b}_{m-1|m-1}(\mathbf{u})$.

If no target is present in a cell of the received data, the measured amplitude is due to background clutter-plus-thermal noise. This interference is formulated using the compound Gaussian model [97], which comprises a speckle component for modelling the small ripples on top of larger waves. These larger waves are then modelled by a slowly varying texture which can be assumed constant over a typical radar dwell period. The interference PDF can be represented by a K-distribution, where the shape parameter ν is replaced by an effective shape, $\tilde{\nu}$. As described in Chapter 2, by matching the moments of the PDF with and without thermal noise, the following relationship can be obtained:

$$\tilde{\nu} = \nu\,(1 + 1/\mathcal{C})^2 \tag{8.24}$$

where $\mathcal{C}$ is the clutter-to-noise ratio (CNR). For the remainder of this section, $\tilde{\nu}$ is used as the shape parameter without explicitly defining the CNR. The compound PDF model is given by

$$P_r(r) = \int_0^\infty P_{r|\tau}(r|\tau)P_\tau(\tau)d\tau \tag{8.25}$$

where the speckle amplitude, r, is Rayleigh distributed,

$$P_{r|\tau}(r|\tau) = \frac{2r}{\tau}\exp\left(-\frac{r^2}{\tau}\right) \tag{8.26}$$

with mean $\langle r\rangle = \sqrt{\pi\tau}/2$. For a K-distribution, the mean square, $\langle r^2\rangle = \tau$, is treated as a gamma distributed random variable,

$$P_\tau(\tau) = \frac{\tau^{\tilde{\nu}-1}}{b^{\tilde{\nu}}\Gamma(\tilde{\nu})}\exp\left(-\frac{\tau}{b}\right) \tag{8.27}$$

where $\Gamma(\cdot)$ denotes the gamma function, $\tilde{\nu} > 0$ is the shape parameter and $b = \langle\tau\rangle/\tilde{\nu}$ is the scale. The final expression for the K-distribution amplitude PDF is then [98]

$$P_r(r|\tilde{\nu}, b) = \frac{4r^{\tilde{\nu}}}{\sqrt{b}^{\tilde{\nu}+1}\,\Gamma(\tilde{\nu})}\mathcal{K}_{\tilde{\nu}-1}\left(\frac{2r}{\sqrt{b}}\right) \tag{8.28}$$

where $\mathcal{K}_n(\cdot)$ refers to the modified Bessel function of the second kind. Now suppose a Swerling type 1 target is present with power A^2. In many scenarios, the target's signature will spread to more than one range or azimuth cell. If there are K range and N azimuth bins, then the data for each scan can be indexed by (k, n) where $k = 1, \ldots, K$ and $n = 1, \ldots, N$. For the (k, n)th cell, the target signature can be related to the mean target power, A_m^2, via the (intensity) point spread function, $S_\mathrm{I}(\cdot)$,

$$S_\mathrm{I}^{(k,n)}(\mathbf{u}_m) = \frac{A_m^2}{2\pi w_x w_y}\exp\left(-\frac{(k-\mathrm{x}_m)^2}{2w_x^2} - \frac{(n-\mathrm{y}_m)^2}{2w_y^2}\right) \tag{8.29}$$

where w_x and w_y determine the spread in range and azimuth, respectively. The speckle amplitude in the cell of the map is Rayleigh distributed as in (8.26), but with the parameter, $\tau + S_\mathrm{I}$,

$$P_{r|\tau}\left(r|\tau + S_\mathrm{I}^{(k,n)}(\mathbf{u}_m)\right) = \frac{2r}{\tau + S_\mathrm{I}^{(k,n)}(\mathbf{u}_m)}\exp\left(-\frac{r^2}{\tau + S_\mathrm{I}^{(k,n)}(\mathbf{u}_m)}\right) \tag{8.30}$$

where τ is gamma distributed as in (8.27). In this case, the amplitude PDF does not have an analytic expression and must be computed numerically. It is given by [99]

$$P_{r|A}\left(r|S_{\mathrm{I}}^{(k,n)}(\mathbf{u}_m),\tilde{\nu},b\right)=\int_0^\infty P_{r|\tau}\left(r|\tau+S_{\mathrm{I}}^{(k,n)}(\mathbf{u}_m)\right)P_\tau\left(\tau|\tilde{\nu},b\right)d\tau$$

$$=\frac{2r}{b^{\tilde{\nu}}\Gamma(\tilde{\nu})}\int_0^\infty\frac{\tau^{\tilde{\nu}-1}}{\tau+S_{\mathrm{I}}^{(k,n)}(\mathbf{u}_m)}\exp\left(-\frac{r^2}{\tau+S_{\mathrm{I}}^{(k,n)}(\mathbf{u}_m)}-\frac{\tau}{b}\right)d\tau. \tag{8.31}$$

In summary, if no target is present in cell (k,n) of the measurement map at time index m, the amplitude PDF is given by

$$P_0^{(k,n)}(r)=P_r(r|\tilde{\nu},b) \tag{8.32}$$

while if a target in state $\mathbf{u}_m$ is contributing power $S_{\mathrm{I}}^{(k,n)}(\mathbf{u}_m)$, the amplitude PDF is given by

$$P_1^{(k,n)}\left(r|\mathbf{u}_m\right)=P_{r|A}\left(r|S_{\mathrm{I}}^{(k,n)}(\mathbf{u}_m),\tilde{\nu},b\right). \tag{8.33}$$

8.4.2 The Bernoulli track before detect filter

The Bernoulli filter is the optimal recursive Bayesian estimator for an appearing/disappearing target [86]. It propagates two quantities over time. The first is the posterior probability of target presence, $\tilde{q}_{m|m}=P(\varepsilon_m=1|\mathbf{r}_{1:m})$ where $\mathbf{r}_{1:m}\equiv\mathbf{r}_1,\ldots,\mathbf{r}_m$ is the sequence of range/azimuth amplitude maps from m previous scans. The second quantity is the posterior spatial probability distribution conditioned on target existence, $\tilde{s}_{m|m}(\mathbf{u})=P(\mathbf{u}_m|\mathbf{r}_{1:m})$. If at scan $m-1$, both $\tilde{q}_{m-1|m-1}$ and $\tilde{s}_{m-1|m-1}(\mathbf{u})$ are available, then the recursive computation of $\tilde{q}_{m|m}$ and $\tilde{s}_{m|m}(\mathbf{u})$ can be carried out in two steps, prediction and update. The prediction equations of the Bernoulli filter are as follows [86]:

$$\tilde{q}_{m|m-1}=P_{\mathrm{b}}\left(1-\tilde{q}_{m-1|m-1}\right)+P_{\mathrm{s}}\,\tilde{q}_{m-1|m-1}, \tag{8.34}$$

$$\tilde{s}_{m|m-1}(\mathbf{u})=\frac{P_{\mathrm{b}}\left(1-\tilde{q}_{m-1|m-1}\right)\int\vartheta_{m|m-1}(\mathbf{u}|\mathbf{u}')\,\tilde{b}_{m-1|m-1}(\mathbf{u}')d\mathbf{u}'}{\tilde{q}_{m|m-1}}+\frac{P_{\mathrm{s}}\,\tilde{q}_{m-1|m-1}\int\vartheta_{m|m-1}(\mathbf{u}|\mathbf{u}')\,\tilde{s}_{m-1|m-1}(\mathbf{u}')d\mathbf{u}'}{\tilde{q}_{m|m-1}} \tag{8.35}$$

where the target birth PDF models the state of the target if it had appeared at time index $m-1$. The update equations are given by

$$\tilde{q}_{m|m} = \frac{\tilde{q}_{m|m-1} \displaystyle\int \mathcal{L}(\mathbf{r}_m|\mathbf{u})\,\tilde{s}_{m|m-1}(\mathbf{u})\,d\mathbf{u}}{1 - \tilde{q}_{m|m-1} + \tilde{q}_{m|m-1} \displaystyle\int \mathcal{L}(\mathbf{r}_m|\mathbf{u})\,\tilde{s}_{m|m-1}(\mathbf{u})\,d\mathbf{u}}, \tag{8.36}$$

$$\tilde{s}_{m|m}(\mathbf{u}) = \frac{\mathcal{L}(\mathbf{r}_m|\mathbf{u})\,\tilde{s}_{m|m-1}(\mathbf{u})}{\displaystyle\int \mathcal{L}(\mathbf{r}_m|\mathbf{u})\,\tilde{s}_{m|m-1}(\mathbf{u})\,d\mathbf{u}} \tag{8.37}$$

where $\mathcal{L}(\mathbf{r}_m|\mathbf{u})$ is the likelihood ratio defined as

$$\mathcal{L}(\mathbf{r}_m|\mathbf{u}) = \frac{\prod_k \prod_n P_1^{(k,n)}(r_m|\mathbf{u})}{\prod_k \prod_n P_0^{(k,n)}(r_m)}. \tag{8.38}$$

Ideally this ratio would be determined at every value of $\mathbf{r}_m$, but this would be too computationally expensive. Instead, a 2D window is defined around the target range/Doppler location of $\mathbf{u}$ with the result being the ratio of products of $P_1^{(k,n)}(\cdot)$ and $P_0^{(k,n)}(\cdot)$, respectively. The likelihood ratio plays a crucial role in the Bernoulli filter for TBD. As discussed in [100], when the clutter is spiky and has small $\tilde{\nu}$, it is conservative in the sense that it 'reacts' less dramatically to large amplitude values. Also, the parameters $\tilde{\nu}$ and b are necessary for the computation of the likelihood ratio and must be estimated from the data. The filter reports a track only if $\tilde{q}_{m|m} > \gamma_t$, where γ_t is a parameter referred to as the reporting threshold.

Prediction and update of the probability of target presence, $\tilde{q}_{m|m}$, are straightforward using the analytic expressions in (8.34) and (8.36). To represent the posterior spatial density $\tilde{s}_{m|m}(\mathbf{u})$ and then predict and update as new data becomes available, the sequential Monte Carlo method [84] can be used with the Bernoulli TBD filter implemented as a particle filter [86, Sec.VIII.A]. The particles representing the birth PDF, $\tilde{b}_{m-1|m-1}(\mathbf{u})$, are created by letting $\tilde{b}_{m-1|m-1}(\mathbf{u})$ be approximated by $N_t \gg 1$ weighted particles $\{\tilde{w}_{m-1}^{(i)}, \mathbf{u}_{m-1}^{(i)}\}_{i=1}^{N_t}$, where $\mathbf{u}_{m-1}^{(i)}$ is the state of the ith random sample (particle) from $\tilde{b}_{m-1|m-1}(\mathbf{u})$ and the weight, $\tilde{w}_{m-1}^{(i)} > 0$.

An adaptive measurement driven birth PDF is also implemented based on [101]. Thus in the (x, y) subspace of the state space, $\tilde{b}_{m-1|m-1}(\mathbf{u})$ is designed to cover the cells of the range/Doppler map, $\mathbf{r}_{m-1}$, whose amplitude is above a certain threshold and indicates the region where targets can potentially appear. The position of particles $\mathbf{u}_{m-1}^{(i)}$, for $i = 1, \ldots, N_t$, is hence a random draw from this region with equal probability. In the $(\dot{x}, \dot{y})$ subspace, a uniform distribution is adopted from $-v_{max}$ to $+v_{max}$, where v_{max} is the maximum expected target speed. In the subspace of the mean target power,

A^2, the birth density is a gamma distribution with parameters (ν_t, b_t), chosen so that it covers the SIR values of interest. The weights of birth particles are uniform, i.e. $\tilde{w}_{m-1}^{(i)} = 1/N_t$, for $i = 1, \ldots, N_t$.

8.4.3 Simulation details

Consider a generic scanning radar with $K = 128$ range bins and $N = 64$ azimuth cells. The sea clutter is modelled with a K-distribution and due to fluctuations in the swell, the mean clutter level fluctuates both spatially and temporally over a time scale of seconds [97]. Figure 8.23 shows a simulated range/azimuth map of the sea clutter with unity mean and shape parameter, $\tilde{\nu} = 3$. Due to the short timescale, the autocorrelation function is only represented in the range direction with a negative exponential function, $\rho_{spat}(R) = \exp(-R/R_L)$ where R is the range bin and the correlation coefficient is $R_L = 4$.

The following parameters expressed in arbitrary units are used in the simulation: sampling interval, $T_{sc} = 1$, range and azimuth cell sizes are 1×1 units, process noise parameters, $\omega_s = 0.002$, $\omega_i = 0$, the correlation coefficient, $R_L = 4$ and the target range and azimuth standard deviations are $w_x = w_y = 1$. For the Bernoulli filter, $P_b = 0.01, P_s = 0.98$ and the number of particles, $N_t = 5{,}000$. The birth density parameters are chosen to cover the expected range of target amplitudes and velocities, with $\nu_t = 4.5$, $b_t = 1.7$ and $v_{max} = 0.6$. Estimation of the K-distribution parameters are based on the current and the previous scan (i.e. $\mathbf{r}_m$ and $\mathbf{r}_{m-1}$) using the method of moments estimator (see Chapter 2). This results in 2,048 uncorrelated samples for each scan which is intentionally set small to highlight any performance degradation in the TBD algorithm.

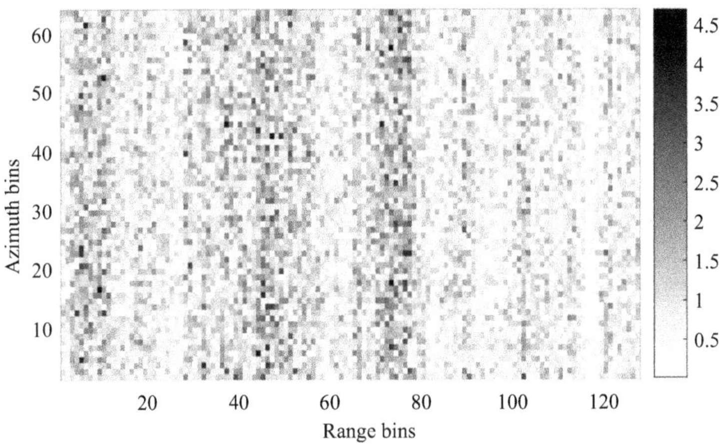

Figure 8.23 Simulated range/azimuth map, $\mathbf{r}_m$, with unity mean and $\tilde{\nu} = 3$

8.4.4 Detection and tracking performance

The first result looks at the probability of false-track declaration as a function of threshold γ_t. This is done in the absence of a target (interference-only case) and for a range of shape parameters, $\tilde{\nu}$. Figure 8.24 shows the estimated probabilities, obtained numerically by running the TBD algorithm over 3,000 scans with independent sea clutter realisations. From this result, the smaller the shape parameter $\tilde{\nu}$, the higher the probability of false-track declaration. This makes sense as smaller $\tilde{\nu}$ values imply spikier clutter (i.e. the K-distribution has a longer tail) and more frequently resembles a target return. Second, there appears to be a sudden change in behaviour between $\tilde{\nu} = 1$ and 1.5, suggesting that the TBD algorithm is not able to work efficiently in an extremely spiky sea clutter environment. Ideally, the reporting threshold should be selected so that the probability of declaring a false track is below a certain user-specified value depending on the estimated shape parameter $\tilde{\nu}$. For the remainder of the simulation results, it is set to $\gamma_t = 0.7$.

Joint detection and tracking performance are measured using the mean optimal sub-pattern assignment (OSPA) metric [102] which requires an order and cut-off to be defined. This metric penalises both the error in the cardinality estimate (number of targets) and the error in the target position, but does not consider the target velocity and mean power in the performance assessment. For these results, the order is set to 2, so the metric is an ℓ_2 norm measure, while the cut-off which expresses the penalty assigned to the cardinality error in terms of the number of cells is set to 30. The mean OSPA metric is estimated by averaging over 100 Monte Carlo runs, each containing a sequence of 40 scans.

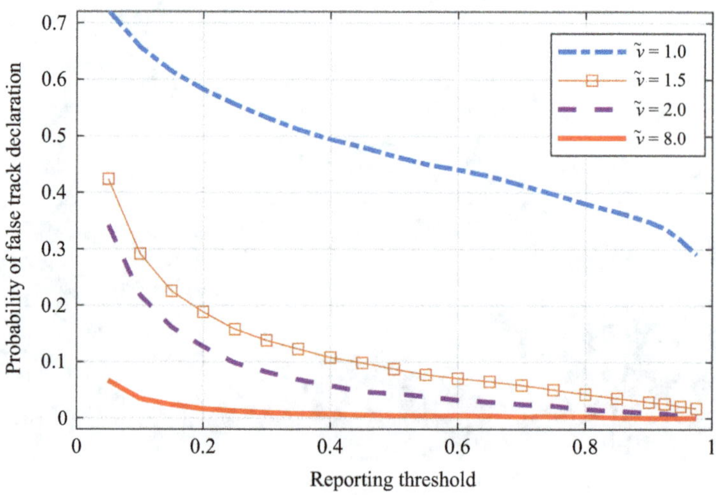

Figure 8.24 Probability of false-track declaration as a function of γ_t for $\tilde{\nu} = 1, 1.5, 2, 8$ [95]

Two versions of the Bernoulli particle filter are compared. 'Bern-PF-1' knows exactly the true values of the K-distribution parameters, while 'Bern-PF-2' estimates them from the data. These filters are compared to a conventional tracking algorithm which uses estimated point measurements (detections). For this purpose a Bernoulli filter for point measurements, 'Bern-conv', is implemented as a Gaussian sum filter [86, Sec.VIII-B] with the point measurements extracted from each scan using a CA-CFAR detection algorithm. The threshold multiplier is computed for each scan so that the probability of false alarm is 10^{-3} and the probability of detection is fixed at $P_{\mathrm{d}} = 0.9$. Finally, the P_{b} and P_{s} parameters are the same as those used in the Bernoulli TBD.

Figure 8.25 shows the rate of false-track declarations in the absence of a target for shape values, $\tilde{\nu} = 2$ and $\tilde{\nu} = 8$. The low values for each algorithm indicate small errors in the cardinality estimate and hence a low false-track declaration rate. Furthermore, in agreement with Figure 8.24, the performance is slightly better for $\tilde{\nu} = 8$ than for $\tilde{\nu} = 2$.

When considering the TBD performance with targets, the initial state vector is randomly generated, making sure the target remains in the surveillance volume with the initial velocity set to 0.45 and 0.25 for the x and y directions, respectively, and the mean power scaled to achieve SIR values of 3 and 8 dB. The resulting mean OSPA results are shown in Figures 8.26 and 8.27, respectively, for shape values of $\tilde{\nu} = 2$ and $\tilde{\nu} = 8$.

For the lower SIR results in Figure 8.26, the mean OSPA error at small m is large due to the cardinality error which reflects a delay in detecting and subsequently tracking the target. After about 20 scans, both of the Bernoulli TBD filters reach steady-state with the mean OSPA error approaching a level of 4. This high value of mean OSPA is due to the occasional track breaks, caused by oscillations of the target amplitude. Both Bernoulli TBD filters are characterised by similar performance,

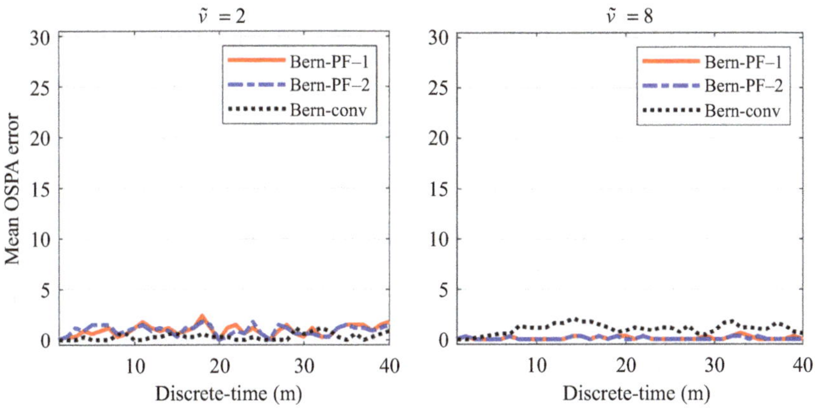

Figure 8.25 Mean OPSA in the absence of a target (interference only): left $\tilde{\nu} = 2$, right $\tilde{\nu} = 8$ [95]

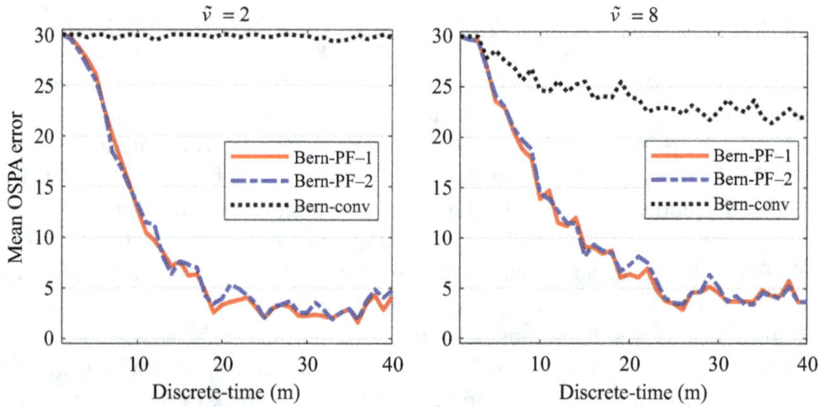

Figure 8.26 Mean OPSA for the target-plus-interference case with SIR = 3 dB: left $\tilde{\nu} = 2$, right $\tilde{\nu} = 8$ [95]

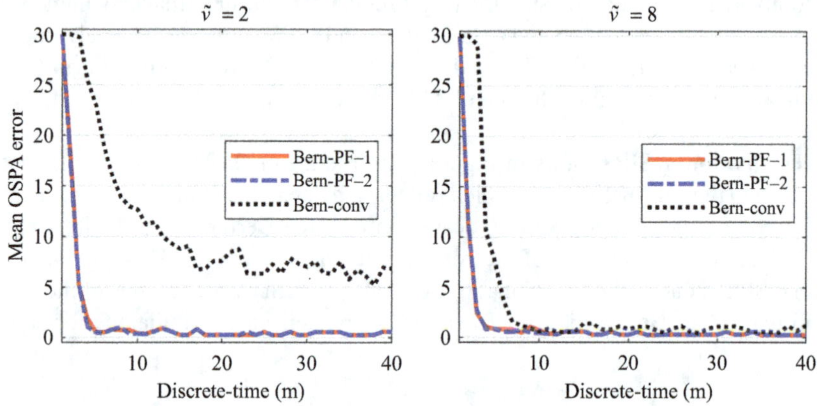

Figure 8.27 Mean OPSA for the target-plus-interference case with SIR = 8 dB: left $\tilde{\nu} = 2$, right $\tilde{\nu} = 8$ [95]

indicating an important observation that the loss of performance is negligible in the absence of knowledge of the K-distribution parameters. Finally, it is important to note that the conventional approach is unable to detect/track the target at such a low SIR, although its performance is slightly better at $\tilde{\nu} = 8$ than at $\tilde{\nu} = 2$.

In Figure 8.27, the results for a stronger target SIR demonstrate improved detection and tracking performance for all three algorithms. The two Bernoulli TBD filters need only three scans to establish and track the target with sub-cell resolution accuracy. Again, there is no significant difference between the Bernoulli TBD filters, which is a remarkable result. Finally, with an SIR of 8 dB, even the conventional Bernoulli filter can track the target, although at $\tilde{\nu} = 2$ its performance is still significantly worse than that of the TBD algorithms.

8.4.5 Summary and further developments

The work presented here has been extended in a number of further publications. The first looks at the case of multiple targets [103] with the multiple target Bernoulli TBD algorithm developed for both closely and well separated targets. In both scenarios, the extended TBD filter shows good performance, characterised by a small number of false tracks and successful detection and tracking of targets with SIRs of 5 dB or more.

The computational demand of TBD algorithms is significant. For the Bernoulli TBD filter, the main contributors are the sequential Monte Carlo estimation method and the numerical integration of (8.31). Techniques to speed up these two aspects will be key to successful real-time implementation of this algorithm. A variation of the Bernoulli TBD filter designed to reduce the computational Burdon was presented in [104]. This hybrid approach takes the advantage of the amplitude information but operates in the framework of conventional point-measurement Bayesian tracking. The resulting filter uses the Rao–Blackwell decomposition to estimate the target amplitude, while the target position and velocity are estimated using a Gaussian mixture filter. Numerical results show that the statistical and computational efficiency of the hybrid filter is reduced when compared to the Bernoulli TBD but are greater than the conventional point measurement tracker.

Another extension involves the exploitation of Doppler [105,106]. To enable the use of Doppler, the radar must operate with a slower scan rate and hence two new Bernoulli TBD filters were developed under this operational context. The first approach in [105] extracts the Doppler information as a point measurement and applies the TBD paradigm to the range/azimuth data, while the second approach [106] processes the full three-dimensional (3D) data cube (range/azimuth/Doppler). Numerical results indicate that the 3D Bernoulli TBD performs remarkably better than its counterpart developed for the fast scanning mode, when considering long exposure targets at low SIRs.

References

[1] W. L. van Rossum and M. Caro Cuenca, "Behavior subtraction applied to radar," *International Radar Conference*, 2014, pp. 1–6.

[2] M. Pan, J. Chen, S. Wang, and Z. Dong, "A novel approach for marine small target detection based on deep learning," *IEEE International Conference on Signal and Image Processing*, 2019, pp. 395–399.

[3] X. Mou, X. Chen, J. Guan, B. Chen, and Y. Dong, "Marine target detection based on improved faster R-CNN for navigation radar PPI images," *International Conference on Control, Automation and Information Sciences*, 2019, pp. 1–5.

[4] N. Ehara, I. Sasase, and S. Mori, "Weak radar signal detection based on wavelet transform," *Acoustics, Speech, and Signal Processing, 1994. ICASSP-94., 1994 IEEE International Conference on*, vol. ii, Apr. 1994, pp. II/377–II/380.

[5] N. Wang, Y. Zhang, and S. Wu, "Radar waveform design and target detection using wavelets," *CIE International Conference on Radar*, 2001, pp. 506–509.
[6] S. Zhang, J. Fan, L. Shou, and J. Dong, "A detection method of radar signal by wavelet transforms," *International Conference on Fuzzy Systems and Knowledge Discovery*, vol. 2, Aug. 2007, pp. 710–714.
[7] G. Davidson and H. D. Griffiths, "Wavelet detection of low observable targets within sea clutter," *IEEE International Radar Conference*, 2002, pp. 238–242.
[8] F. Jangal, "Wavelet contribution to remote sensing of the sea and target detection for a high-frequency surface wave radar," *IEEE Geoscience and Remote Sensing Letters*, vol. 5, no. 3, pp. 552–556, 2008.
[9] J. Torres, A. Vega-Corona, S. Torres, and D. Andina, "Chirp detection through discrete wavelet transform," *Proceedings of the World Scientific and Engineering Academy*, 2002, pp. 1–5.
[10] S. Mallat, *A wavelet tour of signal processing*, 2nd ed. Academic Press, Cambridge, MA, 1999.
[11] J. E. Ball and A. Tolley, "Low SNR radar signal detection using the continuous wavelet transform (CWT) and a Morlet wavelet," *IEEE Radar Conference*, 2008, pp. 1–6.
[12] V. Duk, B. Ng, and L. Rosenberg, "The potential of 2D wavelet transforms for target detection in sea clutter," *International Radar Conference*, 2015, pp. 0901–0906.
[13] V. Duk, L. Rosenberg, and B. Ng, "Target detection in sea clutter using stationary wavelet transforms," *IEEE Transactions on Aerospace and Electronic Systems*, vol. 53, no. 3, pp. 1136–1146, 2017.
[14] V. Duk, L. Rosenberg, B. Ng, M. Ritchie, R. Palamà, and H. Griffiths, "Target detection in bistatic radar sea clutter using stationary wavelet transforms," *International Radar Conference*, 2017, pp. 1–6.
[15] S. G. Mallat, "A theory for multiresolution signal decomposition: The wavelet representation," *Pattern Analysis and Machine Intelligence, IEEE Transactions on*, vol. 11, no. 7, pp. 674–693, 1989.
[16] I. Daubechies, *Ten lectures on wavelets*. SIAM, Philadelphia, PA, 1992.
[17] M. Vetterli and C. Herley, "Wavelets and filter banks: Theory and design," *IEEE Transactions on Signal Processing*, vol. 40, no. 9, pp. 2207–2232, 1992.
[18] J. C. Goswami and A. K. Chan, *Fundamentals of wavelets: Theory, algorithms, and applications*. Wiley, Hoboken, NJ, 2011.
[19] G. P. Nason and B. W. Silverman, "The stationary wavelet transform and some statistical applications," *Wavelets and Statistics*. Springer-Verlag, Berlin, 1995, pp. 281–300.
[20] J.-C. Pesquet, H. Krim, H. Carfantan, and J. G. Proakis, "Estimation of noisy signals using time-invariant wavelet packets," *Conference Record of The Twenty-Seventh Asilomar Conference on Signals, Systems and Computers*, vol. 1, 1993, pp. 31–34.
[21] J.-L. Starck, J. Fadili, and F. Murtagh, "The undecimated wavelet decomposition and its reconstruction," *IEEE Transactions on Image Processing*, vol. 16, no. 2, pp. 297–309, 2007.

[22] J. Pesquet, H. Krim, and H. Carfantan, "Time-invariant orthonormal wavelet representations," *IEEE Transactions on Signal Processing*, vol. 44, no. 8, pp. 1964–1970, 1996.

[23] G. Calvagno, M. Ermani, R. Rinaldo, and F. Sartoretto, "A multiresolution approach to spike detection in EEG," *IEEE International Conference on Acoustics, Speech, and Signal Processing*, vol. 6, 2000, pp. 3582–3585.

[24] N. Ahuja, S. Lertrattanapanich, and N. Bose, "Properties determining choice of mother wavelet," *IEE Proceedings on Vision, Image and Signal Processing*, vol. 152, no. 5, pp. 659–664, 2005.

[25] C. Torrence and G. P. Compo, "A practical guide to wavelet analysis," *Bulletin of the American Meteorological Society*, vol. 79, pp. 61–78, 1998.

[26] R. W. Lindsay, D. B. Percival, and D. A. Rothrock, "The discrete wavelet transform and the scale analysis of the surface properties of sea ice," *IEEE Transactions on Geoscience and Remote Sensing*, vol. 34, no. 3, pp. 771–787, 1996.

[27] C. Shannon, "A mathematical theory of communication," *Bell System Technical Journal*, vol. 27, 1948.

[28] K. F. Wallis, "A note on the calculation of entropy from histograms," *Review Literature and Arts of the Americas*, 2006.

[29] B. Harris, "Entropy," *Encyclopedia of Statistical Sciences*, 2006.

[30] T. M. Cover and J. A. Thomas, *Elements of information theory*. Wiley-Interscience, Hoboken, NJ, 2006, Wiley Series in Telecommunications and Signal Processing.

[31] M. V. Wickerhauser, *Adapted wavelet analysis from theory to software*, AK Peters/CRC Press, Natick, MA, 1994.

[32] S.-J. Wei, X.-L. Zhang, S. Jun, and G. Xiang, "Sparse reconstruction for SAR imaging based on compressed sensing," *Progress in Electromagnetics Research*, vol. 109, pp. 63–81, 2010.

[33] S. I. Kelly and M. E. Davies, "RFI suppression and sparse image formation for UWB SAR," *International Radar Symposium*, vol. 2, 2013, pp. 655–660.

[34] D. Wu, M. Yaghoobi, and M. Davies, "A new approach to moving targets and background separation in multi-channel SAR," *IEEE Radar Conference*, 2016, pp. 1–4.

[35] L. Rosenberg, V. Duk, and B. Ng, "Detection in sea clutter using sparse signal separation," *IEEE Transactions on Aerospace and Electronic Systems*, vol. 56, no. 6, pp. 4384–4394, 2020.

[36] V. Duk, B. Ng, and L. Rosenberg, "Adaptive regularisation for radar sea clutter signal separation using a sparse-based method," *International Radar Conference*, 2017, pp. 1–6.

[37] I. W. Selesnick, "Resonance-based signal decomposition: A new sparsity-enabled signal analysis method," *Signal Processing*, vol. 91, no. 12, pp. 2793–2809, 2011.

[38] I. W. Selesnick, "Sparse signal representations using the tunable Q-factor wavelet transform," *SPIE Conference on Wavelets and Sparsity XIV*, vol. 8138, 2011, p. 81381U.

[39] L. Rosenberg and B. Ng, "Sparse signal separation methods for target detection in sea clutter," *IEEE Radar Conference*, 2018, pp. 110–115.

[40] M. A. T. Figueiredo, J. M. Bioucas-Dias, and M. V. Afonso, "Fast frame based image deconvolution using variable splitting and constrained optimisation," *IEEE Workshop on Statistical Signal Processing*, 2009, pp. 109–112.

[41] B. Ng, L. Rosenberg, and P. Berry, "Comparison of sparse signal separation algorithms for maritime radar target detection," *International Radar Conference*, 2018, pp. 1–6.

[42] I. Selesnick, "TQWT toolbox guide," Polytechnic Institute of New York University, User Guide, October 2011.

[43] M. V. Afonso, J. M. Bioucas-Dias, and M. A. T. Figueiredo, "Fast image recovery using variable splitting and constrained optimization," *IEEE Transactions on Image Processing*, vol. 19, no. 9, pp. 2345–2356, 2010.

[44] M. Farshchian and I. Selesnick, "Application of a sparse time-frequency technique for target with oscillatory fluctuations," *International Waveform and Diversity Conference*, Jan. 2012, pp. 191–196.

[45] I. Selesnick, "Wavelet transform with tunable Q-factor," *IEEE Transactions on Signal Processing*, vol. 59, no. 8, pp. 3560–3575, 2011.

[46] S. T. N. Nguyen and W. A. Al-Ashwal, "Sea clutter mitigation using resonance-based signal decomposition," *IEEE Geoscience and Remote Sensing Letters*, vol. 12, no. 11, pp. 2257–2261, 2015.

[47] B. Ng, L. Rosenberg, and S. T. N. Nguyen, "Target detection in sea clutter using resonance based signal decomposition," *IEEE Radar Conference*, 2016, pp. 1–6.

[48] J. Starck, F. Murtagh, and J. M. Fadili, *Sparse image and signal processing: Wavelets, curvelets, morphological diversity*. Cambridge University Press, Cambridge, 2010.

[49] M. Farshchian, "Target extraction and imaging of maritime targets in the sea clutter spectrum using sparse separation," *IEEE Transactions on Geoscience and Remote Sensing*, vol. 14, no. 2, pp. 232–236, 2017.

[50] L. Anitori, M. Maleki, R. G. Baraniuk, and P. Hoogeboom, "Design and analysis of compressed sensing radar detectors," *IEEE Transactions on Signal Processing*, vol. 61, no. 4, pp. 813–827, 2013.

[51] H.-K. Wong, E. Aboutanios, and L. Rosenberg, "Multi-channel maritime radar target detection using morphological component analysis," *International Radar Conference*, 2020, pp. 1–6.

[52] H.-K. Wong, E. Aboutanios, and L. Rosenberg, "Improved target detection in spiky sea clutter using sparse signal separation," *IEEE Radar Conference*, 2021, pp. 1–6.

[53] J. A. Tropp and A. C. Gilbert, "Signal recovery from random measurements via orthogonal matching pursuit," *IEEE Transactions on Information Theory*, vol. 53, no. 12, pp. 4655–4666, 2007.

[54] Y. C. Eldar, P. Kuppinger, and H. Bolcskei, "Block-sparse signals: Uncertainty relations and efficient recovery," *IEEE Transactions on Signal Processing*, vol. 58, no. 6, pp. 3042–3054, 2010.

[55] R. Rubinstein, M. Zibulevsky, and M. Elad, "Efficient implementation of the K-SVD algorithm using batch orthogonal matching pursuit," *CS Technion*, vol. 40, no. 8, pp. 1–15, 2008.

[56] E. Van Den Berg and M. P. Friedlander, "Probing the Pareto frontier for basis pursuit solutions," *SIAM Journal on Scientific Computing*, vol. 31, no. 2, pp. 890–912, 2008.

[57] R. Tibshirani, "Regression shrinkage and selection via the LASSO," *Journal of the Royal Statistical Society. Series B (Methodological)*, pp. 267–288, 1996.

[58] B. Efron, T. Hastie, I. Johnstone, *et al.*, "Least angle regression," *The Annals of statistics*, vol. 32, no. 2, pp. 407–499, 2004.

[59] E. J. Candes, J. K. Romberg, and T. Tao, "Stable signal recovery from incomplete and inaccurate measurements," *Communications on Pure and Applied Mathematics*, vol. 59, no. 8, pp. 1207–1223, 2006.

[60] R. Rubinstein, A. M. Bruckstein, and M. Elad, "Dictionaries for sparse representation modeling," *Proceedings of the IEEE*, vol. 98, no. 6, pp. 1045–1057, 2010.

[61] F. Giovanneschi and M. A. González-Huici, "A preliminary analysis of a sparse reconstruction based classification method applied to GPR data," *International Workshop Advanced Ground Penetrating Radar*, 2015, pp. 1–4.

[62] M. Aharon, M. Elad, and A. Bruckstein, "K-SVD: An algorithm for designing overcomplete dictionaries for sparse representation," *IEEE Transactions on Signal Processing*, vol. 54, no. 11, pp. 4311–22, 2006.

[63] M. Elad, B. Matalon, and M. Zibulevsky, "Image denoising with shrinkage and redundant representations," *Computer Vision and Pattern Recognition, 2006 IEEE Computer Society Conference on*, vol. 2. IEEE, 2006, pp. 1924–1931.

[64] J. Mairal, F. Bach, J. Ponce, and G. Sapiro, "Online dictionary learning for sparse coding," *International Conference on Machine Learning*, 2009, pp. 689–696.

[65] F. Giovanneschi, K. V. Mishra, M. A. Gonzalez-Huici, Y. C. Eldar, and J. H. Ender, "Dictionary learning for adaptive GPR landmine classification," *IEEE Transactions on Geoscience and Remote Sensing*, vol. 57, no. 12, pp. 10 036–10 055, 2019.

[66] F. Giovanneschi, L. Rosenberg, and D. Cristallini, "Online dictionary learning techniques for sea clutter suppression," *IEEE Radar Conference*, 2020, pp. 1–6.

[67] M. R. Osborne, B. Presnell, and B. A. Turlach, "A new approach to variable selection in least squares problems," *IMA Journal of Numerical Analysis*, vol. 20, no. 3, pp. 389–403, 2000.

[68] S. Blackman and R. Popoli, *Design and analysis of modern tracking systems*. Artech House, Norwood, MA, 1999.

[69] Y. Barniv, "Dynamic programming solution for detecting dim moving targets," *IEEE Transactions on Aerospace and Electronic Systems*, vol. 21, no. 1, pp. 144–156, 1985.

[70] J. Kramer and W. Reid, "Track-before-detect processing for an airborne type radar," *Radar Conference, 1990., Record of the IEEE 1990 International*, 1990, pp. 422–427.

[71] J. Arnold, S. Shaw, and H. Pasternack, "Efficient target tracking using dynamic programming," *IEEE Transactions on Aerospace and Electronic Systems*, vol. 29, no. 1, pp. 44–56, 1993.

[72] S. M. Tonissen and R. J. Evans, "Performance of dynamic programming techniques for track-before-detect," *IEEE Transactions on Aerospace and Electronic Systems*, vol. 32, no. 4, pp. 1440–1451, 1996.

[73] L. A. Johnston and V. Krishnamurthy, "Performance analysis of a dynamic programming track before detect algorithm," *IEEE Transactions on Aerospace and Electronic Systems*, vol. 38, no. 1, pp. 228–242, 2002.

[74] S. Buzzi, M. Lops, and L. Venturino, "Track-before-detect procedures for early detection of moving target from airborne radars," *IEEE Transactions on Aerospace and Electronic Systems*, vol. 41, no. 3, pp. 937–954, 2005.

[75] D. Orlando, L. Venturino, M. Lops, and G. Ricci, "Track-before-detect strategies for STAP radars," *IEEE Transactions on Signal Processing*, vol. 58, no. 2, pp. 933–938, 2010.

[76] W. Yi, H. Jiang, T. Kirubarajan, L. Kong, and X. Yang, "Track-before-detect strategies for radar detection in G0-distributed clutter," *IEEE Trans on Aerospace and Electronic Systems*, vol. 53, no. 5, pp. 2516–2533, 2017.

[77] J. Gao, J. Du, and W. Wang, "Radar detection of fluctuating targets under heavy-tailed clutter using track-before-detect," *Sensors*, vol. 18, no. 7, p. 2241, 2018.

[78] D. Orlando, G. Ricci, and Y. Bar-Sharlom, "Track-before-detect algorithms for targets with kinematic constraints," *IEEE Transactions on Aerospace and Electronic Systems*, vol. 47, no. 3, pp. 1837–1849, 2011.

[79] F. Ehlers, D. Orlando, and G. Ricci, "A batch tracking algorithm for multistatic sonars," *IET Radar, Sonar, Navigation*, vol. 6, no. 8, pp. 746–752, 2012.

[80] S. J. Davey and H. X. Gaetjens, *Track-before-detect using expectation maximisation: The histogram probabilistic multi-hypothesis tracker: Theory and applications*. Springer, New York, 2018.

[81] R. L. Streit, "Tracking on intensity-modulated data streams," Naval Undersea Warfare Center Division, Rhode Island, Technical Report NUWC-NPT 11, 221, May 2000.

[82] R. L. Streit, M. L. Graham, and M. J. Walsh, "Multitarget tracking of distributed targets using histogram-PMHT," *Digital Signal Processing*, vol. 12, no. 2, pp. 394 – 404, 2002.

[83] D. J. Salmond and H. Birch, "A particle filter for track-before-detect," *American Control Conference*, vol. 5, 2001, pp. 3755–3760.

[84] B. Ristic, S. Arulampalam, and N. Gordon, *Beyond the Kalman filter: Particle filters for tracking applications*. Artech House, Norwood, MA, 2004.

[85] M. G. S. Bruno, "Bayesian methods for multi-aspect target tracking in image sequences," *IEEE Transactions on Signal Processing*, vol. 52, no. 7, pp. 1848–1861, 2004.

[86] B. Ristic, B.-T. Vo, B.-N. Vo, and A. Farina, "A tutorial on Bernoulli filters: Theory, implementation and applications," *IEEE Transactions on Signal Processing*, vol. 61, no. 13, pp. 3406–3430, 2013.

[87] F. Papi, V. Kyovtorov, R. Giuliani, F. Oliveri, and D. Tarchi, "Bernoulli filter for track-before-detect using MIMO radar," *IEEE Signal Processing Letters*, vol. 21, no. 9, pp. 1145–1149, 2014.

[88] B.-N. Vo, B.-T. Vo, N.-T. Pham, and D. Suter, "Joint detection and estimation of multiple objects from image observations," *IEEE Transactions on Signal Processing*, vol. 58, no. 10, pp. 5129–5141, 2010.

[89] F. Papi and D. Y. Kim, "A particle multi-target tracker for superpositional measurements using labeled random finite sets," *IEEE Transactions on Signal Processing*, vol. 63, no. 16, pp. 4348–4358, 2015.

[90] D. Y. Kim, "Multi-target track before detect with labeled random finite set and adaptive correlation filtering," *IEEE International Conference on Control, Automation and Information Sciences*, 2017, pp. 44–49.

[91] A. F. Garcia-Fernandez, "Track-before-detect labeled multi-Bernoulli particle filter with label switching," *IEEE Transactions on Aerospace and Electronic Systems*, vol. 52, no. 5, pp. 2123–2138, 2016.

[92] S. J. Davey, M. G. Rutten, and B. Cheung, "A comparison of detection performance for several track-before-detect algorithms," *EURASIP Journal on Advances in Signal Processing*, vol. 2008, p. 41, 2008.

[93] M. I. Skolnik, *Radar handbook*, 3rd ed. McGraw-Hill, New York, 2008.

[94] D. Y. Kim, B. Ristic, X. Wang, L. Rosenberg, J. Williams, and S. Davey, "A comparative study of track-before-detect algorithms in radar sea clutter," *International Radar Conference*, Toulon, France, 2019, pp. 1–6.

[95] B. Ristic, L. Rosenberg, D. Y. Kim, X. Wang, and J. Williams, "Bernoulli track-before-detect filter for maritime radar," *IET Radar Sonar and Navigation*, vol. 14, pp. 356–363, 2020.

[96] Y. Bar-Shalom, X. R. Li, and T. Kirubarajan, *Estimation with applications to tracking and navigation*. John Wiley & Sons, Hoboken, NJ, 2001.

[97] K. D. Ward, R. J. A. Tough, and S. Watts, *Sea clutter: Scattering, the K-distribution and radar performance*, 2nd ed. The Institute of Engineering Technology, London, 2013.

[98] P. Z. Peebles, *Radar principles*. Wiley, Hoboken, NJ, 1998.

[99] E. Brekke, O. Hallingstad, and J. Glattetre, "Tracking small targets in heavy-tailed clutter using amplitude information," *IEEE Journal of Oceanic Engineering*, vol. 35, no. 2, pp. 314–329, 2010.

[100] B. Ristic, L. Rosenberg, D. Y. Kim, X. Wang, and J. Williams, "Bernoulli filter for track-before-detect: Swerling-1 target in K-distributed clutter," *International Radar Conference*, Toulon, France, 2019, pp. 1–5.

[101] B. Ristic, D. Clark, B.-N. Vo, and B.-T. Vo, "Adaptive target birth intensity for PHD and CPHD filters," *IEEE Transactions on Aerospace and Electronic Systems*, vol. 48, no. 2, pp. 1656–1668, 2012.

[102] D. Schuhmacher, B.-T. Vo, and B.-N. Vo, "A consistent metric for performance evaluation of multi-object filters," *IEEE Transactions on Signal Processing*, vol. 56, no. 8, pp. 3447–3457, 2008.

[103] D. Y. Kim, B. Ristic, R. Guan, and L. Rosenberg, "A Bernoulli track-before-detect filter for interacting targets in maritime radar environment," *IEEE Transactions on Aerospace and Electronic Systems*, vol. 57, pp. 1981–1991, 2021.

[104] B. Ristic, L. Rosenberg, D. Y. Kim, and R. Guan, "Bernoulli filter for tracking maritime targets using point measurements with amplitude," *Signal Processing*, vol. 181, no. 107919, 2021.

[105] D. Y. Kim, B. Ristic, L. Rosenberg, R. Guan, and R. Evans, "Exploiting Doppler in Bernoulli track-before-detect," *IEEE Radar Conference*, 2021, pp. 1–6.

[106] B. Ristic, D. Y. Kim, L. Rosenberg, and R. Guan, "Exploiting Doppler in Bernoulli track-before-detect for a scanning maritime radar," *IEEE Transactions on Aerospace and Electronic Systems*, 2021.

Appendix A
Goodness of fit metrics

Assessing the accuracy of a probability density function (PDF) model is important to determine their suitability for representing fluctuations of the sea clutter amplitude. In the statistical literature, there are many tests and measurement techniques for assessing models. Three common ones are summarised here and include the chi-square (CS) and Kolmogorov–Smirnov (KS) tests and the Bhattacharyya distance (BD) [1]. These are all based on measuring differences between histograms of the model and the data, by considering the entire distribution fit. However, for many applications the goal is an accurate fit to the distribution tail, and these tests require some modifications. The final goodness of fit metric is the threshold error (TE) which specifically focusses on quantifying the accuracy of the model fit in the distribution tail.

A.1 Chi-square test

Pearson's chi-square (CS) goodness of fit test [2] is a measure of how well a proposed distribution model matches the data. It is implemented by dividing the range of intensity values, z into K_{int} equal intervals with the modelled PDF, $P_{\text{mod}}(\cdot)$ having N_1 samples and the empirical PDF, $P_{\text{dat}}(\cdot)$ having N_2 samples,

$$D_{\text{CS}} = \sum_{k=1}^{K_{\text{int}}} \frac{(P_{\text{dat}}(z_k)N_2\Delta_z - P_{\text{mod}}(z_k)N_1\Delta_z)^2}{P_{\text{mod}}(z_k)N_1\Delta_z} \tag{A.1}$$

where z_k is the kth sample and Δ_z is the intensity spacing between samples. The lower value of D_{CS} implies a better match between the data and the model being considered. A probability value (p-value) is typically used to determine if the null hypothesis should be accepted (i.e. the model matches the data if the p-value is above 0.05).

To better focus on the distribution tail, a modified CS test can be introduced [3] so that only data samples with complementary cumulative distribution function (CCDF) less than $\bar{F}_{\zeta 0}$ are considered. This is equivalent to restricting the intensity values to be greater than ζ_0 and hence the summation range becomes $k = k_{\zeta 0}, \ldots, K_{\text{int}}$ where $k_{\zeta 0}$ is the first bin for where $z_k \geq \zeta_0$.

A.2 Kolmogorov–Smirnov test

The KS test [4] measures the maximum absolute error between the CDFs of the model, $F_{\text{mod}}(\cdot)$ and data $F_{\text{dat}}(\cdot)$. It is given by

$$D_{\text{KS}} = \max_{z} \left(|F_{\text{mod}}(z_k) - F_{\text{dat}}(z_k)| \right), \quad k = 1, \ldots, K_{\text{int}}. \tag{A.2}$$

Similar to the CS test, a lower value of D_{KS} implies a better match between the data and the model being considered. Similar to the CS test, a p-value can be used to determine if the null hypothesis should be accepted. To better focus on the tail, the intensity range can be reduced to $k = k_{\zeta 0}, \ldots, K_{\text{int}}$ where $z_k \geq \zeta_0$ [5].

A.3 Bhattacharyya distance

The BD measures the similarity between two distributions [6]. It is given by

$$D_{\text{BD}} = -\ln \left(\sum_{k=1}^{K_{\text{int}}} \sqrt{P_{\text{mod}}(z_k) P_{\text{dat}}(z_k)} \right) \tag{A.3}$$

and varies between 0 and ∞, where equal distributions have a distance measure of 0. The values of D_{BD} are often very small and can be represented in dBs. A good fit can be declared if the BD is below a specified value, such as −20 dB [7].

To better focus on the tail of the distribution, the intensity range can be reduced. In this case, there needs to be appropriate normalising terms to ensure both PDFs maintain a unity sum. The modified Bhattacharyya distance (MBD) is then given by

$$D_{\text{MBD}} = -\ln \left(\sum_{k=k_{\zeta 0}}^{K_{\text{int}}} \sqrt{\frac{P_{\text{mod}}(z_k)}{\sum_{k_1=k_{\zeta 0}}^{K_{\text{int}}} P_{\text{mod}}(z_{k_1})} \frac{P_{\text{dat}}(z_k)}{\sum_{k_2=k_{\zeta 0}}^{K_{\text{int}}} P_{\text{dat}}(z_{k_2})}} \right). \tag{A.4}$$

A.4 Threshold error

The TE focusses on the tail of the distribution by considering the threshold difference between two CCDFs. The CCDF is also known as the probability of false alarm in a detection scheme and is typically measured in the log domain. The TE is determined by measuring the difference between the CCDFs of the model and data at a fixed CCDF value. If the threshold values at the desired CCDF levels are z_1 and z_2, then

$$D_{\text{TE}} = |z_1 - z_2|. \tag{A.5}$$

A good fit can be declared if the TE is below a specified value, such as 1 dB [7].

References

[1] D. H. Kil and F. B. Shin, *Pattern recognition and prediction with applications to signal processing*. Springer-Verlag New York, Inc., New York, 1998.

[2] K. Pearson, "On the criterion that a given system of deviations from the probable in the case of a correlated system of variables is such that it can be reasonably supposed to have arisen from random sampling," *Philosophical Magazine*, vol. 50, no. 302, pp. 157–175, 1900.

[3] H. C. Chan, "Radar sea-clutter at low grazing angles," *IEE Proceedings*, vol. 137, no. 2, pp. 102–112, 1990.

[4] Encyclopedia of mathematics, "Kolmogorov-Smirnov test." [Online]. Available: http://www.encyclopediaofmath.org/index.php?title=Kolmogorov-Smirnov_test&oldid=22659

[5] A. Farina, F. Gini, M. V. Greco, and L. Verrazzani, "High resolution sea clutter data: Statistical analysis of recorded live data," *IEEE Proceedings on Radar, Sonar and Navigation*, vol. 144, no. 3, pp. 121–130, 1997.

[6] A. Bhattacharyya, "On a measure of divergence between two statistical populations defined by their probability distributions," *Bulletin of the Calcutta Mathematical Society*, vol. 35, pp. 99–109, 1943.

[7] L. Rosenberg and V. Duk, "Land clutter statistics from an airborne passive bistatic radar," *IEEE Transactions on Geoscience and Remote Sensing*, 2021.

Appendix B
Bimodal parameters for the evolving Doppler spectrum

In Chapter 2, the evolving Doppler spectrum model was proposed as a mixture of two Gaussian-shaped spectra designed to model the Bragg and fast components,

$$G_{\mathrm{bi}}(f) = \frac{\alpha\tau}{\sqrt{2\pi}w_1}\exp\left(-\frac{(f-m_{\mathrm{f}_1}(\tau))^2}{2w_1^2}\right) + \frac{(1-\alpha)\tau}{\sqrt{2\pi}w_2}\exp\left(-\frac{(f-m_{\mathrm{f}_2}(\tau))^2}{2w_2^2}\right) \tag{B.1}$$

where α is a weighting factor and w_1 and w_2 are the spectrum widths, which can be modelled as random variables. The mean Doppler frequency is related to the normalised mean speckle power, $\bar{\tau} = \tau/\langle\tau\rangle$ with

$$m_{\mathrm{f}_1}(\tau) = \begin{cases} A_{\mathrm{c}} + B_{\mathrm{c}}\bar{\tau} + f_{\mathrm{bi}}, & \bar{\tau} \le \tau_{\mathrm{bi}} \\ A_{\mathrm{c}} + B_{\mathrm{c}}\tau_{\mathrm{bi}} + f_{\mathrm{bi}}, & \bar{\tau} > \tau_{\mathrm{bi}} \end{cases} \tag{B.2}$$

$$m_{\mathrm{f}_2}(\tau) = A_{\mathrm{c}} + B_{\mathrm{c}}\bar{\tau} + f_{\mathrm{bi}}. \tag{B.3}$$

where f_{bi} is a zero mean Gaussian random variable with standard deviation σ_{bi}, which represents the noisy nature of the fit between the mean Doppler and local intensity. Dependent on the value of α, this model can have the effect of reducing the rate of increase of the mean Doppler shift with intensity and broadening the spectrum if $\bar{\tau}$ exceeds a threshold τ_{bi}.

The parameters of this model can be represented using the generic model presented in Chapter 4. Examples of the data and model fits are given in Figures B.1–B.5 with the model coefficients from [1] given in Tables B.1–B.3.* These results show very similar trends to those discussed in Chapter 4 with some parameters being wind driven, while others having peaks in the upswell direction. Note that for the weighting value α shown in Figure B.4, many of the estimated values have an extreme value of 1. Therefore to improve the parametric modelling, all bimodal model fits which produce a value of 0.8 or greater have been set to 1. Finally, the associated extended Doppler model coefficients from [2] are given in Table B.4.

*Note the values for σ_{bi} have been corrected since the original publication in [1].

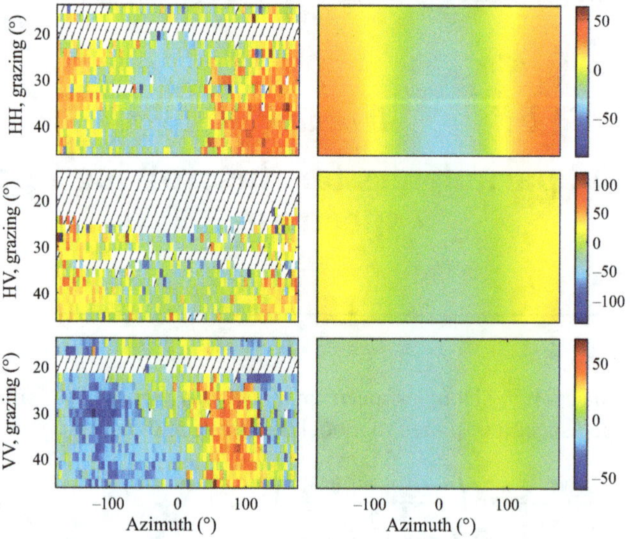

Figure B.1 *Variation of bimodal parameter, A_c, with azimuth and grazing angle. Data on left, model on right. Top plot: HH, middle plot: HV, bottom plot: VV. The wind speed and wave height are 10.2 m/s and 1.21 m, respectively [1]*

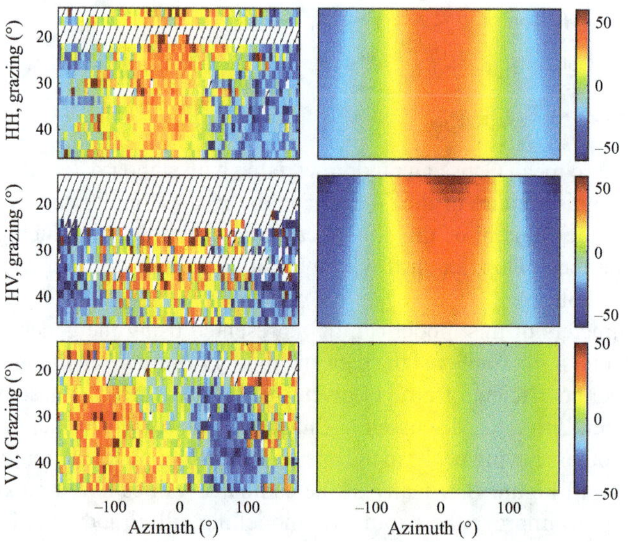

Figure B.2 *Variation of bimodal parameter, B_c, with azimuth and grazing angle. Data on left, model on right. Top plot: HH, middle plot: HV, bottom plot: VV. The wind speed and wave height are 10.2 m/s and 1.21 m, respectively [1]*

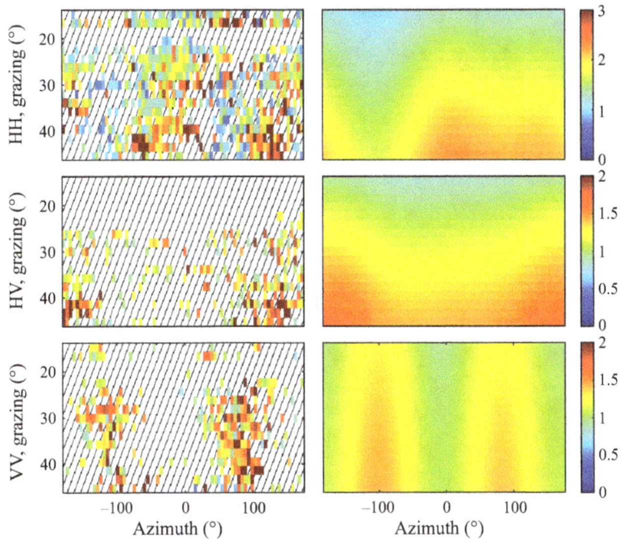

Figure B.3 *Variation of bimodal parameter,* τ_{bi}*, with azimuth and grazing angle. Data on left, model on right. Top plot: HH, middle plot: HV, bottom plot: VV. The wind speed and wave height are 10.2 m/s and 1.21 m, respectively [1]*

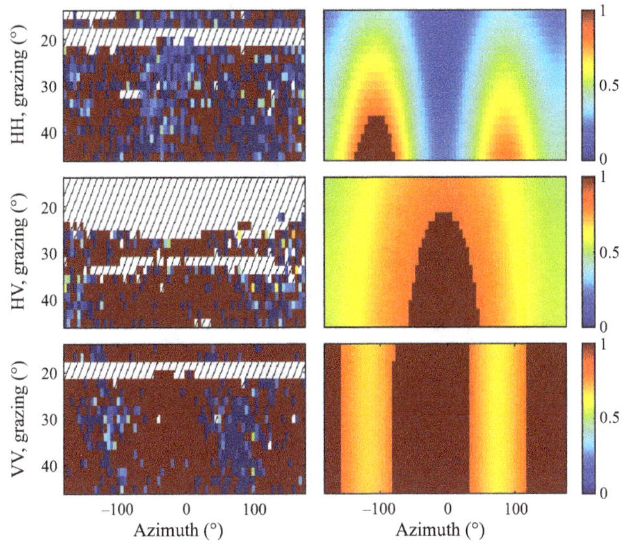

Figure B.4 *Variation of bimodal parameter,* α*, with azimuth and grazing angle. Data on left, model on right. Top plot: HH, middle plot: HV, bottom plot: VV. The wind speed and wave height are 10.2 m/s and 1.21 m, respectively [1]*

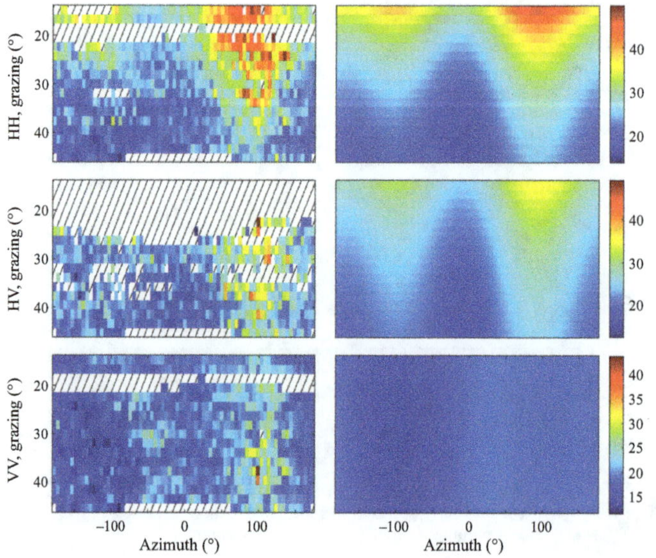

Figure B.5 Variation of bimodal parameter, σ_{bi}, with azimuth and grazing angle. Data on left, model on right. Top plot: HH, middle plot: HV, bottom plot: VV. The wind speed and wave height are 10.2 m/s and 1.21 m, respectively [1]

Table B.1 Evolving Doppler model bimodal coefficients for A_c and B_c

Pol.		A_c			B_c		
		f_{gen}	g_{gen}	h_{gen}	f_{gen}	g_{gen}	h_{gen}
HH	γ_{gen}	−2.38	2.77	0.044	−1.88	1.44	0.059
	a_{gen}	127.9	−125.1	−0.46	−1.98	5.62	−0.018
	b_{gen}	−262.3	225.7	3.43	148.9	−103.6	−5.03
	c_{gen}	126.1	−123.6	−1.29	−85.09	81.82	0.58
	d_{gen}	6.81	−15.26	3.69	−28.09	38.42	−4.21
	e_{gen}	6.71	−4.54	−0.38	4.50	−4.30	−0.26
HV	γ_{gen}	5.85	−3.73	−0.54	2.65	−2.95	−0.094
	a_{gen}	−18.1	19.57	−0.045	5.03	−2.05	−0.69
	b_{gen}	−40.31	24.39	−2.56	147.3	−101.3	−1.34
	c_{gen}	18.09	−16.47	−0.23	−47.33	45.75	−0.92
	d_{gen}	−37.14	35.17	−0.43	49.71	−42.13	−1.14
	e_{gen}	11.2	−10.07	−0.094	−21.95	21.32	−0.76
VV	γ_{gen}	−3.01	3.86	−0.23	−2.79	3.22	−0.10
	a_{gen}	0.24	−0.14	0.028	7.45	−6.87	1.22
	b_{gen}	−8.63	−13.65	9.85	16.37	6.29	−9.86
	c_{gen}	−4.33	5.26	−2.18	1.29	−2.37	2.23
	d_{gen}	0.72	−2.75	3.85	−2.90	5.56	−3.97
	e_{gen}	−2.08	3.93	−1.43	0.13	−1.15	1.19

Table B.2 Evolving Doppler model bimodal coefficients for τ_{bi} and α

		τ_{bi}			α		
Pol.		f_{gen}	g_{gen}	h_{gen}	f_{gen}	g_{gen}	h_{gen}
HH	γ_{gen}	5.27	–4.48	–0.11	7.41	–5.96	–0.28
	a_{gen}	–0.55	2.15	0.027	–0.21	0.65	0.0025
	b_{gen}	–0.24	0.40	0.035	0.15	–0.25	–0.0098
	c_{gen}	0.42	–0.45	0.045	0.25	–0.37	–0.024
	d_{gen}	–0.21	0.45	0	0.069	–0.12	–0.0028
	e_{gen}	–0.024	0.0016	0.024	–0.034	0.13	–0.066
HV	γ_{gen}	0.41	0.25	–0.084	–1.96	2.03	–0.0048
	a_{gen}	0.59	0.56	0.030	0.62	0.11	–0.018
	b_{gen}	0.096	–0.22	0.020	0.019	0.15	–0.021
	c_{gen}	0.026	0.11	–0.045	0.48	–0.54	0.026
	d_{gen}	0.017	–0.031	0.012	0.76	–0.72	–0.026
	e_{gen}	–0.31	0.35	–0.019	0.095	–0.11	0.0021
VV	γ_{gen}	0.031	0.22	–0.065	0.097	–0.080	–0.0039
	a_{gen}	1.00	0.11	0.031	0.89	0.18	–0.090
	b_{gen}	0.10	–0.11	–0.0095	0.021	–0.06	0.025
	c_{gen}	0.23	–0.31	–0.030	–0.043	0.016	0.059
	d_{gen}	0.25	–0.40	0.053	0.059	–0.089	–0.0090
	e_{gen}	0.13	–0.16	–0.0087	–0.038	–0.0058	0.061

Table B.3 Evolving Doppler model bimodal coefficients for σ_{bi}

		σ_{bi}		
Pol.		f_{gen}	g_{gen}	h_{gen}
HH	γ_{gen}	–1.051	0.25	0.12
	a_{gen}	27.37	–5.67	1.43
	b_{gen}	3.89	–3.84	0.67
	c_{gen}	1.84	–3.96	–0.19
	d_{gen}	–9.40	11.21	0.80
	e_{gen}	3.48	–3.75	–0.29
HV	γ_{gen}	0.16	–0.56	0.062
	a_{gen}	24.72	–4.52	1.28
	b_{gen}	3.79	–4.98	0.85
	c_{gen}	–2.30	0.22	–0.24
	d_{gen}	–4.72	3.73	1.43
	e_{gen}	11.54	–11.49	–0.32
VV	γ_{gen}	–0.21	0.22	0.0037
	a_{gen}	10.05	1.40	2.46
	b_{gen}	1.16	–0.77	0.63
	c_{gen}	–1.99	3.38	–0.20
	d_{gen}	–2.00	3.19	0.41
	e_{gen}	2.21	–4.03	0.65

Table B.4 Extended Doppler model bimodal coefficients

Pol.		A_{bi}	B_{bi}	α	τ_{bi}	σ_{bi}
HH	a_{ext}	−15.62	101.60	−0.0003	−1.98	−23.19
	b_{ext}	18.01	−23.50	0.012	0.18	1.34
	c_{ext}	0.056	−0.053	−0.55	−0.51	−0.57
	d_{ext}	0	−92.02	−99.96	7.49	42.96
	e_{ext}	0	−0.049	−2.12	−0.40	−0.20
HV	a_{ext}	−58.93	34.04	−0.21	−1.11	−9.53
	b_{ext}	−14.77	0	0.0062	0.086	0.52
	c_{ext}	−0.061	0	−1.020	−0.72	−0.77
	d_{ext}	100	−100	1.42	9.028	25.35
	e_{ext}	−0.074	−0.21	−0.53	−0.61	−0.30
VV	a_{ext}	−69.38	100.51	0	−8.28	−35.46
	b_{ext}	58.57	−76.02	0	3.17	1.60
	c_{ext}	−0.010	−0.034	0	−0.091	−0.58
	d_{ext}	11.13	−27.24	0	5.48	37.93
	e_{ext}	−0.031	−0.074	0	−0.043	−0.047

References

[1] S. Watts, L. Rosenberg, S. Bocquet, and M. Ritchie, "The Doppler spectra of medium grazing angle sea clutter; part 2: Exploiting the models," *IET Radar Sonar and Navigation*, vol. 10, no. 1, pp. 32–42, 2016.

[2] L. Rosenberg, "Parametric modelling of sea clutter Doppler spectra," *IEEE Transactions on Geoscience and Remote Sensing*, 2021.

Appendix C
Atmospheric and precipitation losses

As the radar signal propagates through the atmosphere, there is attenuation due to molecular absorption of oxygen and water vapour, as well as potential attenuation due to precipitation or fog. The two-way atmospheric loss, L_a, can be determined over the range in km, R_{km}, by summing twice the one-way attenuation per km due to clear air, $l_{c,dB}$, rain, $l_{r,dB}$, snow, $l_{s,dB}$, and fog, $l_{f,dB}$ [1]:

$$L_a = 2R_{km}\left(l_{c,dB} + l_{r,dB} + l_{s,dB} + l_{f,dB}\right). \tag{C.1}$$

The one-way attenuation per km in clear air is given by

$$l_{c,dB} = a_o + \frac{a_v a_w a_h}{100} \tag{C.2}$$

where a_w is the water content, a_h is the relative humidity and a_o and a_v are the oxygen and water absorptions factors, respectively, with values given in Table C.1. At a given temperature in °C, $T_{a,C}$, the water content in g/m^3 can be approximated by

$$a_w \approx 5.25 + 0.13T_{a,C} + 0.023T_{a,C}^2. \tag{C.3}$$

As an example, consider an X-band radar operating at sea level with a temperature of 18°C and a relative humidity of 70%. In clear air, the one-way attenuation would be 0.0106 dB/km.

The attenuation in rain depends on the rain rate, ρ_r, in mm/h. Typical values are 0.5 mm/h (drizzle), 1 mm/h (light rain), 4 mm/h (moderate rain) or 16 mm/h

Table C.1 Molecular absorption at sea level (dB/km) one-way

Frequency (GHz)	**3**	**9**	**15**
Oxygen, a_o	0.0066	0.007	0.008
Water absorption (g/m^3), a_v	0.000025	0.00034	0.0016

Table C.2 Coefficients for the Briggs rain model in (C.4) for a temperature of 18°C

Frequency (GHz)	3	9	15
a_r	–3.523	–2.062	–1.403
b_r	0.884	1.247	1.182

(heavy rain). In [1], a simple model has been developed by Briggs for the one-way attenuation per km based on measured data by Kingsley and Quegan [2],

$$l_{r,dB} = 10^{a_r + b_r \log(\rho_r + 0.00001)} \tag{C.4}$$

where the coefficients, a_r, b_r at 18°C are given in Table C.2. For an X-band radar in heavy rain with a temperature of 18°C, the one-way attenuation would be 0.28 dB/km. At 0°C and 40°C, the attenuation is about 15% less and typically horizontal polarisations have a 10% higher attenuation than vertical.

In general, snow storms have a lower rain rate, ρ_r, than rain storms. A model for the one-way attenuation per km has been suggested by Blake [3]:

$$l_{s,dB} = 34.9 + 10^{-12}\frac{\rho_r^{1.6}}{\lambda^4} + 22 \times 10^{-6}\frac{\rho_r}{\lambda} \tag{C.5}$$

where λ is the radar wavelength.

There are two types of fog that impact on radio propagation. The first is radiation fog that occurs at night under clear skies and light wind. As heat is radiated from the land, it cools the surface and the air mass just above it. This condenses the moisture into a fog which may spill offshore. Advection fog is the more common sea-fog that occurs when relatively warm, moist air encounters a cold surface such as the sea. Dense fog can cause a similar level of attenuation as moderate rain. Barton and Leonov [4] have proposed models that determine an equivalent rain rate based on the visibility, V_f. For advection fog,

$$\rho_r = 0.00269/V_f^{1.48} \tag{C.6}$$

and for radiation fog,

$$\rho_r = 0.00339/V_f^{1.65}. \tag{C.7}$$

The one-way attenuation per km, $l_{f,dB}$, can then be determined using (C.4). In thick advection fog, the visibility could be as small as 0.2 km, giving a rain rate of 0.2913 mm/h. At an X-band, this gives a one-way loss of 0.00186 dB/km.

References

[1] J. N. Briggs, *Target detection by marine radar*. The Institution of Electrical Engineers, London, 2004.
[2] S. P. Kingsley and S. Quegan, *Understanding radar systems*. McGraw-Hill, New York, 1992.
[3] L. V. Blake, "A guide to pulse radar maximum range calculations," US Naval Research Laboratory, Tech. Rep., 1969.
[4] D. K. Barton and S. A. Leonov, *Radio technology encyclopedia*. Artech House, Norwood, MA, 1996.

Index

adaptive DPCA algorithm 258
adaptive matched filter (AMF) 258
airborne maritime surveillance radars 2, 21, 246, 257, 258
airborne scanning radar 164
amplitude distributions 27–42
  distribution models 31–9
    K+Rayleigh distribution 35
    KA-distribution 37–8
    K-distribution 33–5
    KK-distribution 37
    log–normal distribution 32–3
    Pareto distribution 36–7
    tri-modal discrete (3MD) distribution 38–9
    Weibull distribution 33
  effective shape relationships 42
  example fits 39–42
anomalous propagation 17, 20–1, 109, 112, 195
atmospheric and precipitation losses 339–40
autocorrelation function (ACF) 152, 154, 199
autoregressive (AR) process 264

Bernoulli track before detect filter 315–17, 319–20
Bhattacharyya distance (BD) 182, 329–30
binary scan-to-scan processing 246
bistatic Doppler shift 67, 71
bistatic geometry 68–71, 126
bistatic mean reflectivity 122–7
  in-plane scattering 122–6
  out-of-plane scattering 126–7
bistatic measurements, 11, 73, 78
bistatic radars, 11, 67, 102
bistatic range resolution 67–9, 87, 90, 92, 95, 97, 99, 101
bistatic sea clutter 67–102
  active bistatic radar 71–83
    amplitude statistics 77–8
    correlation 78–80
    Doppler spectra 80–2
    in-plane reflectivity 74–5
    out-of-plane reflectivity 75–7
    sea clutter measurements 72–4
    sea spikes 82–3
  passive bistatic radar 84–102
    data processing 89–91
    delay/Doppler amplitude statistics 91–3
    Doppler spectra 98–101
    sea clutter measurements 84–9
    time domain amplitude statistics and correlation 93–8
Bragg scattering 8, 29, 43, 47–9, 52–3, 98, 172
Briggs rain model 340

cell-averaging (CA)-CFAR 197, 224, 240, 249, 289, 319
censored mean (CM)-CFAR 243
chi-square (CS) test 40, 329
chi-square distribution 202
clutter characteristics 5–6
clutter range profile (CRP) 240–1, 289, 301
coherent amplitude statistics 49–52
coherent detection 257–67
  coherent detection strategies 260

Doppler spectrum aliasing 265–7
model-based detection 264–5
multichannel radar 258–60
post-Doppler detection 263
post-Doppler whitening 263–4
pre-Doppler whitening 260–3
coherent processing 194, 210–18, 257
adaptive processing 213–18
Doppler filtering 210–13
coherent signal processing 210, 236
compound G2 model 152, 175–8, 185
compressive sensing 291, 304
constant false alarm rate (CFAR) 194, 235, 281
continuous wavelet transform (CWT) 282
contour integration 207, 209
CSIR X-band data set 10, 45, 53

Dalle Mese Giuli (DMG) approximation 209
data sets 9–10
CSIR X-band data set 10
Ingara X-band data set 9–10
detection in heterogeneous environments 267–75
fixed scale hybrid 269
single data set detector 267–9
texture estimation and normalisation hybrid 270
variable scale hybrid 269
dictionary learning 304–11
classical DL algorithms 305
dictionary learning techniques 305–6
performance evaluation 306–9
discrete wavelet transform (DWT) domain 282
displaced phase centre array (DPCA) 258
Domville bistatic mean reflectivity model 123–5
Doppler centroid 80, 81
Doppler characteristics 44–58
coherent amplitude statistics 49–52
mean Doppler spectra 52–3
time and range varying spectra 53–8
Doppler model bimodal coefficients, extended 338
Doppler processing 9, 235–6, 246, 263, 265
Doppler spectrum 6, 132–43
evolving Doppler spectrum model 137–43
mean Doppler spectrum models 132–6
temporal decorrelation model 136
Doppler spectrum model, evolving 136–43, 98
Douglas sea state 4–5
drop off mini-batch online dictionary learning (DOMINODL) algorithm 305–7, 309, 311
DSTG continuous mean backscatter model 119–22
DSTG continuous shape model 129–32
coefficients 130

electromagnetic (EM) modelling 20, 107
electromagnetic scattering models 7–8
empirical modelling 8–9
end-to-end radar performance, 218–29
discussion 229
radar performance examples 218–25
simulated performance prediction 225–9
entropy 287
entropy-SWT (eSWT) detector 289
evolving Doppler spectrum 157–71
frequency domain simulation 157–60
multiple phase centres 161–3
scanning radar scenario 163–5
simulation examples 166–71
time domain simulation 160–1
extended Doppler model fits 144
extended Doppler model linear coefficients 145

first-order small perturbation models 24
first-order small slope approximation (SSA) 24
Fourier synthesis (FS) 153
frequency agility, effects of 9

Gaussian distribution 5, 15, 49, 111, 159, 176, 237
generalised inner product (GIP) 267
generalised likelihood ratio test (GLRT) 258
geometrical optics (GO) 24
geometric mean (GM) detector 249
geometric optics-small slope approximation (GO-SSA) models 20, 25
Georgia Institute of Technology (GIT) model 109–11
goodness of fit metrics 329–30
  Bhattacharyya distance 330
  Chi-square test 329
  Kolmogorov–Smirnov test 330
  threshold error 330
GO-SSA mean backscatter models 24–7
greater of (GO)-CFAR 243

histogram probabilistic multi-hypothesis tracker (H-PMHT) algorithm 312
hybrid mean backscatter model 111–12

Imaging Radar Systems Group (IRSG) model 115–18
Ingara L-band data set 306–7
Ingara X-band data set 39–40, 152, 171, 226, 257, 284, 300
in-plane reflectivity measurements 74
in-plane scattering 73, 122–6
IRSG mean backscatter model coefficients 116
iterative shrinking thresholding algorithm (ISTA) 292

K+Rayleigh distribution 35
KA-distribution 37–8
K-distribution 33–5, 156, 174, 182, 184
K-distribution shape 127–32
  DSTG continuous shape model 129–32
  low grazing angle models 127–8
KK-distribution 37
K-means singular value decomposition (KSVD) 305, 307, 309
Kolmogorov–Smirnov (KS) test 40, 329–30

least absolute shrinkage and selection operator (LASSO) 305
least-angle regression (LARS) 305, 309
log–normal distribution 32
low grazing angle models 127–8

Marcum model 200
maritime target detection techniques 235–74
  coherent detection 257–67
    coherent detection strategies 260
    Doppler spectrum aliasing 265–7
    model-based detection 264–5
    multichannel radar 258–60
    post-Doppler detection 263
    post-Doppler whitening 263–4
    pre-Doppler whitening 260–3
  detection in heterogeneous environments 267–75
  non-coherent detection 236–48
    alternative methods 247–8
    CA-CFAR performance in spatially correlated clutter 243–4
    clutter statistics 237–40
    constant false alarm rate detectors 240–3
    scan-to-scan processing 244–7
  Pareto CFAR detectors 248–57
    type I complete sufficient statistics 250–1
    type I transformed CFAR 249–50

type II Bayesian CFAR 252–3
type II transformed CFAR 251–2
Masuko models 114–15
mean backscatter 17–27
anomalous propagation 20–1
GO-SSA mean backscatter models 24–7
multipath scattering 21–4
mean Doppler spectrum models 132–6
mean optimal sub-pattern assignment (OSPA) metric 318
memoryless non-linear transform (MNLT) 153–7
modified Bhattacharyya distance (MBD) 330
modified chi-square (MCS) test 41–2
modified Kolmogorov–Smirnov (MKS) test 42
monostatic mean backscatter 107–22
DSTG continuous mean backscatter model 119–22
Georgia Institute of Technology (GIT) model 109–11
hybrid model 111–12
Imaging Radar Systems Group (IRSG) model 115–18
Masuko models 114–15
Naval Research Laboratory model 118–19
Royal Radar Establishment model 108–9
Sittrop's model 109
Technology Service Corporation (TSC) model 112–14
monostatic sea clutter 15–60
amplitude distributions 27–42
distribution models 31–9
effective shape relationships 42
example fits 39–42
Doppler characteristics 44–58
coherent amplitude statistics 49–52
mean Doppler spectra 52–3
time and range varying spectra 53–8
mean backscatter 17–27
anomalous propagation 20–1
GO-SSA mean backscatter models 24–7
multipath scattering 21–4
sea-spike characteristics 42–4
texture correlation 58–60
Monte Carlo simulations 225, 228, 253, 289, 300
morphological component analysis (MCA) 291
MQ-4C Triton unmanned aerial system (UAS) 2
multipath factor 22
multipath scattering 21–4
multiple phase centres 161–3
multi-resolution analysis 283

NATO anti-air warfare systems (NAAWS) model 111
Naval Research Laboratory (NRL) model 118–19
NetRAD clutter measurement campaigns 74
non-coherent detection 236–48
alternative methods 247–8
CA-CFAR performance in spatially correlated clutter 243–4
clutter statistics 237–40
constant false alarm rate detectors 240–3
scan-to-scan processing 244–7
non-coherent processing 194, 197–209, 236, 257
compound clutter 203–4
constant target 200–2
correlated returns 205–9
fluctuating target 202–3
normalised adaptive matched filter (NAMF) 214–15, 258, 304, 309
normalised sample covariance matrix (NSCM) 225, 262
NRL mean backscatter model coefficients 119

one-way propagation factor 22–3
online dictionary learning (ODL) algorithm 305–7, 309
Optus-D3 bistatic collection geometry 89
Optus-D3 Passive Seas data set 88–9
Optus-D3 spectrograms 95
ordered statistic (OS)-CFAR 243, 249
orthogonal matching pursuit (OMP) 305
out-of-plane scattering 126–7

P-8A Poseidon aircraft 2
parametric modelling 107–47
  bistatic mean reflectivity 122–7
    in-plane scattering 122–6
    out-of-plane scattering 126–7
  Doppler spectrum 132–43
    evolving Doppler spectrum model 136–43
    mean Doppler spectrum models 132–6
    temporal decorrelation model 136
  K-distribution shape 127–32
    DSTG continuous shape model 129–32
    low grazing angle models 127–8
  monostatic mean backscatter 107–22
    DSTG continuous mean backscatter model 119–22
    Georgia Institute of Technology (GIT) model 109–11
    hybrid model 111–12
    Imaging Radar Systems Group (IRSG) model 115–18
    Masuko models 114–15
    Naval Research Laboratory model 118–19
    Royal Radar Establishment model 108–9
    Sittrop's model 109
    Technology Service Corporation (TSC) model 112–14
  spatial correlation 143–7
Pareto CFAR detectors 248–57
  type I complete sufficient statistics 250–1
  type II Bayesian CFAR 252–3
  type II transformed CFAR 251–2
  type I transformed CFAR 249–50
Pareto distribution 36–7, 152, 156, 240, 252
passive bistatic radar 84–102
  data processing 89–91
  delay/Doppler amplitude statistics 91–3
  Doppler spectra 98–101
  sea clutter measurements 84–9
  time domain amplitude statistics and correlation 93–8
Passive Seas trial 84, 88
performance prediction modelling 193–229
  assessing end-to-end radar performance 218–29
    radar performance examples 218–25
    simulated performance prediction 225–9
  coherent processing 210–18
    adaptive processing 213–18
    Doppler filtering 210–13
  non-coherent processing 197–209
    compound clutter 203–4
    constant target 200–2
    correlated returns 205–9
    fluctuating target 202–3
  radar range equation 195–7
physical models 24
point spread function (PSF) 166
pulse chasing 67
pulse jittering 164
pulse-to-pulse frequency agility 29
pulse-to-pulse integration 9, 163–4, 198, 199

radar range equation 195–7
RADARSAT Constellation Mission 3
retrospective processing 247

Rician distribution 200–2
root-mean-square (RMS) error 182
Royal Radar Establishment (RRE) 108
  mean backscatter model coefficients 109

sample covariance matrix (SCM) 225, 261
scan-to-scan integration 198–200
scan-to-scan processing techniques 236
sea clutter modelling 6–9
  electromagnetic scattering models 7–8
  empirical modelling 8–9
sea clutter simulation 151
  evolving Doppler spectrum 157–71
    frequency domain simulation 157–60
    multiple phase centres 161–3
    scanning radar scenario 163–5
    simulation examples 166–71
    time domain simulation 160–1
  simulation algorithms 171–89
    algorithm comparison 178–85
    compound G2 model 175–8
    spatially and temporally limited clutter model 172–5
  uniform Doppler spectrum 152–7
    memoryless non-linear transform 153–7
sea-spike characteristics 42–4
short-time Fourier transforms (STFTs) 281, 293–6
simulation algorithms 171–89
  algorithm comparison 178–85
  compound G2 model 175–8
  spatially and temporally limited clutter model 172–5
single data set (SDS) 267
Sirius-XM data set 84–8, 97
Sittrop's mean backscatter X- and Ku-band model coefficients 110
Sittrop's model 109
smaller of (SO)-CFAR 243
space–time adaptive processing (STAP) algorithms 258
  principle of 259
sparse signal separation 291–304
  detection approaches 296–301
  detection performance 301–4
  short-time Fourier transform 293–6
  summary and further developments 303–4
  tuned Q wavelet transform 292–3
spatial correlation 143–7, 176, 182
spatial correlation model coefficients 146
spatially and temporally limited clutter (STLC) model 152, 172–5 185
spherically invariant random process (SIRP) 50, 153, 210
spherically invariant random vector (SIRV) 153
split augmented Lagrangian shrinkage algorithm (SALSA) 292, 296
Stratton–Chu equations 8
sub-band isolation and reconstruction 283–5
Swerling models 200–3
Stationary wavelet transformation (SWT) signal decomposition 284

target detection, 281–321
  dictionary learning 304–11
  sparse signal separation 291–304
  time/frequency target separation 282–91
  track before detect 311–21
Technology Services Corporation (TSC) model 112–14
temporal decorrelation model 136
texture correlation 11, 58–60
threshold error (TE) 39, 329–30
time/frequency target separation 282–91
  detection performance 289–90
  sub-band isolation and reconstruction 283–5

sub-band selection 285–9
track before detect (TBD) 281, 311–21
Bernoulli track before detect filter 315–17
detection and tracking performance 318–20
mathematical models 312–15
simulation details 317
traditional coherent detection techniques 210
trimmed mean (TM)-CFAR 243
tri-modal discrete (3MD) distribution 38–9
tuned *Q*-wavelet transforms (TQWTs) 281, 292–3
two-dimensional (2D) stationary wavelet transforms (SWT) 282

uniform Doppler spectrum 152–7

Watts and Wicks shape parameter coefficients 128
wavelet transforms (WT) 282
Weibull distribution 33
Weinstock fluctuations 200, 202
WiMAX 84, 87
World Meteorological Organisation, 5

www.ingramcontent.com/pod-product-compliance
Lightning Source LLC
LaVergne TN
LVHW011838200926
848484LV00002B/369

* 9 7 8 1 7 8 5 6 1 9 1 3 7 *